ATOMIC
COLLISIONS IN SOLIDS

Volume 2

ATOMIC
COLLISIONS IN SOLIDS

Volume 2

Edited by

Sheldon Datz, B.R. Appleton, and C.D. Moak

*Oak Ridge National Laboratory
Oak Ridge, Tennessee*

SPRINGER SCIENCE+BUSINESS MEDIA, LLC

Library of Congress Cataloging in Publication Data

International Conference on Atomic Collisions in Solids, 5th, Gatlinburg, Tenn.,
   1973.
   Atomic collisions in solids.

   Includes bibliographical references and index.
   1. Solids—Congresses. 2. Collisions—(Nuclear physics)—Congresses. I. Datz, Sheldon,
ed. II. Appleton, B. R., 1937-   ed. III. Moak, C. D., 1922-   ed. IV.
United States. National Laboratory, Oak Ridge, Tenn. V. Title.
QC176.A1I53 1973   539.7'54   74-26825
ISBN 978-1-4615-8998-3   ISBN 978-1-4615-8996-9 (eBook)
DOI 10.1007/978-1-4615-8996-9

Second half of the Proceedings of the Fifth International Conference on Atomic Collisions in Solids held in Gatlinburg, Tennessee, September 24-28, 1973, sponsored by
Oak Ridge National Laboratory

Perhaps the most controversial aspect of this volume is the number (V) assigned to the conference in this series. Actually, the first conference to be held under the title "Atomic Collisions in Solids" was held at Sussex University in England in 1969 and the second at Gausdal, Norway in 1971, which would logically make the conference held at Gatlinburg, Tennessee, U.S.A. in 1973 the third (III). However, the appearance of the proceedings of the 1971 Gausdal Conference (published by Gordon and Breach) bore the number IV. The reasoning behind this was that, in fact, two previous conferences had been largely dedicated to the same subject area. The first of these was at Aarhus, Denmark in 1965 and the second in 1967 was held in Chalk River, Canada. Hence, the number V for the 1973 meeting.

Actually, the conference can easily be traced back to Paris, France in 1961 when it went under the colorful title of "Le Bombardement Ionique." In 1962 a small conference was held at Oak Ridge, Tennessee, U.S.A. at which the discovery of channeling was first formally annunciated. This was followed by conferences at Chalk River, Canada in 1963 and at Harwell, England in 1964. Moreover, immediately following the Chalk River conference in 1967 there was a conference on higher energy collisions at Brookhaven, New York, U.S.A. Thus, strictly speaking, the Gatlinburg meeting is the tenth (X) in the series.

The accent of the Gatlinburg conference was on primary atomic collision processes, states of ions penetrating solids and their relation to macroscopic effects. Applications of the results to various technologies such as ion implantation and analysis of solid structures have grown large enough to warrant separate conferences.

The effect of the crystal lattice on the motion of penetrating particles (channeling) was first mentioned by M. T. Robinson at the Paris meeting (1961) and the study of these effects grew to the point where the Gausdal 1971 meeting was subtitled "The Physics of Channeling." The phenomenon had been sufficiently well investigated to permit its application to a broad range of problems, principally the location of lattice impurities and defects. At Gatlinburg, papers on channeling were generally restricted to those aspects which related to scattering in solids. Even with this restriction *four of the ten sessions were devoted exclusively to channeling papers and the remaining sessions contained many papers which utilized the technique.*

The increasingly detailed knowledge of the consequences of inelastic atomic collisions and of charge exchange processes in solids are accented in three sessions. All of these observations contribute to a new level of understanding which permits the treatment of such macroscopic effects as stopping power and radiation damage on a more fundamental basis.

CONTENTS OF VOLUME 2

SECTION VII: SURFACE SCATTERING

SECTION IX: CHANNELING

SECTION X: DECHANNELING

CONTENTS OF VOLUME 1

SECTION V: X RAYS

SECTION VI
ELECTRONS, PHOTONS, AND CHANNELING

RADIATIVE ELECTRON CAPTURE AND BREMSSTRAHLUNG

HERBERT W. SCHNOPPER and JOHN P. DELVAILLE
Center for Space Research
Massachusetts Institute of Technology
Cambridge, Massachusetts 02139, U.S.A.

ABSTRACT

The applicability of radiative electron capture (REC) as a means for investigating the electron momentum distribution of target electrons in atomic collisions is discussed. A model based on an impulse approximation is used to predict the position, shape, and cross section for the REC spectral feature. Experimental results obtained for proton, lithium, carbon, and oxygen projectiles incident on various gaseous targets are presented.

Introduction

Of the many channels for x-ray production which are available in the interaction of fast charged particles with target atoms, there are two whose contribution to the experimental data may be easily overlooked or misinterpreted. Radiative electron capture [1] (REC) and bremsstrahlung [2] processes contribute only in a small way to the total production of x rays in heavy-ion collisions, but one of them, REC, has the possibility of becoming a significant tool in the study of the momentum distribution of target electrons. This possibility has only just begun to be explored, and this discussion will outline briefly some of the current thinking on the subject.

We shall deal primarily with fast positive ions which inter- act with outer or weakly bound electrons in solid or gaseous atomic or molecular targets. The interaction may or may not lead to electron capture and, in either case, the interaction may or may not lead to the production of radiation. When REC occurs, it will be argued that the resulting spectrum of x rays is related to the electron momentum distribution of target electrons. Our discussion follows similar arguments put forth to explain the shape of the spectrum obtained from the Compton scattering of x rays [3]. In addition, there is already a fairly complete theory for the non-radiative electron capture [4] (NREC) cross section which, presumably, can be altered to meet the requirements of REC. Our discussion, however, will proceed in a different way.

Radiative Electron Capture

To begin, consider the interaction of a completely stripped projectile of atomic number Z with a free electron gas. It is convenient to view the interaction from the rest frame of the projectile where the total momentum of a target electron is

$$\vec{p}' = \vec{p}_o + \vec{p}, \tag{1}$$

where p is the momentum of the electron in the target frame and $p_o = mv_o$ where v_o is the projectile velocity in the laboratory frame. If the electron is captured into a shell with binding energy $-E_f$, then a photon of energy $h\nu$ is emitted where

$$h\nu = p'^2/2m + E_f = T' + E_f, \tag{2}$$

where T' is the total kinetic energy of the electron in the projectile frame.

$$T' = (p_o^2 + p^2)/2m + \vec{p}_o \cdot \vec{p}/m = T_r + T + 2\sqrt{TT_r}\cos\theta \tag{3}$$

where $T_r = p_o^2/2m$ and $T = p^2/2m$ is the kinetic energy of the electron in the target frame.

Bethe and Salpeter [5] give an expression for the cross section $\sigma_R(T')$ for the emission of recombination radiation during the process of radiative electron capture by a hydrogen-like ion. This cross section together with dn'/dT', the number of electrons with energy T' in the interval dT', determine the REC yield $Y(h\nu)$

$$Y(h\nu) = N_p \, GP(h\nu)\varepsilon(h\nu)\sigma_R(T')dn'/dT'$$

$$= N_p n_e \, GP(h\nu)\varepsilon(h\nu)\sigma_R(T') \, P(T'), \tag{4}$$

where N_p is the number of incident particles, n_e is the number of electrons per cm^3, G is a geometrical factor which includes the solid angle subtended by and the length of beam seen by the detector [6], P(hν) is the energy dependent part of the G factor, ε(hν) is the energy dependent efficiency of the detector system, and P(T) is the probability of finding an electron with kinetic energy T in dT.

The procedure to be followed here is to consider an isotropic distribution of electron momenta in the target frame and
to transform that distribution to the laboratory frame in order
to arrive at dn'/dT'. Although non-iosotropic distributions lead
to similar results, the analysis is more complex and will not be
treated here. Small relativistic effects will also be ignored.

Since the total number of electrons in either frame is an
invariant, it follows that

$$\left(\frac{d^3N'}{dT'd\Omega'dV'} \right) dT'd\Omega'dV' = \left(\frac{d^3N}{dTd\Omega dV} \right) dTd\Omega dV, \tag{5}$$

where N, T, Ω and V are the number, kinetic energy, solid angle
and volume in the projectile (primed) and target (unprimed) frames,
respectively. Since dV' = dV, Eq. 5 can be expressed in terms of
electron number densities n' and n

$$\left(\frac{d^2n'}{dT'd\Omega'} \right) dT'd\Omega' = \left(\frac{d^2n}{dTd\Omega} \right) dTd\Omega. \tag{6}$$

With an isotropic distribution in the target frame this becomes

$$\left(\frac{d^2n'}{dT'd\Omega'} \right) dT'd\Omega' = \frac{1}{4\pi} \left(\frac{dn}{dT} \right) dTd\Omega$$

$$= \frac{n_e}{2} P(T) \; dTd(\cos\theta). \tag{7}$$

From Eq. 3

$$dT' = 2 \sqrt{TT_r} \; d(\cos\theta). \tag{8}$$

Thus,

$$\left(\frac{d^2n'}{dT'd\Omega'} \right) dT'd\Omega' = \frac{n_e}{4 \sqrt{TT_r}} P(T)dTdT'. \tag{9}$$

Integrating both sides yields

$$P(T') = \frac{1}{4\sqrt{T_r}} \int_{LL}^{UL} P(T) \frac{dT}{\sqrt{T}} \tag{10}$$

The upper and lower limits need some discussion. For a given T',
i.e., a given hν, there is a well defined range of T which can
contribute to the integral. In addition, there is a range of T

intrinsic to the target. The interrelationship of both of these ranges defines the limits of the integral. For a given T', the largest and smallest values of T which can contribute to the integral are:

$$\sqrt{T_u} = \sqrt{T'} + \overline{T_r} \quad \theta = 180°$$

$$\sqrt{T_\ell} = |\sqrt{T'} - \sqrt{T_r}| \quad \theta = 0°, \tag{11}$$

If the range of T is $T_L < T < T_U$ then

$$UL = T_U \qquad\qquad \sqrt{T'} > \sqrt{T_U} - \sqrt{T_r}$$

$$= (\sqrt{T'} + \sqrt{T_r})^2 \quad \sqrt{T'} < \sqrt{T_U} - \sqrt{T_r}$$

$$LL = (\sqrt{T'} - \sqrt{T_r})^2 \begin{cases} \sqrt{T'} > \sqrt{T_L} + \sqrt{T_r} > \sqrt{T_r} \\[2mm] \sqrt{T'} < \sqrt{T_r} - \sqrt{T_L} < \sqrt{T_r} \end{cases}$$

$$= T_L \qquad\qquad \text{otherwise} \tag{12}$$

In the projectile frame, the energy range is $T'_L < T' < T'_U$. Thus,

$$\sqrt{T'_U} = \sqrt{T_U} + \sqrt{T_r}$$

$$\sqrt{T'_L} = \begin{cases} \sqrt{T_r} - \sqrt{T_U}, \sqrt{T_r} > \sqrt{T_U} \\[2mm] 0 \qquad, \sqrt{T_L} < \sqrt{T_r} < \sqrt{T_U} \\[2mm] \sqrt{T_L} - \sqrt{T_r}, \sqrt{T_r} < \sqrt{T_L} \,. \end{cases} \tag{13}$$

Eq. 10 defines the REC line profile and its derivative has special properties which are most apparent if the right hand side is expressed in terms of the electron momentum. Noting that for an isotropic distribution $P(T)dT = 4\pi P(p)p^2 dp$, $P(T')$ becomes an

$$P(T') = \frac{2\pi m}{P_o} \int_{P_{LL}}^{P_{UL}} P(p)\,p\,dp \tag{14}$$

where, as before, the photon energy dependence is introduced in the limits of the integral. For most cases dealt with here, the upper limit will be a finite constant or infinity. Thus,

$$\frac{dP(T')}{dT'} = \frac{dP(T')}{d(h\nu)} = -\frac{2\pi m}{p_0} P(p_{LL})p_{LL}\frac{dp_{LL}}{d(h\nu)}$$

$$= \frac{-\pi m \sqrt{2m}}{p_0\sqrt{h\nu-E_f}} P(p_{LL})p_{LL}. \tag{15}$$

The spherically symmetric momentum distribution $|\chi(p)|^2$ is obtained immediately from Eq. 15.

$$|\chi(p_{LL})|^2 = 4\pi p_{LL}^2 P(p_{LL})$$

$$= (2/m)^{3/2}\sqrt{h\nu-E_f}p_0 p_{LL} [dP(T')/d(h\nu)]. \tag{16}$$

This is similar to the results which are obtained from an analysis of the Compton profile [3].

In the case of the collision with an atom, the situation is complicated by the fact that the electron is no longer free, but instead is bound to the target atom with an energy $-E_i$. Instead of Eq. 2, the energy released during capture should be

$$h\nu = E_f - E_i + T_r. \tag{17}$$

This condition destroys the simplicity of the discussion leading to Eq. 4 and, unless certain conditions are met, the target binding can only be dealt with in a complex manner. In the Compton case [3], the simplifying condition is

$$E_i (\hbar^2 k^2/2m) \ll 1, \tag{18}$$

where k is the change in photon wave number, i.e., that the binding energy of the target electron is much smaller than the recoil energy acquired by the electron. A similar condition is stated here without proof for REC:

$$E_i/T' \ll 1, \tag{19}$$

i.e., the binding energy of the target electron is much smaller than its kinetic energy relative to the projectile. Put another way, the velocity of the projectile is much larger than the orbital velocity of the electron and the time of collision will be short compared with the orbital period of the electron. Satisfying either Eq. 18 or 19, known as the impulse approximation, has the same interpretation for Compton scattering and REC. The potential energy of the electron in the target is not really neglected. It is assumed to remain constant throughout the interaction which leads to photon emission and, therefore, cancels out of the photon energy calculation. In either case, the binding <u>does</u> serve to provide a distribution of momentum in the initial state and to an accuracy of $(E_i/T')^2$ Eq. 14 can still be used to describe the REC process. Now, $P(\vec{p})$ is given by

$$P(\vec{p}) = \frac{1}{(2\pi)^3} \left| \int e^{-i\vec{p}\cdot\vec{r}} \, \phi(\vec{r}) \, d^3r \right|^2$$

$$= |\psi(\vec{p})|^2, \tag{20}$$

where $\phi(r)$ is the initial state wave function. Within this approximation, the REC process can be thought of as a free-bound transition, and we intend to use this theoretical framework to explain and interpret the experimental REC results.

Several simple spherically symmetric cases can be illustrated immediately. These are summarized in Table I. In all cases the peak of the $P(T')$ distribution occurs at $h\nu = T_r + E_f$. The energy dependence of $\sigma_R(T')$ may cause the peak to shift slightly. The full width at half maximum of the $P(T')$ distribution is also given since it is a convenient parameter which can be derived from experimental data obtained with an Si(Li) x-ray detector.

Bremsstrahlung

What happens to those electrons which are unbound from the target and captured in a continuum state on the projectile? An electron of initial momentum p_i in the target frame is accelerated to a final state momentum p_f. The maximum value for p_f is mv_o which occurs when the electron is captured at the bottom of the projectile continuum. Within the impulse approximation, a photon will be emitted whose energy is

$$h\nu = (\vec{p}_i + \vec{p}_f)^2/2m = p_i^2/2m + p_f^2/2m + \vec{p}_i\cdot\vec{p}_f/m. \tag{21}$$

A bremsstrahlung spectrum is expected since the final state is in the projectile continuum. REC and bremsstrahlung processes are therefore closely related, the only difference being the final state. When viewed from the rest frame of the projectile, the bremsstrahlung process represents, within the impulse approximation, a free-free transition in the projectile continuum.

In the lowest order of approximation, the electron may be taken to be at rest in the target and it has, therefore, a relative kinetic energy $T_r = p_o^2/2m = (m/M)T_p$ where m and M are the masses of the

Table I. Examples of Transformed Energy Distributions*

	Target Frame $P(T)$	Projectile Frame $P(T')$	FWHM	Remarks
Delta Function	$\delta(T^*)$	$\begin{cases} 1, & T_r - T^* < T < T_r + T^* \\ 0, & \text{otherwise} \end{cases}$	$4\sqrt{T_r T^*}$	$T_r > T^*$
Rect Function	$\begin{cases} 1, & T^* - a/2 < T < T^* + a/2 \\ 0, & \text{otherwise} \end{cases}$	$\sqrt{T_U} - \lvert\sqrt{T'} - \sqrt{T_r}\rvert$ (special case)	$2\sqrt{T_r T_U}$	$T_r \gg T_U$ $a = 2T^* = T_U$
Fermi Distribution	$\begin{cases} \sqrt{T}, & T < \varepsilon_f \\ 0, & T > \varepsilon_f \end{cases}$	$\varepsilon_f - (\sqrt{T'} - \sqrt{T_r})^2$	$2\sqrt{2 T_r \varepsilon_f}$	$T_r > \varepsilon_f$
Maxwell Distribution	$\sqrt{T}\, e^{-\beta T}$	$\exp[-\beta(T' + T_r)] \cdot \sinh(2\beta\sqrt{T' T_r})$	$4\left(\dfrac{T_r \ln 2}{\beta}\right)^{1/2}$	$\dfrac{1}{\beta} = kT_k$
1s Atomic Wave Function	$\dfrac{\sqrt{T}}{(1+\beta T)^4}$	$[1 + \beta(\sqrt{T'} - \sqrt{T_r})^2]^{-3}$	$4\left(\dfrac{.26 T_r}{\beta}\right)^{1/2}$	$\dfrac{1}{\beta} = E_i$
2p Atomic Wave Function	$\dfrac{T^{3/2}}{(1+\beta T)^6}$	$\dfrac{1}{4y^4} - \dfrac{1}{5y^5}$ where $y \equiv 1 + \beta(\sqrt{T'} - \sqrt{T_r})^2$	$4\left(\dfrac{.46 T_r}{\beta}\right)^{1/2}$	

*Unnormalized

electron and projectile, respectively, and T_p is the projectile kinetic energy as measured in the laboratory. The end point of the continuum will occur when the electron is at rest in the projectile frame in the final state and corresponds to a photon energy of T_r. Photons of lower energy correspond to final states which lie higher in the continuum.

This approach which ignores the momentum distribution in the target is justified since the expected continuum is broad compared with $P(T')$, the transformed kinetic energy distribution of weakly bound outer electrons. Calculations involving deeply bound electrons, however, must include the target momentum distribution.

If the model presented here is correct, the experimental yields can be predicted by applying the conventional formalism for the production of bremsstrahlung. Some care must be used in choosing the appropriate frame of reference. According to Bethe and Salpeter [7], the differential cross section for bremsstrahlung production per unit photon energy in the energy range of interest is

$$\frac{d\sigma}{d(h\nu)} = \frac{\sigma_o \alpha Z^2 mc^2}{\pi \, T_r h\nu} \ln[(\sqrt{T_r} + \sqrt{T_r - h_\nu})/(\sqrt{T_r} - \sqrt{T_r - h\nu})], \quad (22)$$

where σ_o is the Thompson cross section (.665 barn), α is the fine structure constant (1/137), Z is the projectile charge and $h\nu$ is the photon energy.

In some cases it is more convenient to compare the experimental and theoretical values for the integrated yield of x rays above some particular photon energy $h\nu_1$. This procedure has the advantage of removing the need for the small systematic corrections to the data for Doppler shift and instrumental broadening.

The integrated yield is

$$Y(>h\nu_1) = \frac{N_p n_e G}{4\pi} \int_{h\nu_1}^{T_r} \frac{d\sigma}{d(h\nu)} P(h\nu) \epsilon(h\nu) d(h\nu)$$

$$= \frac{N_p n_e G}{4\pi} \Sigma(>h\nu_1), \quad (23)$$

where $\Sigma(>h\nu_1)$ is the integrated cross section.

For sufficiently small solid angles, $P(h\nu)$ represents the energy dependent part of the G factor and is obtained as follows. The angular distribution of the emitted radiation is dipole near the high energy photon cutoff and tends toward a more isotropic distribution at lower energies. Neglecting retardation effects and assuming a linear dependence of the polarization [7] on photon energy for $h\nu > T_r/8$, the differential cross section at 90° in the lab frame is given by

$$\left| \frac{d^3\sigma}{d(h\nu)d\Omega} \right|_{90°} = \frac{d\sigma}{d(h\nu)} \frac{P(h\nu)}{4\pi} \,, \tag{24}$$

where $P(h\nu) \simeq 21\, T_r/(22T_r - 8\, h\nu)$.

Experiments

Experiments have been performed with projectile beams of protons and lithium, carbon and oxygen ions incident on a variety of gaseous targets. The choice of gas targets was made in order to test the basic assumptions in both the REC and bremsstrahlung models since accurate ground state wave functions are available.

All data were taken at the Tandem Van de Graaff facility at Brookhaven National Laboratory. Protons were accelerated in a tandem MP and the heavy ions in a three-staged MP configuration. For the proton runs, the beam enters and exits an electrically isolated gas cell through 0.0025 mm nickel windows and is detected in a Faraday cup. Special care is taken to eliminate spurious electron currents produced by proton interactions in the window. When heavy ion projectiles are used, a differentially pumped gas cell is necessary since nickel windows are no longer feasible. A thin mylar window allows the x rays to escape the cell at 90° to the beam axis. Gas pressures ranged between 50 and 100 Torr, and in some cases small amounts of impurities such as argon were present.

The detection system is a well calibrated, high resolution, 30 mm^2 Si(Li) crystal with a 0.025 mm beryllium entrance window and standard electronics including pulse pileup rejection. Care was taken to shield the detector from scattered particles.

X rays are viewed at 90° to the direction of the incoming projectile since this procedure minimizes the Doppler shift correction. Also, in some cases, a thin polystyrene film is used to reduce the counting rate at low photon energies. This procedure has only a small effect on the experimental results presented here.

Experimental Results

Bremsstrahlung data are discussed first. Data for protons at 10 MeV are shown in Fig. 1(a). In addition to the bremsstrahlung continuum, the characteristic x-ray line produced by K-shell excitation of the argon impurity in the gas is seen.

Also shown on Fig. 1(a) is a theoretical yield $Y(h\nu)$ based on conventional bremsstrahlung formalism [7] including polarization effects and the energy dependent efficiency of the detector system.

$$Y(h\nu) = \frac{N_p\, n_e\, G}{4\pi} \frac{d\sigma}{d(h\nu)} P(h\nu)\varepsilon(h\nu) \,. \tag{25}$$

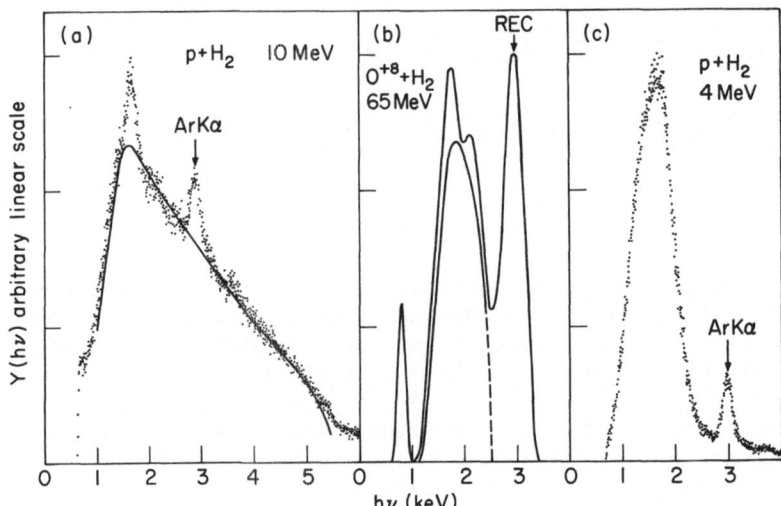

Fig. 1. Experimental x-ray yield curves. In (a) a theoretical
curve which does not include instrumental broadening is normalized
to the data (about a 10% correction) at 3 keV. In (b) the broad
curve between 1 and 2.5 keV is obtained by stripping the lines
from the smoothed oxygen data and is to be compared with the proton data (c) since the particle velocities are approximately the
same. In all cases the cutoff below about 1.5 keV is an artifact
introduced by the lack of transmission of the 0.025 mm beryllium
detector window. In (b) a mylar window in the gas cell introduces
additional cutoff below 1.5 keV. The sharp feature below 1 keV is
attributable to characteristic emission lines which occur after
NREC to high lying orbitals. Other unidentified features in the
spectrum are as yet unexplained.

In addition to having the proper high energy cutoff, $T_r =$
5.45 keV, the shape of the yield function is also in excellent
quantitative agreement with the data. Direct proton and knock-
on-electron-bremsstrahlung contributions are probably present as
part of the underlying background in Fig. 1(a). However, the end
points for these continua are expected to occur at T_p and $4T_r$,
respectively, and the x-ray yields are much reduced.

It is of interest to compare the theoretical and experimental
curves for the integrated yield $\Sigma(>1$ keV$)$ as calculated from Eq. 23
compared with the experimental yield for protons up to 14 MeV. No
adjustable parameters have been used, and the agreement is excellent (see Fig. 2).

Bremsstrahlung is also expected for the case of a fast, highly
stripped, heavy ion projectile. Fig. 1(b) shows the experimental

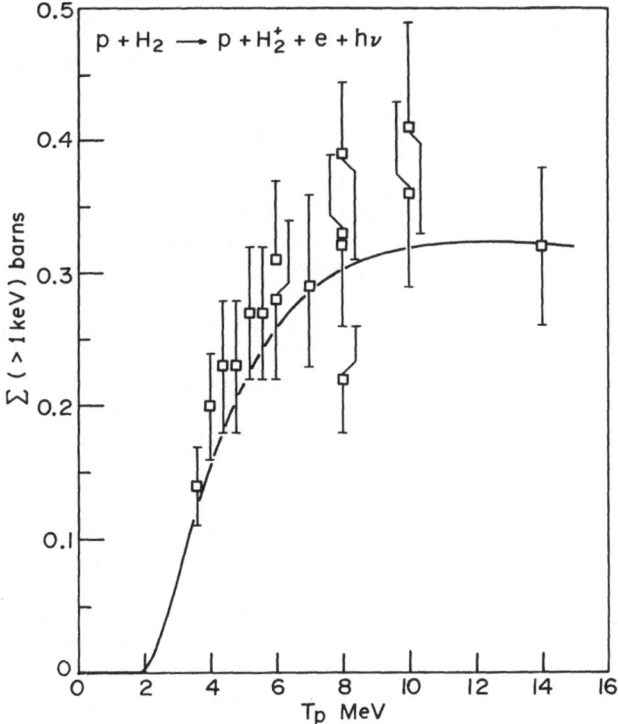

Fig. 2. A comparison of the theoretical (solid curve) and experimental (open squares) values for the integrated cross section for bremsstrahlung above 1 keV. No adjustable parameters are used. The ±20% error bars represent estimates of systematic errors only.

results for the case of 65 MeV O^{+8} incident on hydrogen. In addition to the bremsstrahlung continuum, there is a fairly sharp feature labelled REC. Our model predicts that the REC peak should occur at an energy approximately equal to T_r plus the binding energy of the shell into which the electron is captured. The K-shell binding energies for O^{+7} and O^{+8} are 740 eV and 870 eV, respectively, and, from an extenison of the results of Macdonald and Martin [8], it is expected that the equilibrium distribution of projectile charge states in this experiment is predominately O^{+7}. The measured REC peak occurs about 740 eV above the 2.2 keV end point of the bremsstrahlung continuum in agreement with the requirements of the model.

For a fixed projectile velocity the cross section for production of a continuum photon of energy $h\nu$ should increase as Z^2, where Z is the charge of the incoming projectile. Apart from being scaled by a factor of Z^2, the shape of the continuum spectrum should be independent of the choice of projectile. That this is in fact the case is seen in Figs. 1(b) and (c).

Experimental data obtained from collisions of beams of lithium, carbon and oxygen with various gaseous targets are shown in Figs. 3-5. A continuous line has been drawn through the smoothed data points. Except for lithium, the statistical error is too small to be shown. If the REC data are examined in some detail, it is clear that there exists a good quantitative agreement with the major predictions of the model outlined above.

Lithium (Fig. 3): Only one small set of data was obtained. The results are dominated by background and only a small bremsstrahlung spectrum is seen. No REC peak is observed. The end point for the bremsstrahlung spectrum is expected at about 3.1 keV in agreement with the data.

Carbon (Fig. 3): The results are somewhat richer in detail. A resolved REC peak is observed for 45 MeV projectiles, but the data are still dominated by bremsstrahlung. Continuum cutoffs are expected at about 1.5 and 2.0 keV for the 34 and 45 MeV data, respectively. Binding energies for the C^{+6} and C^{+5} ions are about 490 and 395 eV, respectively.

Oxygen (Figs. 4, 5): A very good set of data have been obtained, and they will be discussed in detail. The peak of the REC profile occurs at $h\nu = T_r + E_f$, and the data can be compared with this prediction using E_f = 870 and 742 eV for O^{+8} and O^{+7}, respectively. The comparison for the 65 MeV cases is shown in Fig. 5 as the closely spaced pair of lines drawn through the REC peak, and labelled +7 and +8.

In all cases, a sharp feature is present at about 815 eV labelled O. The detection system is almost cut off at this energy; the combination of the 0.0025 cm beryllium detector window and the 200μM gas cell window leads to a transmission of only about 1/200 of the incident flux. We identify this feature with a radiative transition to the ground state of an outer shell electron with principal quantum n ≥ 4 in the O VIII system. The excited state is obtained either through electron shake-up from the ground state of an incident O^{+7} ion or an NREC on a bare oxygen nucleus where the electron goes directly or by a cascade to the appropriate outer shell. No significant contribution is seen at about 775 eV since the transmission for the transition from the n=3 shell is even further reduced. Poor transmission also rules out a large contribution from transitions in the O VII system since the series limit occurs at about 740 eV. The line labelled C corresponds to the expected high energy cut-off of the bremsstrahlung continuum (excluding the effects of instrumental broadening and initial state momentum distribution).

It seems reasonable to expect REC features for processes in which an electron is captured in the n=2 shell of an ion which is initially in the ground state of O^{+8}, O^{+7} or O^{+6}. These ions have L-shell binding energies of about 220, 175, and 135 eV, respectively. Contributions from REC to the L shells of O^{+6}, O^{+7}, and O^{+8} will appear as unresolved structures around the lines labelled +6 and +8 just to the right of C in Fig. 5. REC to higher lying

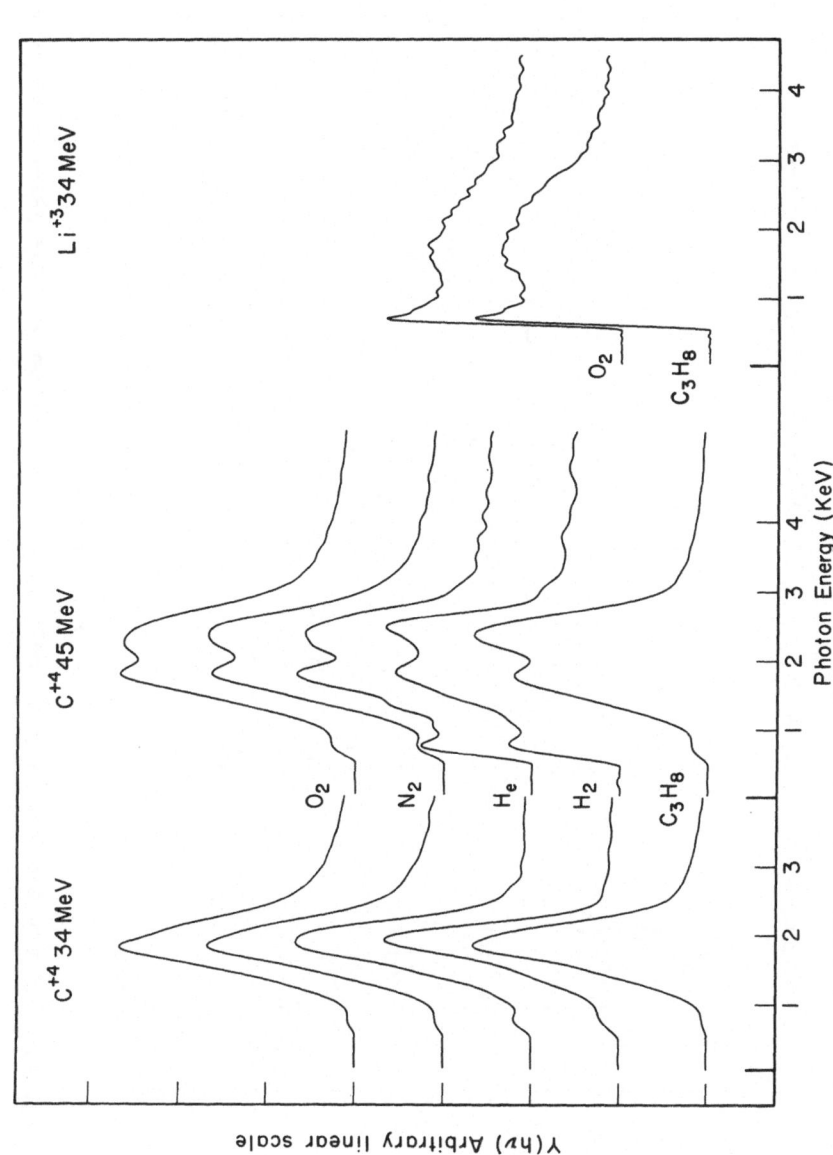

Fig. 3. X-ray yields for lithium and carbon projectiles incident on various gaseous targets.

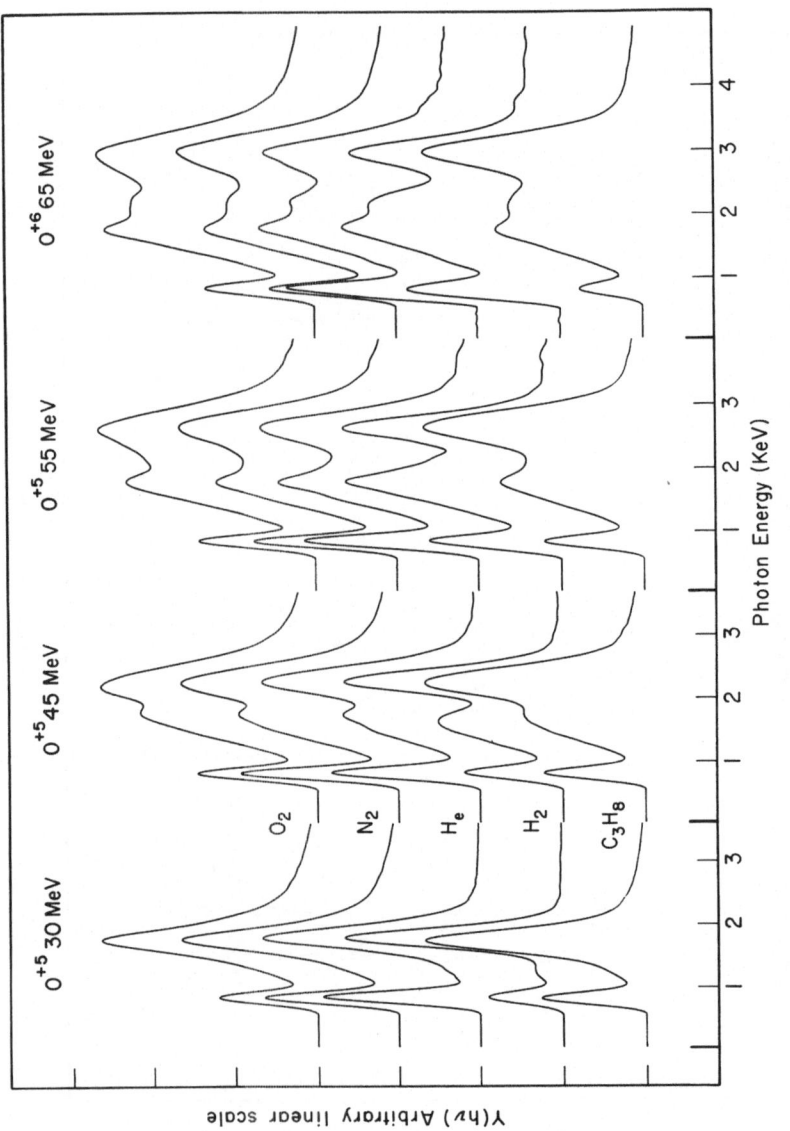

Fig. 4. X-ray yield for oxygen projectiles incident on various gaseous targets.

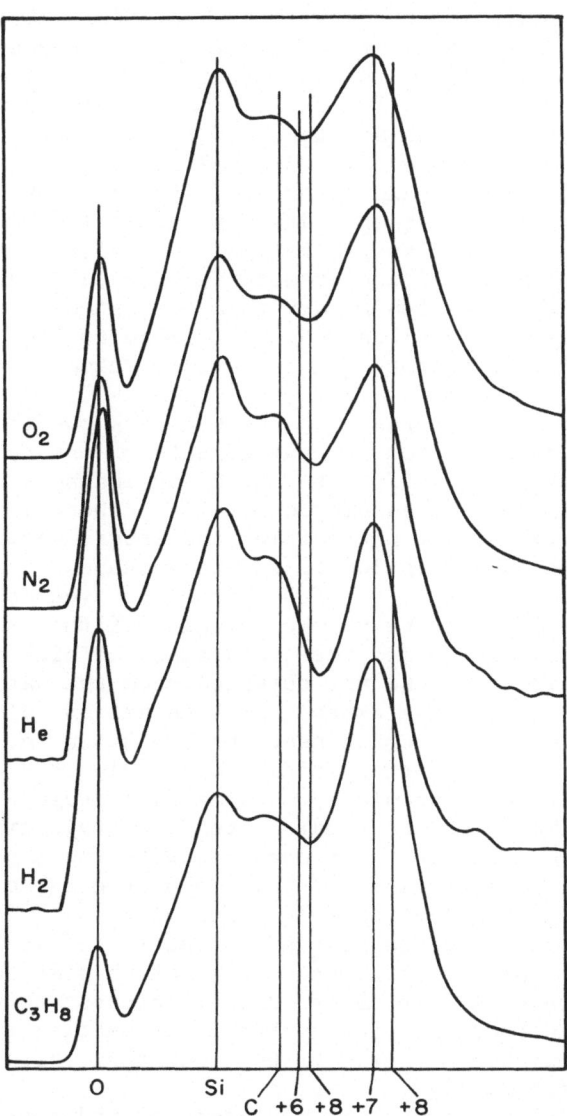

Fig. 5. Data for O^{+6} at 65 MeV showing the expected positions for K- and L-shell REC, the bremsstrahlung cutoff, and characteristic radiation from the projectile and the detector.

orbitals will occur at lower energies than for the L shell with the
series limit coming at C. The whole REC/bremsstrahlung structure
has the appearance of a kind of Raman/Compton spectrum.

Capture of core electrons with high binding energies in the
target will not lead to a sharp REC peak. When the energy dependence of the cross section is taken into account, core electron
capture may be responsible for the low energy broadening seen in
the REC peaks (i.e., at 65 MeV the FWHM for capture from the oxygen
2p shell is 470 eV compared with 2.2 keV from the 1s shell).

All spectra contain additional details, some of which can be
explained. Part of the sharp feature at about 1.6 to 1.8 keV is an
artifact created by the sharply falling transmission of the detector
window and the steeply rising bremsstrahlung yield (except for
oxygen at 30 MeV). The other part is a persistent narrow line of
unexplained origin. It falls, however, close to the silicon Kα
energy. Other sharp features in the spectra have not been identified.

Fig. 6 shows a comparison of the REC peak positions obtained
when the data are corrected for the effects of instrumental broadening with those obtained from the model including a small Doppler
shift correction and the assumption of O^{+7}. There is excellent
agreement for high energy projectiles and progressively poorer
agreement as the energy is lowered. The agreement for the 45 MeV
O_2 point is exceptionally poor for reasons not yet understood.
These results are sensitive to the validity of the impulse approximation which can be applied more accurately at high energies.

Cross Sections: These are obtained from the data with appropriate corrections for the experimental conditions discussed in
Eq. 4. Experimental REC cross sections are shown in Fig. 6 where
they are compared with a theoretical calculation for a free electron
gas. The theoretical calculation assumes a hydrogen-like projectile
and, owing to the number of final states per projectile, should be
divided by a factor of two when compared with the experimental data
for helium-like projectiles. If this done, the agreement with the
hydrogen and helium target data is excellent. Data for propane,
nitrogen and oxygen targets have a distinctly different shape.
This may arise from the difference in electronic structure of the
two groups of targets. It is of further interest to note that the
ratio of the REC cross section to the cross section for the features
labelled O is, within a factor of two, constant for all combinations
of target and projectile.

Widths: The agreement between measured and expected REC line
widths is not as good as for the peak positions and cross sections.
As expected, the measured widths increase approximately as $(T_r)^{1/2}$
but are, in general, a factor of two larger than implied by the
formulae in Table I. This may be due in part to the breakdown of
the impulse approximation and in part to contributions from core
electron capture and smearing from different projectile charge
states.

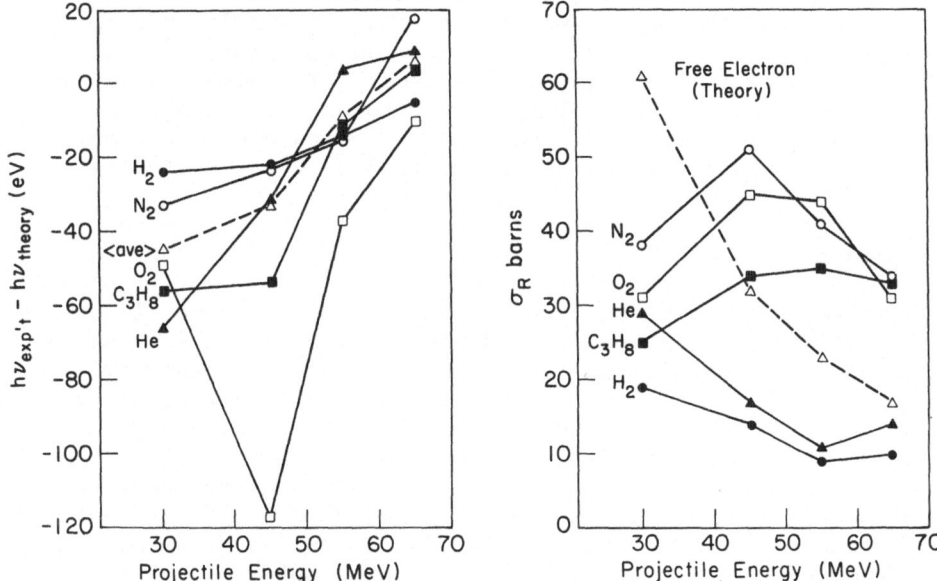

Fig. 6. Left: the deviation of the peak of the REC feature from
the expected position based on O^{+7} capture. Right: REC cross
section versus projectile energy. Experimental data (solid lines)
are compared with a calculation (dashed line) based on free electron theory. Since the experiment is believed to be O^{+7} capture,
the free electron calculation which is based on O^{+8} capture should
be reduced by a factor of two.

It should be emphasized that these width measurement defici-
encies are inherent in the nature of this experiment and should
not be taken as a failure of the model to achieve its objective of
predicting accurately the electron momentum distribution. Contributions to the REC line from specific target electron shells can be
selected by coincidence techniques. Smearing arising from projectile
charge state distribution can be reduced by carefully controlling
the target thickness and by choosing a projectile energy and initial
charge state which minimizes the net probability for charge exchange.
Furthermore, by increasing the resolution of the detector system,
the data will reflect more nearly the true REC spectrum and only
small instrumental corrections will be required.

Conclusion

In this discussion, we have outlined a new method of determining
experimentally the momentum distribution of electrons in atomic,
molecular and solid state systems. Theoretical predictions which

are based on a relatively simple impulse approximation model are
in excellent agreement with most of our data. Where divergences
between experiment and theory occur, they can be understood in
terms of our current experimental techniques. The means to improve
our results are at hand and point the way to our future program.
Ultimately we believe that REC studies will produce electron momentum distributions which rival those obtained by Compton studies.

Acknowledgments

The authors are spokesmen for a large group who have all worked
towards understanding the material discussed in this paper. At
M.I.T. they are K. Kalata and A. R. Sohval; and at Brookhaven,
K. W. Jones and H. E. Wegner. We have also benefited from discus-
sions with P. Morrison, S. Olbert and F. Villars of M.I.T. and A.
A. Dalgarno of Harvard University.

This work was supported in part by the National Aeronautics
and Space Administration under Grant No. NGL 22-009-015, the National
Science Foundation under Grant No. GP-31378, and the U. S. Atomic
Energy Commission under Contract No. AT (11-1) 3069.

The authors also hold guest appointments at Brookhaven National
Laboratory.

References

[1] H. W. Schnopper, H. D. Betz, J. P. Delvaille, K. Kalata, A. R.
    Sohval, K. W. Jones and H. E. Wegner, Phys. Rev. Lett. $\underline{29}$, 898
    (1972).
[2] H. W. Schnopper, J. P. Delvaille, K. Kalata, A. R. Sohval,
    M. Abdulwahab, K. W. Jones and H. E. Wegner, Phys. Lett. $\underline{47A}$,
    61 (1974).
[3] P. Eisenberger and P. M. Platzman, Phys. Rev. $\underline{A2}$, 415 (1970).
[4] V. S. Nikolaev, Soviet Physics Uspekii $\underline{8}$, 269 (1965); Soviet
    Physics JETP $\underline{24}$, 847 (1967).
[5] H. A. Bethe and E. E. Salpeter, Quantum Mechanics of One- and
    Two-Electron Atoms (Academic, New York, 1957), pp. 320-322.
[6] E. A. Silverstein, Nuc. Inst. Meth. $\underline{4}$, 53 (1959).
[7] H. A. Bethe and E. E. Salpeter, $\underline{op}$. $\underline{cit}$. pp. 323-324.
[8] J. R. Macdonald and F. W. Martin, Phys. Rev. $\underline{A4}$, 1965 (1971).

RADIATIVE ELECTRON CAPTURE BY CHANNELED OXYGEN IONS*

B. R. APPLETON, T. S. NOGGLE
Solid State Division, Oak Ridge National Laboratory
Oak Ridge, Tennessee 37830, U. S. A.

C. D. MOAK, J. A. BIGGERSTAFF
Physics Division, Oak Ridge National Laboratory
Oak Ridge, Tennessee 37830, U. S. A.

S. DATZ, H. F. KRAUSE
Chemistry Division, Oak Ridge National Laboratory
Oak Ridge, Tennessee 37830, U. S. A.

and

M. D. BROWN
Kansas State University
Physics Department
Manhattan, Kansas 66502

Introduction

Recently Schnopper et al. [1] reported a new feature in the x-ray spectra observed when solid targets were bombarded with energetic, highly-stripped heavy ions. They observed a broad x-ray band well above the highest energy characteristic x-ray line from the projectile. This new feature was identified as radiative electron capture (REC). The phenomenon of REC becomes most probable when the incident heavy ions are completely stripped, and no outer shell electrons are available to fill K-shell vacancies. Then a free or weakly bound target electron can be captured directly into the K-shell of the moving ion, emitting a photon. The energy of a REC photon resulting from the capture of a free electron by an ion of charge Z, mass M, and energy E, is given by the expression:

$$E_p = Z^2 \, Ry \; + \; \frac{M_e}{M} \; E \tag{1}$$

where Ry is the Rydberg energy and M_e the electron mass. The first
term is the binding energy of an electron in the hydrogen-like ion
and the second the energy of an electron moving at a velocity equivalent to that of the ion. As noted by Schnopper [1],the actual width
of the observed REC line is affected by a number of things. In
measurements utilizing heavy ion beams incident on amorphous solids,
the heavy ions will have a distribution of charge states inside the
solids and each charge state can result in a REC photon of a different
energy as can be seen from eq. (1). In most cases the differences
in energy are the order of the resolution of the detecting systems,
thus the line can be shifted in energy as well as broadened. The
intrinsic width of a REC line arising from a single charge state
should reflect the velocity distribution of the target electrons
captured by the moving ion. Even the small energies of the free
electrons of the target (~5 ev) contribute significantly to the REC
line width when transformed to the rest frame of the moving projectile.
It was this latter exciting possibility of utilizing the REC process
to measure electron density distributions in solids that prompted us
to initiate the experiments to be discussed here. As we shall see,
the unique constraints imposed on well-channeled heavy ions makes it
possible to study the REC process under well defined conditions
where it should be possible to extract these widths in an unambi-
guous manner.

Experimental

     We have investigated the REC phenomenon for 17.8 to 40.0
MeV ^{16}O ions channeled through thin Ag single crystals along the
[110] axial direction. The path length through the crystal in
this orientation was 0.80 um. The experiemental arrangement is the
same as that described in a companion paper of these Proceedings [2]
with the exception that a Si(Li) x-ray detector with a 0.5 mil Be
window was located at 90° to the Ag single crystal. With the incident beam parallel to the [110] axis the crystal was tilted at 45°
to both the incident beam and the Si(Li) x-ray detector. The x-ray
detector viewed the side of the foil from which the ions emerged,
had a solid acceptance angle of 1.6×10^{-2} steradians, and an
energy resolution of 200 eV FWHM. A beam monitor was used for
relative normalization of the various measurements but no absolute
yield measurements were made.

Measurements

     When the Ag single crystal was oriented with the incident
oxygen ions parallel to the [110] axis, the yield of characteristic
x rays from the Ag target and O ions was reduced to 4% of that

observed when the beam was incident on the crystal in a random orientation. This can be understood on the basis of the channeling
phenomenon whereby 96% of the ions are channeled and never approach
closer than 0.2Å to Ag atoms and thus do not excite characteristic
x-rays. The remaining 4% of the incident ions are deflected through
too large an angle to be channeled on entering the crystal (so-called
minimum yield fraction X m) or escape the channel in traversing the
crystal. These ions encounter the Ag atoms at random and have a
normal probability for exciting characteristic x rays.

One important consequence of the channeling phenomenon is that
oxygen ions channeled in Ag (like oxygen ions channeled in Au [3]) have
"frozen" charge states. This can be deduced from charge state fractions
like those shown in Fig. 1 for 27.8 MeV ^{16}O ions channeled through
a Ag single crystal along the [110] axis. Unlike the equilibrium
charge distribution observed when the beam was incident in a random
direction in the crystal (dashed curve) the charge distributions
for channeled ions of various incident charge states are far from
equilibrium. Actual measurements discussed in a separate paper [2]
show that both capture and loss cross sections in the channel are
~10^{-19} cm^2 and a significant fraction of ^{16}O ions incident with
charges 6+, 7+ and 8+ essentially traverse the crystal without changing charge.

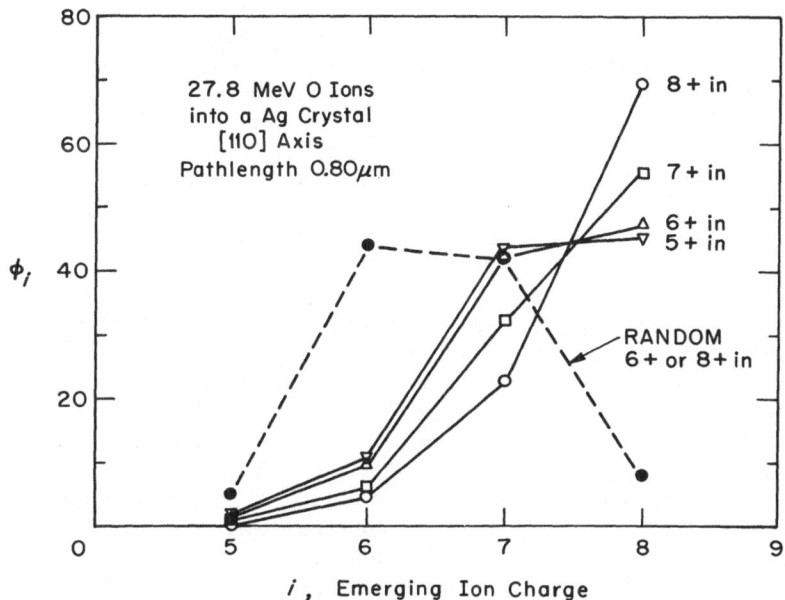

Fig. 1. Measured emerging charge state distribution for 27.8 MeV
ions transmitted parallel to the [110] axial direction of a 0.8$_{\mu m}$
thick Ag single crystal for various input charge states (solid
lines) and in a random orientation of the crystal (dashed lines).

A set of x-ray spectra typical of those obtained in this experiment is shown in Fig. 2. The open circles show the x-ray distribution
observed when 27.8 MeV ^{16}O ions were incident in a random direction
of the Ag single crystal. This random spectrum was the same regard-
less of the input charge state of the ions. The other spectrum
plotted in Fig. 2 (solid circles) was obtained for 27.8 MeV O^{8+}
ions incident parallel to the [110] direction of a 0.80 um thick Ag
single crystal. Both spectra have been corrected for the Be window
absorption. For comparison purposes the random spectrum has been
multiplied by the minimum yield fraction X_m = 0.04. The arrows
on the figure above about 2.5 keV identify the expected energies
for the characteristic x rays from the Ag target atoms. The broad
x-ray band peaked at ~1.7 keV, clearly visible in the channeling
spectrum is attributed to REC photons from the ^{16}O ions. The
arrow marked E_p is the expected energy for a REC photon-arising for
capture of a free electron by a 27.8 MeV O^{8+} ion. Since these
REC photons arise from capture by the ^{16}O ions of weakly bound Ag
electrons they should be present in the random spectrum as well
as the channeling case. The reason for the prominence of the REC
band in the channeling spectrum is that the "background" due to
characteristic Ag x rays excited by the oxygen ions is reduced
to only 0.04 times what it is in the random situation, while the
REC process itself is undiminished since the channeled O^{8+} ions
interact freely with the conduction electrons in the Ag crystal.
Additionally, as we have already discussed, the majority of the ions
contributing to this REC band are in a single charge state, 8^+. If
we assume that the normalized random spectrum represents the background distribution beneath the REC peak we can substract the two
spectra as drawn in Fig. 2 and obtain a width and relative yield
for the REC band. Table 1 gives the measured energies F_p and widths
determined in this manner at three separate incident energies all
for O^{8+} ions incident to the [110] axial direction of Ag. The
widths have the resolution of the detecting system subtracted out.
Also given are the calculated values of E_p using the formula from
Bethe and Salpeter[4] for capture of a free electron by O^{8+} ions
E_p (8+). In calculating these values the mean energy of the ions
traversing the Ag single crystal was used. In general, the ions
lose less than 10% of their incident energy. Cross sections for REC
were obtained at each energy by comparing the area within the subtracted REC band to that of the characteristic Ag L x rays in the
same spectrum. The absolute cross sections for the Ag L x rays
were obtained from the work of Nettles et al. [5] who measured the min-
yields for 12 - 50 MeV ^{16}O ions in Ag. Having measured the minimum yield fraction X_m with the aid of the beam monitor system we
were thus able to obtain absolute cross-sections for the REC process.
The cross sections extracted in this manner, σ_{REC}, are shown plotted
in Fig. 3 versus E the energy of the transmitted ^{16}O ions. Also
shown are the theoretical cross-sections as calculated from Bethe and
Salpeter [4], σ_{BS}, and from the recent theory by Ritchie and Neelavathi
of ORNL. Both theoretical calculations assume one electron per atom.

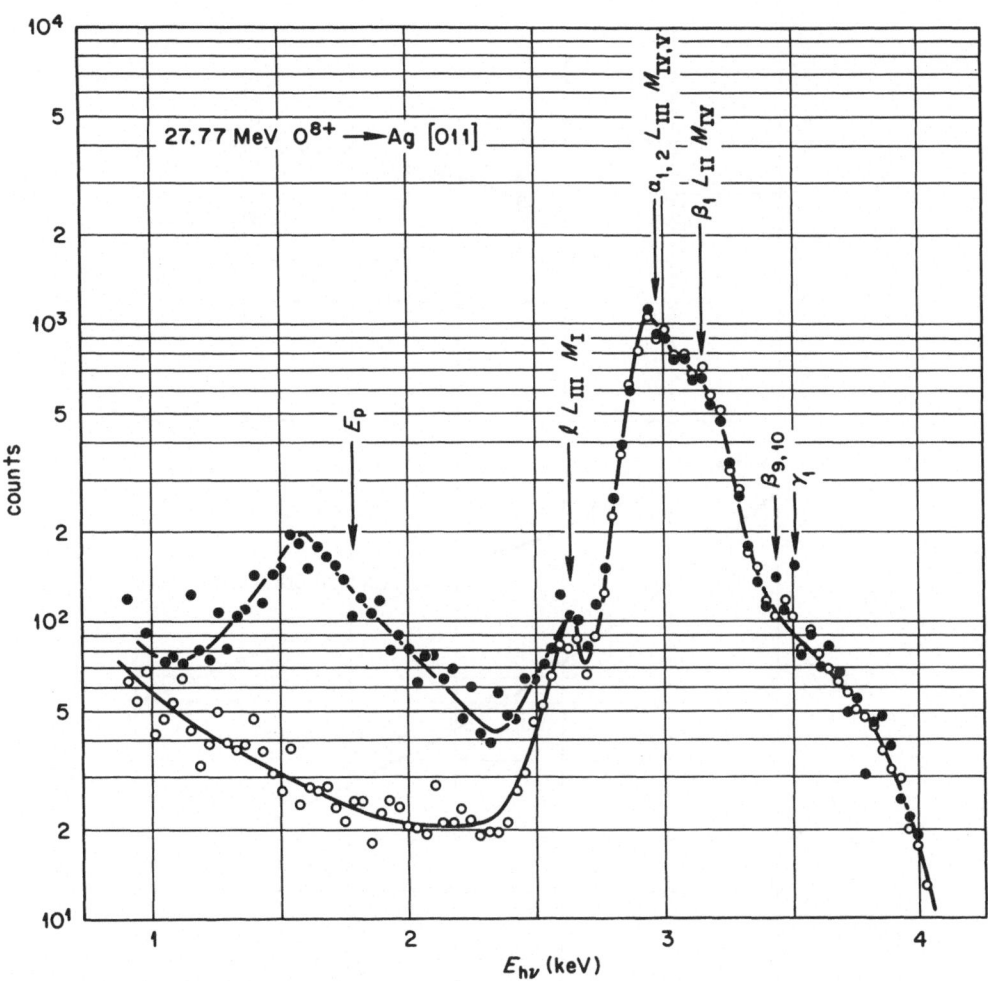

Fig. 2. Normalized x-ray spectra for 27.8 MeV O^{8+} ions incident
parallel to the [110] axis of a Ag single crystal (solid circles)
and in a random direction (open circles).

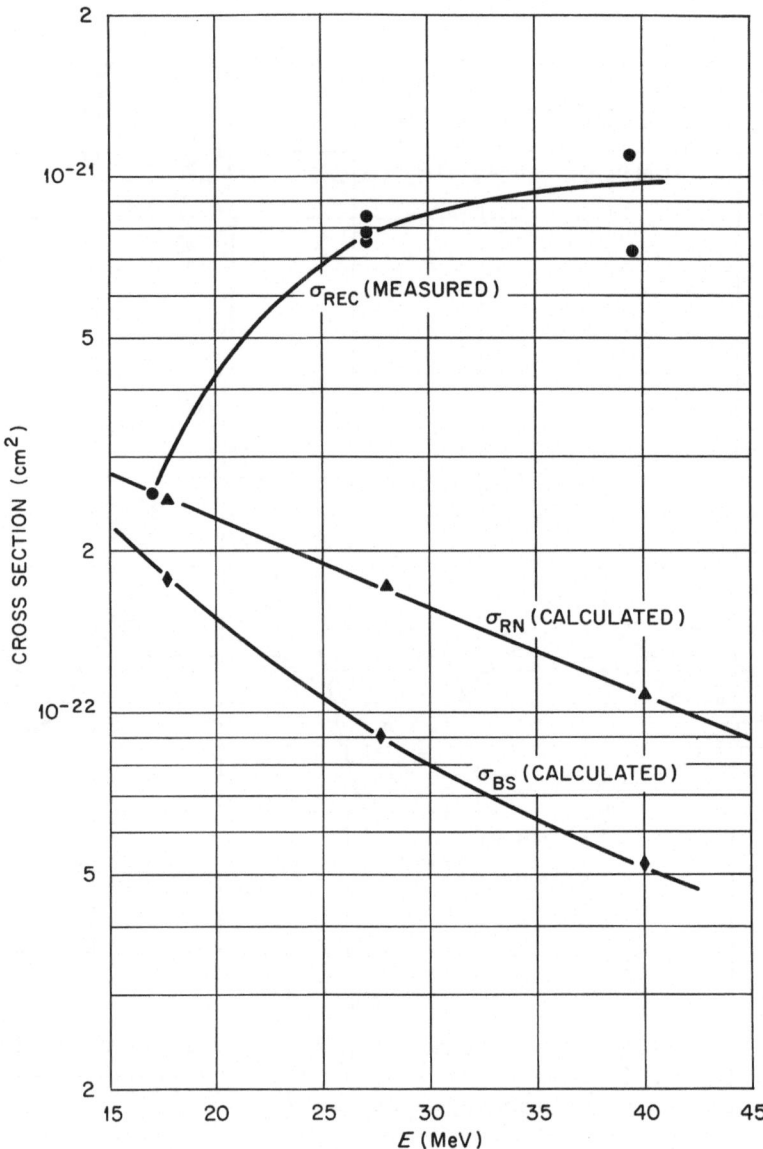

Fig. 3. Measured and calculated cross-sections for REC process as a function of energy.

TABLE 1

E_1 (MeV)	E_p (keV)	MEASURED (FWHM) (eV)	CALCULATED E_p (8+)
17.8	1.30	600	1.449
27.8	1.61	445	1.79
40.0	2.00	1020	2.232

Conclusion

The preliminary measurements discussed here do not yield
definitive results but rather are intended to demonstrate the use
of the channeling technique as a tool for studying the REC process
in the solids under unique well-defined conditions. A complete
paper will soon be published, which includes more detailed measurements and analysis as well as the theoretical calculations of
Ritchie and Neelavathi. Nevertheless, several qualitative conclusions can be made regarding the present results.

1. The energies of the measured REC lines in Table 1 occur at
    consistently lower energies than the calculated values. (The
    reason for this was not understood at the time of the conference
    but subsequently this question has been resolved). †

2. We observed that the most probable energy of the REC bands was
    unchanged regardless of whether 0^{6+} or 0^{8+} ions at a given en-
    ergy were channeled along the [110] axis. This is consistent
    with the observed charge-changing behavior of these ions.[2]
    Some small changes in the widths of the lines are suggested by
    our preliminary measurements.

3. There appears to be structure in some of the measured REC lines
    particularly at 17.8 MeV. This may result because lower charge
    states are populated more in a channeled 0^{8+} beam at lower en-
    ergies. At higher energies the channeled ions may be capturing
    target electrons which are more tightly bound than conduction
    electrons.

4. The shape and magnitude of the measured REC cross-section in
    Fig. 3, deviate from the calculated ones. The trend of the
    measured cross-section curve may be due to some of the effects
    already mentioned, namely, more charge exchange at lower ener-
    gies and the likelihood that REC occurs for other than just
    conduction electrons. These two effects should be separable
    if one makes use of the strong charge state dependence of the
    energy loss for these channeled ions [2]. Slight differences at
    a particular incident energy are observed in REC for different
    input charge states of the ions but these differences are con-
    sistent with the small charge exchange expected.

These preliminary results show that the REC process for well-
channeled heavy ions has several distinct advantages for studying
this effect in solids. Since channeled ions have frozen states
one can study the process without the confusion of the distribution in charge states normally experienced in amorphous or polycrystalline solids. This allows one to extract cross-sections
and line widths as a function of a single charge. Such measurements also provide a stringent test of the theoretical calculations.

The ultimate usefulness of this technique as a tool for measuring the velocity distributions of electrons in single crystal channels depends on a more thorough understanding of the theory of the REC process in the environment of a channel and on more detailed measurements.

References

*Research sponsored by the U. S. Atomic Energy Commission under contract with Union Carbide Corporation.
†We have found that the pulse height to energy conversion formula used in calculating the measured energies in Table 1 and in generating the energy scale in Fig. 2 was in error. More nearly correct x-ray energies may be obtained by applying the recipe:

$$E_{new} = 0.963 \ E_{old} + 0.224 \ keV$$

to values presented in these two places. Also note that this error resulted in a small error in the window attenuation correction applied to the data of Fig. 2. The corrected data are now in good agreement with the calculated values given in Table 1.

[1] H. W. Schnopper, Hans D. Betz, J. P. Delvaille, K. Kalata, A. R. Sohval, K. W. Jones, and H. E. Wegner, Phys. Rev. Lett. 29, 898 (1972).

[2] S. Datz, B. R. Appleton, J. A. Biggerstaff, M. D. Brown, H. R. Fraus, C. D. Moak, and T. S. Noggle, Proceedings of the Fifth International Conference on Atomic Collisions in Solids.

[3] S. Datz, F. W. Martin, C. D. Moak, B. R. Appleton, and L. B. Bridwell, "Radiation Effects", 12, 163 (1972).

[4] H. A. Bethe, and E. E. Salpeter, "Quantum Mechanics of One- and Two-Electron Atoms" (Academic Press, New York, 1957) p. 320-322.

[5] P. H. Nettles, G. A. Bissinger, S. M. Shafroth and A. W. Waltner, Proceedings of the International Conference on Inner Shell Ionization Phenomena, Atlanta, GA (North-Holland--to be published).

THE INFLUENCE OF CHANNELING ON THE SHAPE AND INTENSITY OF DOPPLER-BROADENED SPECTRAL LINES PRODUCED BY LIGHT ION BOMBARDMENT OF METALS

C. KERKDIJK, C. SMITS, D. R. OLANDER[*] and F. W. SARIS

FOM-Institute of Atomic and Molecular Physics
Kruislaan 407, Amsterdam, The Netherlands

ABSTRACT

An interesting consequence of the bombardment of solids by ions or atoms with keV energy is the visible light emitted by the backscattered particles. The observed spectral lines are broadened in a manner consistent with Doppler shifting of the radiation, which arises because the backscattered atoms possess a distribution of energies and directions with respect to the detector. Here we report on calculations of the shape of the Doppler-shifted Balmer β-line and comparison with experimental results for proton bombardment of a Cu target in the energy region from 5 - 100 keV.

In this calculation three independent parameters must be considered:
- the energy distribution of backscattered projectiles,
- the energy dependence on the neutralization probability in an excited state upon emerging from the Cu surface,
- the probability for non-radiative decay via an interaction with the metal surface.

It will be shown how the use of the channeling phenomenon in monocrystalline Cu helps verifying the relative importance of these three parameters on the actual shape and intensity of the spectral lines.

---

[*]Permanent address: University of California, Berkeley, California, U.S.A.

OPTICAL LINE AND BROAD-BAND EMISSION FROM ION-BOMBARDED TARGETS*

W. F. VAN DER WEG** and E. LUGUJJO***
California Institute of Technology
Pasadena, California 91109

ABSTRACT

A variety of solid targets have been bombarded with 40 keV Ar^+ ions. The resulting optical emission was studied with a grating monochromator. The observed spectra consist mainly of lines from sputtered target particles. For several elements also a continuum was present in the spectrum. When metals with filled or nearly filled d-shells were used as a target, the radiation intensity showed a transient behavior as a function of the time elapsed since the beam is switched on. A high initial intensity in these cases will be related to emission from a partly oxidized surface, since it has been observed before that an oxidized surface emits more light than a clean one, probably due to the absence of radiationless de-excitations near oxide surfaces.

Introduction

The interaction of fast ion beams with solid surfaces generally results in the emission of electrons, atoms and ions in many charge states. Also the emission of electromagnetic radiation is commonly observed. Various types of radiation may be observed during ionic bombardment of solids, examples of which are luminescence phenomena, caused by excited electrons in the target material and emission from excited sputtered and scattered particles. A review of the different processes leading to the emission of radiation during ion bombardment has been given by Parilis [1]. Recently, the emission of x rays during beam-solid interactions has been given great attention [2].

Relatively few experiments have been performed in the visible and near U.V. region of the spectrum [3-6]. Some general observations, however, can be made from the available material. The observed spectrum in the 2000 to 7000Å wavelength range consists of a series of sharp lines, in some cases also a broad continuum of radiation is observed [7]. The lines can easily be identified as resulting from excited atoms and ions of both beam [8] and target [5,6] species. The fact that these lines are narrow and have the same wavelengths as found in emission from gas discharges, shows that this radiation is emitted by atoms which are essentially free, i.e., outside the target. If the radiation were emitted in or very close to the surface of the solid, the lines would be displaced and widened, because of the considerable broadening and shift of the involved energy levels, which takes place under the influence of the potential between particle and solid [9].

The observation that the wavelength of a line is not influenced by the presence of the solid makes it possible to use the line spectrum for the elemental analysis of solid surfaces [4]. The intensity, however, of the emission can be drastically modified by the presence of a solid surface. It has been shown [10,11] that near a surface radiationless de-excitation processes take place, which compete with photon emission and reduce markedly the number of excited particles. These processes are essentially resonance type processes of electrons which tunnel from the excited atoms into the solid [9]. Since the probability for these tunneling processes depends strongly on the electron state density in the solid, it is expected that the state of the surface (chemical composition, degree of oxidation) influences the number of excited particles emittted from the surface. It has been observed [6,10] that oxidized metal surfaces produce a much higher line intensity in the spectrum than clean ones. This can be explained by the inhibition of radiationless de-excitation by the low electron state density or by the presence of large band gaps in the oxide.

The present study was undertaken in order to elucidate the influence of the state of the surface on the intensity of photon emission. This was done by recording intensity changes in the line spectra, which were observed after starting the ion bombardment of various metal samples. A number of metals, semiconductors and insulators were bombarded with 40 keV Ar$^+$ ions under different vacuum conditions. A decrease in optical line intensity during the initial period of bombardment is observed for some metals; this effect is attributed to cleaning of the sample by the sputtering process. The cleaning process appears to be strongly influenced by the number of d-shell electrons in the metals. Analogous changes have been observed in the number of sputtered ions, directly after the starting of the bombardment. Especially since these ions are used for secondary ion mass spectroscopy studies of solid surfaces, it is of importance to characterize these transient phenomena.

In the course of this work also information about continuum radiation in the spectra of several metals was obtained. This will be briefly mentioned and more extensively published elsewhere.

Experimental

Polycrystalline samples of the following materials: Al, Ti, Cr, Fe, Ni, Cu, Zr, Mo, Pd, Ag, Ta, W, Pt, Au and SiO_2, and single crystalline samples of Si, Ge and Al_2O_3 were used as targets. Cleaning of the samples was performed by etching and rinsing in dionized water. In some cases additional cleaning in the vacuum system of the accelerator was performed by low energy ion bombardment. This was done with a 2 keV Ar^+ ion gun, at a current density of approximately 50 $\mu A/cm^2$ over periods between 10 min and 1 hr.

The samples were bombarded at an angle of 45° in the 100 keV accelerator of the California Institute of Technology. In all cases Ar^+ ions of 40 keV were used. Beam currents were of the order of 1 μA on an area of 0.2 cm^2. Adequate secondary electron suppression was provided. The vacuum in the target box was maintained by a turbomolecular pump, the final pressure in the system was 6 x 10^{-7} torr. A liquid nitrogen cold finger was mounted near the target position.

An image of the beam spot area on the target was focused by a quartz lens on the entrance slit of a monochromator which was situated perpendicular to the ion beam. A Jarrell-Ash 50 cm focal length grating monochromator of the Ebert type was used. A cooled photo-multiplier on the output slit recorded the photons. The output pulses were, after amplification, fed into a scaler and a ratemeter , which was connected to a recorder.

Results

Spectra

The general features of the spectra obtained bombarding a solid target with 40 keV Ar^+ ions can be seen in Fig. 1. As an example we show the spectra for Si, Pd and W. The lines in these spectra which have no special designation are all atomic lines of the target material. In the silicon spectra we also see emission from sputtered singly charged (marked II) and doubly charged (III) ions. Furthermore, contamination of the surface with hydrocarbons shows up as emission from a CH molecule around 4314Å. In the Pd case, also strong aluminum lines are present, which are caused by a part of the beam striking the (Al) target holder during that run. In the W case a notable continuum is present in the spectrum. We found no strong Ar lines in these spectra.

Transient behavior of line intensities

It was noticed that the intensity of many lines showed a strong variation after the beginning of the ion bombardment. Therefore we studied this phenomenon more systematically in the following way. The line intensity was recorded as a function of time t after

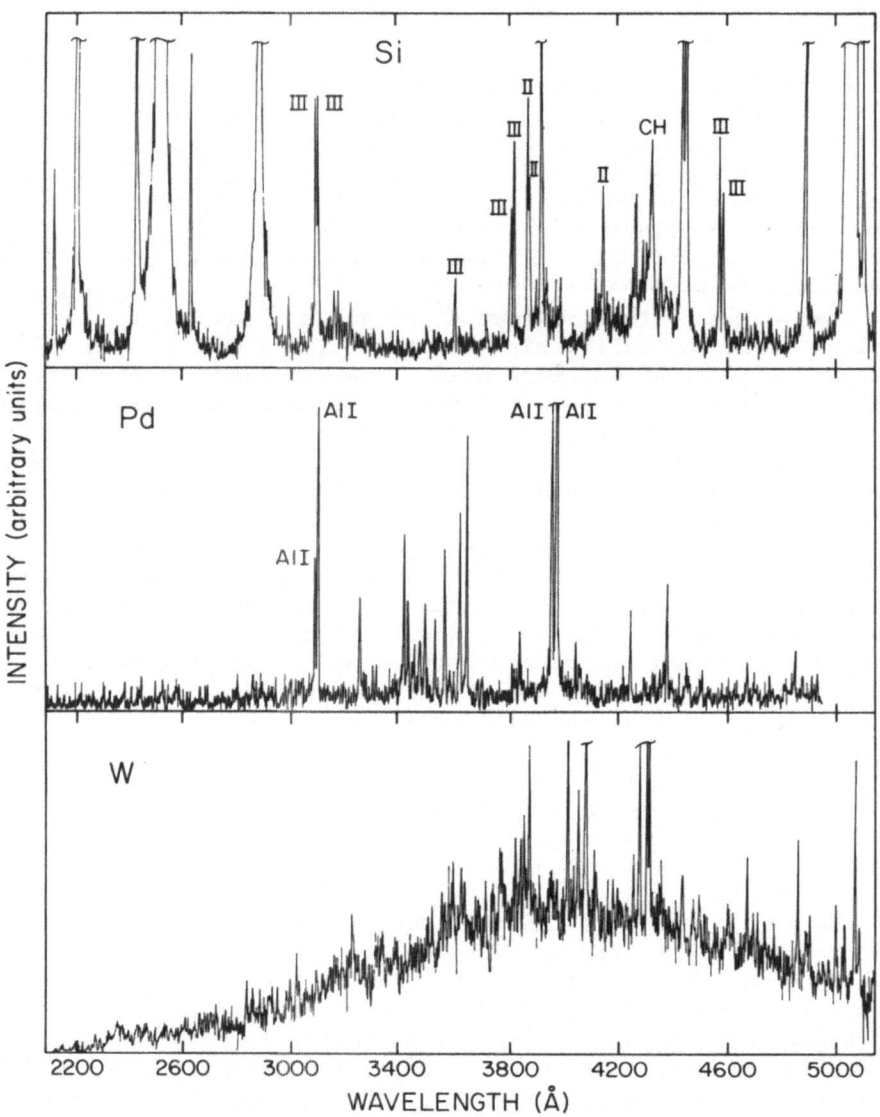

Fig. 1. Spectra obtained for 40 keV Ar⁺ bombardment of Si, Pd and
W targets. The unmarked lines are atomic lines from the target
material. In the Si spectrum also lines from singly charged
(marked II) and doubly charged (III) ions are present.

switching on the beam. After a time which was necessary to reach
an equilibrium intensity, the beam was interrupted for a time τ,
after which the same procedure was repeated. An example is given
in Fig. 2, where the intensity of the Cu I, λ = 3247Å line from
the Cu spectrum is shown as a function of time following different

periods τ without a beam. We denote the intensity of radiation for such a line, t seconds after the beam was put on the target, when it had been interrupted previously for τ seconds, by R(t,τ). We made the following observations:
  a) The equilibrium value R(∞,τ) does not depend on τ.
  b) The intial value R(o,τ) can be up to 6 times larger than the equilibrium value R(∞,τ).
  c) The halfwidth of the line intensity above the equilibrium value, $t_{1/2}$, is larger for smaller beam currents. The value of $t_{1/2}$ for 1 μA beam is 3 seconds.
  d) R(o,τ) initially increases with τ, reaches a saturation value for τ of the order of 1 minute.
  e) The equilibrium intensity R(∞,τ) is strongly affected by the pressure in the target chamber, it decreases when the pressure is lower.
These observations apply to a case like Cu, where a strong transient effect is observed. There are, however, metals which do not show a transient effect at all. As an example, a line in the W spectrum is indicated in Fig. 2.

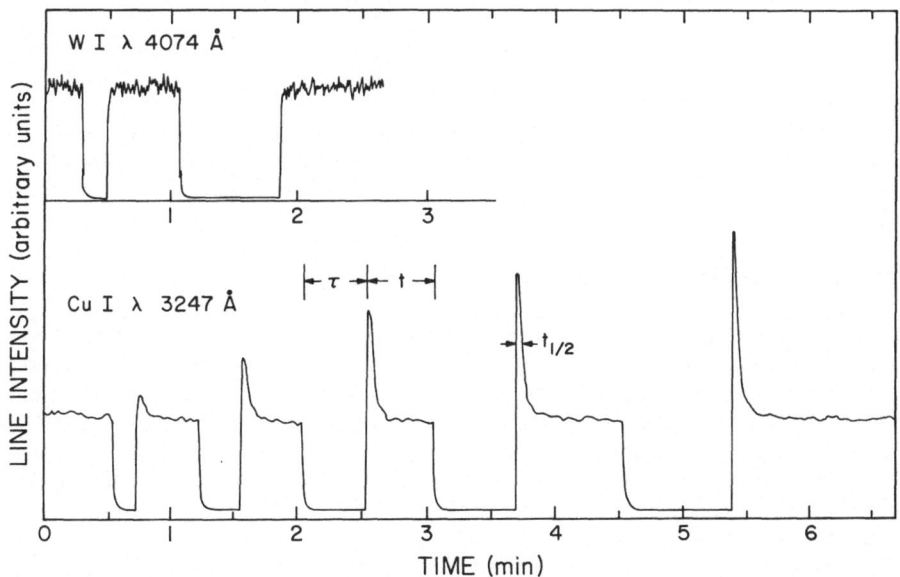

Fig. 2. Intensity of a W I and a Cu I line as a function of time t after an interruption of the beam for a time τ.

Table I

Relation of transient effect and continuum radiation to electron
configuration.

Element	Electron Configuration	Transient Effect	Continuum Radiation
Si	$3p^2$	Yes	Small
Ti	$3d^2 4s^2$	No	Yes
Cr	$3d^5 4s^1$	No	Yes
Fe	$3d^6 4s^2$	Small	Small
Ni	$3d^8 4s^2$	Yes	No
Cu	$3d^{10} 4s^1$	Yes	No
Zr	$4d^2 5s^2$	No	Yes
Mo	$4d^5 5s^1$	Small	Yes
Pd	$4d^{10}$	Yes	No
Ag	$5s^1$	Yes	No
Ta	$5d^3 6s^2$	No	Yes
W	$5d^4 6s^2$	No	Yes
Pt	$5d^9 6s^1$	Yes	No

In Table I we list the various elements which we studied, and
indicate which of them show transient effects. The insulators
SiO_2 and Al_2O_3 showed no transient effect. We also noted that the
elements which show a transient effect do not have a background in
the spectrum, and vice versa. (A possible correlation between
d-shell occupancy and a continuum in the spectrum was first pointed
out by L. Feldman.)

Pressure dependence of line intensities

During the measurement of the transient behavior a dependence
of line intensity on the pressure in the target chamber was found.
In order to determine which of the components in the residual gas
was responsible for this behavior, the target chamber was filled
with nitrogen to a pressure of 1×10^{-5} torr. During this operation
the line intensities remained constant. However, filling the chamber with oxygen to the same pressure caused a strong increase in
the line intensities.

In Fig. 3 this pressure dependence is shown for an atomic
and for an ionic line in the Si spectrum. One notes that the line
intensities increase slowly after a sudden change of the pressure.
After the lines had reached an equilibrium value at the high O_2

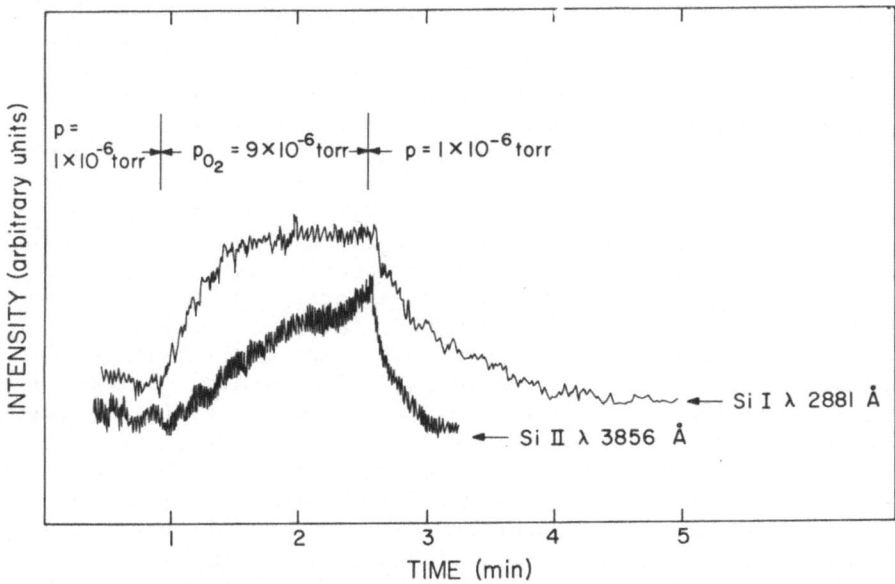

Fig. 3. Intensity of an atomic (Si I) and an ionic (Si II) line as a function of time for different pressures in the target chamber.

pressure, the oxygen supply was shut off. Both lines are seen to decrease to their original value. The growth and decay times, however, differ strongly for these lines. The pressure dependence was measured for several lines in the Si, Cu, Al and W spectra. The lines of the Al and W spectrum, which show no transient effect (see Table I) also do not exhibit a dependence on oxygen pressure. A pressure dependence does exist for the lines in the Cu spectrum and, as noted before, for atomic and ionic lines in the Si spectrum. It was observed that the growth rate of atomic lines was faster than that for lines from singly charged ions, which in turn was faster than the growth of a line from doubly ionized silicon. On the other hand, the decrease of the atomic lines took the longest times.

Discussion

     Both the observed transient effects and the pressure dependence of the line intensities are obviously related to the number of oxygen atoms which is present on the target surface. Since we are not working under ultra high vacuum conditions there is always oxygen present on the surface. This degree of coverage depends on the residual pressure, the sticking probability for oxygen, the beam current density and the sputtering rate of argon ions for adsorbed

oxygen. Before we proceed to make a quantitative estimate of the
degree of coverage of our surfaces we will try to interpret the
effect of oxygen adsorption on the optical line intensities.

The lines which we observe in the spectra are emitted by excited sputtered particles of the target material. These particles are
excited and emitted as a result of the collision cascade produced
by the incoming beam ion. They are known to radiate at distances
of a few mm from the target surface [3,11]. These emitting particles have been able to maintain their excitation state over that
distance, after which they de-excite by photon emission. It is
known, however, that near a metal surface other mechanisms of de-
excitation operate, which are far more effective than photon emission. These are resonance or Auger-type electron transitions from
the excited particle into the solid [9]. A schematic picture of
resonance de-excitation near a metal is given in Fig. 4a. Lifetimes
for these transitions are of the order of 10^{-15}s, which is very
short compared to normal optical lifetimes (10^{-8}s). This shows that
these tunneling transitions, at least at distances of the order of
50Å or less dominate over photon emission. If oxygen atoms are adsorbed on a surface, this has the effect of reducing the density of
available electron states on the surface. The details of the

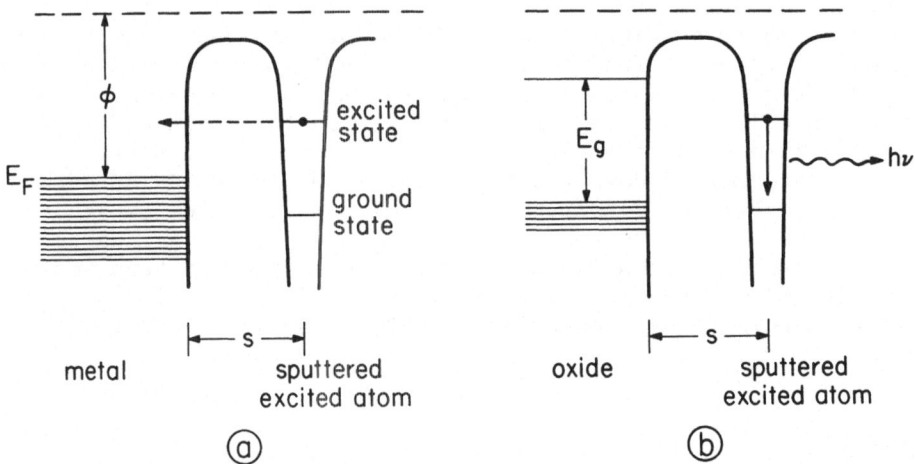

Fig. 4. Schematic potential energy diagram of an excited sputtered
particle at a distance s from a solid. (a) Shows the excited atom
near a metal surface. The metal has a conduction band filled up to
the energy E_F; its work function is denoted by ϕ. A possible electron tunneling process is indicated. (b) Shows the excited atom
near an oxidized surface. A forbidden energy gap E_g exists. Tunneling is impossible and de-exciation takes place through photon
emission.

effects of oxidation on the band structure and the electron conduction are complicated and depend very critically on the type and
stoichiometry of the oxide [12]. Since the structure of the thin
oxides which form on our samples is not known, we can only make
some general observations. In the transition metal oxides, the
valence and conduction band are separated by a large energy gap
(5-10 eV), the only states contributing to conduction would be in
a narrow d-band. Therefore, it is reasonable to assume that in the
case of oxides the electron tunneling processes are largely inhib-
ited, thus causing a relative increase in the probability for de-
excitation by photon emission. This is illustrated in Fig. 4b.

The transient effect, therefore, can be interpreted as a
temporary enhanced probability of radiation due to the accumulation
of oxygen on the surface during the time τ. The sputtering action
of the beam establishes an equilibrium concentration of oxygen,
resulting in a stationary intensity of radiation. The concentration
of oxygen atoms is calculated from:

$$\frac{dN_O}{dt} = -\frac{ISN_O}{N_M} + P\sigma \tag{1}$$

N_O being the density of oxygen on the surface, I the beam current
density, S the sputtering ratio for oxygen atoms, N_M the atomic
surface density of the metal, P the number of oxygen atoms arriving
from the residual gas and σ the sticking probability. The solution
of Eq. 1 is:

$$N_O = \frac{P\sigma N_M}{IS} + C\,\exp\left(-\frac{IS}{N_M}\,t\right) \tag{2}$$

C being an integration constant.

We do not know the exact form of the function which relates
the photon emission probability to the oxygen concentration, but
we can assume that the light intensity increases monotonically with
oxygen concentration. In that case, Eq. 2 represents decay of light
intensity towards an equilibrium value, as observed in Fig. 2.
The calculated decay time constant which is 9 seconds for the case
of 1 µA Ar$^+$ on Cu is comparable to the experimental value of 3
seconds. Furthermore, the observations (a) to (e) noted above
relating to Fig. 2 can be easily inferred from Eq. 2. Equation 2
also predicts that the equilibrium line intensity should depend on
the beam current intensity in a nonlinear way. This is because on
one hand the emission intensity depends on the number of excited
atoms which is proportional to the beam intensity and on the other
hand, the enhancement of the radiation by the oxide formation depends inversely proportional on beam current density I (first right
hand term of Eq. 2). This effect was actually observed in our
experiments.

The observation that the growth and decay rate of a line after pressure changes differ for several lines has an interesting consequence. It could mean that the electron state density or the tunneling probability at different energy levels is affected in a different way by the oxidation process. We observed that the ion lines increase slowly upon oxidation and decrease fast during cleaning with the ion beam. This shows that the oxidation process affects the deeper energy levels (ionic lines) less than the shallower ones (atomic lines).

Table I shows that the occurrence of a transient effect can be correlated with the number of electrons in the d-shell of the element. It appears that elements with filled or nearly filled d-bands show strong transient effects, which are not observed for cases where only a few d-electrons are present.

This behavior can be expected for two reasons. In the first place, it is known that adsorption properties of transition metal surfaces are largely determined by the d-band structure. For instance, the heat of adsorption of hydrogen atoms on metal surfaces was found to decrease with increasing number of d-shell electrons [13]. It can be expected that oxygen behaves in a similar manner. Therefore we expect that at a fixed pressure in the target chamber, the metals with nearly or completely filled d-band are less oxidized than those with fewer d-electrons. The second reason for expecting relatively cleaner surfaces when the d-band is almost filled is related to the dependence of the sputtering ratio on the atomic number of the target material. It was shown [14] that this sputtering rate increases with increasing number of d-shell electrons. Therefore, the metals with almost filled d-bands are cleaned more efficiently under the action of the ion beam. We will give an estimate of the degree of coverage of the surface with the help of Eq. 2. This formula shows that, once equilibrium is attained (large t), the fractional coverage f of the surface by oxygen atoms is given by:

$$f = \left(\frac{N_O}{N_M}\right)_{equilibrium} = \frac{P\sigma}{IS} \tag{3}$$

With the current density used in our experiments, at a pressure of 6×10^{-7} torr, the value of f is equal to $20(\sigma/s)$. Hence, for metals like Fe, Ni, Mo and W, where the sticking probability is nearly unity [15], the sputtering ratio is by far insufficient [16] (S < 5), to establish a value of f smaller than unity. Since, for these metals the current density and pressure used are insufficient to clean the surface, we do not observe transient behavior in these cases. Noble metals like Ag and Pt have small sticking probabilities and high sputtering rates [16] (S > 10), in which case an equilibrium surface f of smaller than unity can be maintained. These surfaces can absorb additional oxygen, when the beam is not on the surface. This additional amount can be removed by sputtering with an ion beam, which results in a transient behavior of the spectral line intensities.

In conclusion it can be remarked that the change of optical emission intensity upon the oxidation of a metal surface may be used to obtain information about the density of electron states near the surface of solids. Since the oxidation affects all the optical lines in the spectrum, it would be possible to probe the electron distribution in the surface region at many energy levels. Also, the enhancement of line intensity by addition of oxygen can be used to make the elemental analysis of surface by ion beam in-duced photon emission more sensitive.

Acknowledgments

We thank J. W. Mayer and M-A. Nicolet for encouragement and helpful discussions. We are much indebted to R. Gorris for skill-ful technical assistance.

References

*Work supported by O.N.R. (L. Cooper).

**Permanent Address: Philips Research Lab., Amsterdam, The Netherlands.

***Supported by the African-American Institute Fellowship.

[1] E. S. Parilis, Atomic Collision Phenomena in Solids, eds. D. W. Palmer, M. W. Thompson and P. D. Townsend, North Holland Publ., 1970, p. 513.

[2] Proc. Int. Conf. on Inner Shell Ionization Phenomena, Atlanta, 1972.

[3] C. Snoek, W. F. van der Weg, P. K. Rol, Physica 30, 341 (1964).

[4] C. W. White, D. C. Simms and N. H. Tolk, Science 177, 481 (1972).

[5] I. Terzic and B. Perovic, Surf. Sci. 21, 86 (1970).

[6] I. S. T. Tsong, Phys. Stat. Sol. (a) 7, 451 (1971).

[7] N. Tolk, C. W. White and P. Sigmund, Bull. Am. Phys. Soc. 18, 686 (1973).

[8] C. Kerkdijk and E. W. Thomas, Physica 63, 577 (1973).

[9] H. D. Hagstrum, Phys. Rev. 96, 336 (1954).

[10] W. F. van der Weg and D. J. Bierman, Physica 44, 206 (1969).

[11] V. V. Gritsyna, T. S. Kiyan, R. Goutte, A. G. Koval and Ya M. Fogel, Bull. Acad. Sci. USSR, Phys. Ser. (English Transl.) 35, 530 (1971).

[12] C. N. R. Rao and G. V. Subba Rao, Phys. Stat. Sol. (a) 1, 597 (1970).

[13] O. Beeck, Disc. Faraday Soc. 8, 118 (1950).

[14] O. Almén, G. Bruce, Trans. 8th Nat. Vacuum Symp. Washington (1961). Pergamon Press Inc., New York, 1962, p. 245.

[15] A. M. Horgan and D. A. King, Surf. Sci. <u>23</u>, 259 (1970).
[16] G. Carter and J. S. Colligon, Ion Bombardment of Solids,
     American Elsevier Publ., New York, 1968, Ch. 7.

REDUCTION OF THE CHARACTERISTIC RADIATION
DUE TO THE CHANNELING EFFECT

N. P. KALASHNIKOV

The Moscow Physical-Engineering Institute
Moscow, U.S.S.R.

ABSTRACT

The channeling effect essentially changes
the electromagnetic processes in a single crystal
when the angle between the momentum of incident
fast nonrelativistic positive charged particle
and crystallographic plane is smaller than the
critical Lindhard channeling angle. The wave
function of the channeling particles appears to
be exponentially small near the atomic rows or
the planes resulting in a substantial reduction
of the probability of the processes characterized
by small impact parameters. The excitation process of the K-electron by the fast positive
ions in a single crystal has been studied when
the momentum of the incident particles is parallel to the crystallographic plane.

When the fast positive charged particles ($pd \gg 1$, where p is
momentum of the incident particles; d, a lattice constant, $\hbar = c$
$= 1$) pass through a single crystal, the space distribution of these
particles is essentially changed by the channeling effect. The
capture of the fast positive charged particles in the channeling
regime takes place when the angle between the momentum of the
incident particle and crystallographic plane or axes, θ_0, is
smaller than the critical Lindhard channeling angle θ_c

$$\theta_0 < \theta_c = \sqrt{\frac{2\pi Z_1 Z_2 e^2}{\kappa d^2 E_{kin}}} \qquad (1)$$

where Z_1e and Z_2e are the atomic charges of the target and incident particle, $\kappa = me^2\sqrt{Z_1^{2/3} + Z_2^{2/3}}$ (reciprocal Thomas–Fermi screening radius of the atomic lattice potential), E_{kin} is kinetic energy of the incident positive charged particle. In this paper we consider the case of the plane channeling effect. If the charged particle enters a single crystal with a small angle ($\theta_0 < \theta_c$) to the crystallographic plane YOZ, the interaction between the positive charged particles and crystal lattice is essentially determined by one dimensional potential [1,2]. But the exact lattice potential $U(\vec{r}) = \Sigma_a U_0(\vec{r} - \vec{R}_a)$ is equal to $U(\vec{r}) = V(x) + W(\vec{r})$, where $V(x) = d^{-3}$.

$$\sum_{S_x=-\infty}^{+\infty} \frac{4\pi Z_1 Z_2 e^2}{(\frac{2\pi}{a} S_x)^2 + \kappa^2} e^{i\frac{2\pi}{a} S_x x} \simeq \sum_n \frac{2\pi Z_1 Z_2 e^2}{\kappa d^2} e^{-\kappa|x+nd|} \quad (2)$$

is the average potential of the crystallographic plane; the potential $W(F)$ takes into account the deviation of the exact potential from the average plane one, i.e., the thermal vibrations and the potential discreteness.

The positive charged particle wave function which satisfies the Schrödinger equation with the periodic potential (2) and the Bloch theorem, can be written in the following form

$$\phi_n(x) = \sum_{\nu=-\infty}^{+\infty} e^{ik\nu d} u_n(x-\nu d), \quad (3)$$

where "k" is the particle quasi-momentum.

$$u_n(x) = \sqrt{\frac{2}{d-2\kappa^{-1}}} \begin{cases} \frac{P_x^{(n)}}{\alpha} \exp[-\alpha(\kappa^{-1}-|x|)] & |x| \le \kappa^{-1} \\ \operatorname{Sin}[P_x^{(n)}(\kappa-x^{-1}+\alpha^{-1})] & -\kappa^{-1} \le x \le d-\kappa^{-1} \end{cases}$$

$$\alpha = \sqrt{M \frac{4\pi Z_1 Z_2 e^2}{\kappa d^2}}$$

$$P_x^{(n)} = \frac{n}{(d-2\kappa^{-1}+2\alpha^{-1})}, \qquad n = 1,2\ldots \quad (4)$$

The total wave function of the particle considered inside a single crystal has the form

$$\Psi_{\vec{p}}(\vec{r}) = \sum_{n,\nu} Q_{n,\nu}(\theta_o) e^{ipZ(1-\frac{P_x^{(n)^2}}{2p^2})} e^{ik\nu d} u_n(x-\nu d) \qquad (5)$$

where p is the incident particle momentum and

$Q_{n\nu}(\theta_o) = \int_o^d d\xi\, e^{ip\theta_o\xi} u_n^*(\xi)$ is the boundary condition at z = o.

This expression (5) immediately exhibits that the establishment of the channeling regime takes place in the crystal thickness L_Z, which is larger than L_0.

$$L_o \simeq 2p[\kappa^2 + p^2\theta_o^2]^{-1} . \qquad (6)$$

The value L_0 characterizes the crystal thickness where essential reconstruction of the particle wave function takes place from the incident plane wave into the channeling wave function, Eqs. (3) – (5).

The analysis $Q_{n\nu}(\theta_o)$ allows one to determine the fraction of the particles which are channeled; thus, for $\theta_o = 0$, one obtains

$$\frac{N_{channel}}{N_o} = 1 - \frac{\kappa}{M} \cdot \frac{\pi}{4Z_1 Z_2 e^2} . \qquad (7)$$

Therefore, the wave function of the channeling particle (3) – (5) is exponentially small near the atomic planes or atomic rows. Consequently, it is of great interest to investigate the probabilities of processes which are characterized by small impact parameters.

We now consider the excitation process of the K-electron by the fast positive charged particles in a single crystal provided that the angle between the momentum of the incident particle and crystallographic plane is smaller than the critical channeling angle (1). The interaction potential between the incident particle and the K-electrons is written as

$$V_{in} = \sum_{a,j} \frac{Z_2 e^2}{|(\vec{r}+\vec{r}_{ja})|} , \qquad (8)$$

where $\vec{r}$ is the particle coordinate, $\vec{r}_{ja}$ is the coordinate "j" electron related to atom "a", and the summation is over all K-electrons and atoms. The crystal thickness L along the incident particle motion direction is much larger than L_0, Eq. (6), that characterizes the thickness for the establishment of the channeling regime. But the upper limit for the crystal thickness L is restricted by the condition for the applicability of perturbation theory to the treatment of the inelastic excitation process of the K-electron in a whole crystal. Using the interaction potential,

Eq. (8), one can obtain the electron transition probability from the ground state "f" of the ionization continuum. Now using the conventional perturbation theory for the inelastic collisions, we consider the matrix element for the electron transition [3]

$$d\delta_{o \to f} = \frac{2}{\pi} \frac{M}{P} (Z_2 e^2) \; N_{tot} \cdot \sum_{j_1, j_2} \frac{d^3 l dS}{l^2 [l_\perp^2 + (1_x - S)^2]}$$

$$\cdot \sum_n |Q_n(\theta_o)|^2 \; F_n(S) <f|e^{-i\vec{l} \vec{r}_{j_1}}|o> [<f|e^{-i\vec{l} \vec{r}_{j_2}}|o>]^*$$

$$\cdot \delta(\frac{\vec{P}^2}{2M} - \frac{(\vec{P} - \vec{l})^2}{2M} - \omega_{fo}). \tag{9}$$

where $N_{tot} = N_x N_y N_z$ is the total number of the crystal atoms, M is the mass of the incident particle, and $\omega_{fo} = E_f - E_o$ is the transition frequency,

$$F_n(s) = \int_o^d dx \; e^{isx} |u_n(x)|^2$$

$$\vec{l}' \equiv \{\vec{l}_i, \; l_x - S\} \; . \tag{10}$$

The minimum momentum transfer in the ionization process of the K-electron is approximately equal to

$$l_{min}^{fo} \simeq \frac{\omega_{fo} M}{P} \tag{11}$$

i.e.,

$$l_{min}^{fo} > \kappa \; . \tag{12}$$

Therefore, in the process considered, the small impact parameters play the essential role. Now the expression (9) can be divided into two parts. The first term of this expression corresponds to the channeling particles, and the second term relates to the non-channeling particles. By using the wave function of the channeling particle, Eqs. (3) - (5), we can obtain the electron transition probability from the ground state to the ionization continuum state with the energy $E_q = \frac{q^2}{2m_e}$. Thus, the interaction with the channel-

ing particles takes the contribution due to the K-electron excitation probability, as is given below,

$$d\delta^{(channel)}_{o \to q} = \frac{2^{11}}{3}\pi \frac{M}{P} (Z_2 e^2)^{-2} \int \frac{d^2 l \exp(-2 l_\perp \kappa^{-1})}{(l_\perp^2 + (p\theta_o)^2)^2} [\frac{2}{d} \frac{(P\theta_o)^2}{l_{11}}]$$

$$N_{tot} \cdot \frac{\nu^6 \exp(-4\nu \text{arcctg}\nu)}{(1+^2)^5 q^5 (1-\exp(-2\pi\nu))}$$

$$q^2 dq \delta(\frac{P^2}{2M} - \frac{(\vec{p}-\vec{l}_\perp)^2}{2M} - (1+\nu^2)\frac{q^2}{2m} , \qquad (13)$$

where $\nu = \dfrac{Z_1 m e^2}{q}$,

and the momentum of the incident particle along the axis OX is equal
to $(p\theta_o)$. By performing the integration in the expression (13),
we obtain

$$\sigma_o(channel) = N_{tot} \frac{2^{19/2} \pi^{3/2}}{3} \frac{1}{m}(\frac{M}{P})^3 (Z_2 e^2) (\frac{\kappa}{l_{min}})^{3/2}$$

$$[\frac{2}{d} + \frac{(P\theta_o)^2}{l_{min}}] \frac{\exp[-2l_{min}\kappa^{-1}-4]}{[l_{min}^2 + (P\theta_o)^2]} , \qquad (14)$$

where $\quad l_{min} = \dfrac{Z_1^2 m e^4 M}{2P}$. $\qquad\qquad\qquad\qquad (15)$

The fraction of the nonchanneling particles is too small, when the
entrance angle is far less than the critical Lindhard channeling
angle

$$\frac{N(nonchannel)}{N_o} \approx \sqrt{\frac{\kappa}{M} \cdot \frac{\pi}{4Z_1 Z_2 e^2}} \qquad (16)$$

and the excitation probability of the inner (K-electron) electrons
by the nonchanneling particles is determined by the excitation
probability in an amorphous medium

$$\sigma(\text{nonchannel}) = \frac{\kappa}{M} \cdot \frac{\pi}{4z_1 z_2 e^2} \; 16\pi \frac{(z_2 e^2 M)^2}{P} \; N_{tot} \; \frac{1}{z_1^2 m^2 e^4}$$

$$(17)$$

where β is a dimensionless parameter which cannot be calculated in the general case [3]. The total excitation probability of the K-electrons is determined by the sum of two terms (15) and (17).

The above expressions (15) and (17) determine the cross section of K-electron excitation when the momentum of the incident particles is parallel to the crystallographic plane. When the entrance angle becomes larger than the critical channeling angle, the cross section of excitation is determined by the excitation probability in an amorphous medium.

References

[1] J. Lindhard, Mat.-Fys. Medd. Dan. Vid. Selsk 34(14) (1965).
[2] N. P. Kalashnikov, "The Proceedings of the Third Soviet Con-
     ference on the Interaction Physics of the Charged Particles
     with Single Crystals," (Moscow) (1972).
[3] L. D. Landau and E. M. Lifshitz, "Quantum Mechanics," Addison-
     Wesley (1965).

ELECTRON CHANNELING IN Si, Ag AND Au CRYSTALS

S. KJAER ANDERSEN, F. BELL,* F. FRANDSEN,
and E. UGGERHØJ
*Institute of Physics, University of Aarhus
Aarhus, Denmark*

ABSTRACT

For the case of Rutherford scattering the rule of
reversibility (reciprocity theorem) has been proved to
hold quantitatively. Thus the enhancement of Rutherford
scattering of an incident beam parallel to a major axis
or plane is equivalent to Kikuchi excess bands normally
observed from crystals a few extinction distances or less
in thickness. The contrast of the excess bands decreases
drastically by increasing the foil thickness due to the
influence of multiple scattering (absorption). It will
be shown that the experimental thickness dependence for
the investigated crystals can be scaled by a mean absorption distance $\ell=(N\sigma)^{-1}$, where N is the density of scatterers and σ the total cross section for incoherent
scattering.

The contrast of the excess bands becomes less than
10 percent for thicknesses larger than 6ℓ. At the same
time Kikuchi defect bands develop in the forward direction.
Experimentally the angular shape of the excess and defect
bands are just reciprocal to each other:

$I_{defect}(\phi) \simeq 1 - \alpha \cdot I_{excess}(\phi)$. According to a recent

theory of Kikuchi defect bands by Thomas and Humphreys
[1], based on the dynamical theory of diffraction, it
can be shown that the above relationship holds theoretically, too, if one assumes that the excess bands are due
to single inelastic scattering.

According to the correspondence principle a classical description of the observed phenomena should be a
rather good approximation if the number of quantum
states is large and if all the Fourier components of the
wave function contribute simultaneously to the observed

quantity.   Such phenomena are the excess yield of
Rutherford scattering, the sum of the elastic Bragg
beam intensities in thin crystals and the reduction of
multiple scattering in forward direction in thick crystals.   For the investigated crystals and energy region
(1-3 MeV) the number of quantum states is only large
for major crystallographic axis.   It turns out that
for this case the angular width of all three quantities scales rather well with Lindhard's critical ψ_1
[2].

References

*Permanent address: Sektion Physik, University of Munich.

[1] L. E. Thomas and C. J. Humphreys, Phys. Stat. Sol. (A)
    3, 599 (1970).
[2] P. Lervig, J. Lindhard, and V. Nielsen, Nucl. Phys.
    A96, 481 (1967).

EMISSION OF ELECTRONS AND POSITRONS FROM CRYSTALS - DIRECTIONAL EFFECTS

A. P. PATHAK

Theoretical Physics Division
A.E.R.E. Harwell, Didcot, Berkshire
England

ABSTRACT

The wave mechanical treatment of directional effects in the emission of the electrons and positrons from crystals as described earlier, has been used to calculate the detailed emission patterns. The Moliere approximation to the Thomas-Fermi potential is used to describe the atomic interaction potential. Single phonon approximation for inelastic phonon scattering and Debye model for lattice vibrations have been used. Within the framework of two beam theory, the effects of phonon absorption are illustrated by comparing the emission pattern for different crystal thickness traversed by the particles. The results are in qualitative agreement with experiments. The angular widths of emission patterns agree with earlier calculations.

Introduction

Long ago, Yoshioka [1] suggested that the absorption effects due to electronic excitation during propagation of charged particles in crystals may be calculated in two beam diffraction theory so that the imaginary part of the phenomenological complex potential could be theoretically calculated. Later, this approach was also used to calculate the absorption due to phonon excitations (Yoshioka and Kainuma [2]). In an earlier work (Pathak and Yussouff [3]) we used this formalism to discuss the directional effects in emission of electrons and positrons from crystals, and calculated analytical expression for phonon absorption using a screened Coulomb

potential. The purpose of this note is to report detailed calculations of emission patterns using a more realistic interatomic potential and to compare these results with earlier calculations and
experiments.

Calculation and Results

Previously it was found (DeWames and Hall [4], Pathak and
Yussouff [3]) that in the two beam approximation, the emitted
intensity of electrons (or positrons) due to an emitter located at
a lattice site in the crystal is given by

$$|\phi_n(t)|^2 = x^2 \exp\{-[1-\varepsilon_h(1+y^2)^{-1/2}]t/\xi_o''\}$$

$$+ (1-x)^2 \exp\{-[1+\varepsilon_h(1+y^2)^{-1/2}]t/\xi_o''\}$$

$$+ 2x(1-x) \exp(-t/\xi_o'') \cos[(1+y^2)^{1/2} t/\xi_h']$$

$$(1)$$

with

$$x = \frac{1}{2}[1 + (y-1)/(y^2+1)^{1/2}]$$

$$y = \zeta_h/2W'_h, \quad [\mathrm{Re}\ W_h = W'_h = (V_h + \mathrm{Re}\ C_{ho})/E_p] \simeq \frac{\mathrm{Sin}2\theta_B}{W_h'}(\theta_B-\theta)$$

$$\varepsilon_h = W_h''/W''_o, \quad [\mathrm{Im}\ W_h = W_h'' = \mathrm{Im}\ C_{ho}/E_p]$$

$$\xi'_h = (k_n W_h')^{-1}; \quad \xi_h'' = -(k_n W_h'')^{-1} \qquad (2)$$

where t is the distance of the emitting atom from the crystal surface, k_n is the wave vector of the emitted particle with energy E_n
and mass m_o and n denotes the initial phonon state of the crystal
at low temperature, and θ_B is the Bragg angle. In eq. (1), $\mathrm{Re}\ C_{ho}$
and $\mathrm{Im}\ C_{ho}$ are real and imaginary parts of C_{ho} given by

$$C_{ho} = -\frac{2m_o}{\hbar^2 V'} \int d\underline{r} \int d\underline{r}' \exp[-i(\underline{K}_h+\underline{k}_n) \cdot \underline{r}+i \underline{k}_n \cdot \underline{r}']$$

$$\sum_{n'\neq n} V_{nn'}(\underline{r})V_{n'n}(\underline{r}')\frac{\exp(i k_n' |\underline{r} - \underline{r}'|)}{4\pi|\underline{r} - \underline{r}'|} \tag{3}$$

where V' is the volume of the crystal, K_h is a reciprocal lattice vector and potential matrix elements are given by

$$V_{nn'}(\underline{r}) = \langle n|V(\underline{r})|n'\rangle = \langle n|\sum_\sigma V_\sigma(\underline{r} - \underline{R}_\sigma)|n'$$

and

$$V_h(n) = \frac{1}{V'} \int V_{nn} (\underline{r}) \exp(-i \underline{K}_h \cdot \underline{r}) d\underline{r} \tag{4}$$

with $\underline{R}_\sigma = \underline{\sigma} + \underline{u}_\sigma$ representing the actual position of the σth atom in the crystal lattice and $\underline{u}_\sigma$ its displacement.

The Moliere approximation to the interaction potential is given by

$$V(r) = Z_1 Z_2 e^2 \sum_\sigma \sum_{i=1}^{3} \alpha_i \frac{e^{-\Lambda_i|\underline{r}-\underline{R}_\sigma|}}{|\underline{r}-\underline{R}_\sigma|} \tag{5}$$

with $\alpha_1 = 0.1$, $\alpha_2 = 0.55$, $\alpha_3 = 0.35$; $\Lambda_1 = 6/a$, $\Lambda_2 = 1.2/a$, $\Lambda_3 = 0.3/a$ and

$$a = 0.47 (Z_1^{2/3} + Z_2^{2/3})^{-1/2} \overset{\circ}{A} .$$

The calculation for the emitted intensity distribution has been made by using potential (5) in the expression (1) for the case of a copper crystal at 20°K. The extinction distance ξ_h' is typically less than 100Å for electron energies above 100 keV and the absorption length ξ_o'' is of order 10^2 times ξ_h'. Thus a collection of emitting atoms spread over several extinction distances from the crystal surfaces will give an intensity which is the average over thickness of equation (1). This average will eliminate the cosine term and in the rest of the expression t will represent average distance of emitters from the crystal surface. The results for 200 keV electrons and positrons emitted from t = 500Å and t = 5×10^5 Å along (100) planes, are shown in Figs. 1 and 2. We see from the calculations that the angular widths of the emission patterns ($\sim 2 \theta_B$, where θ_B is Bragg angle) are in agreement with the calculations of Howie et al. [5] for Ag.

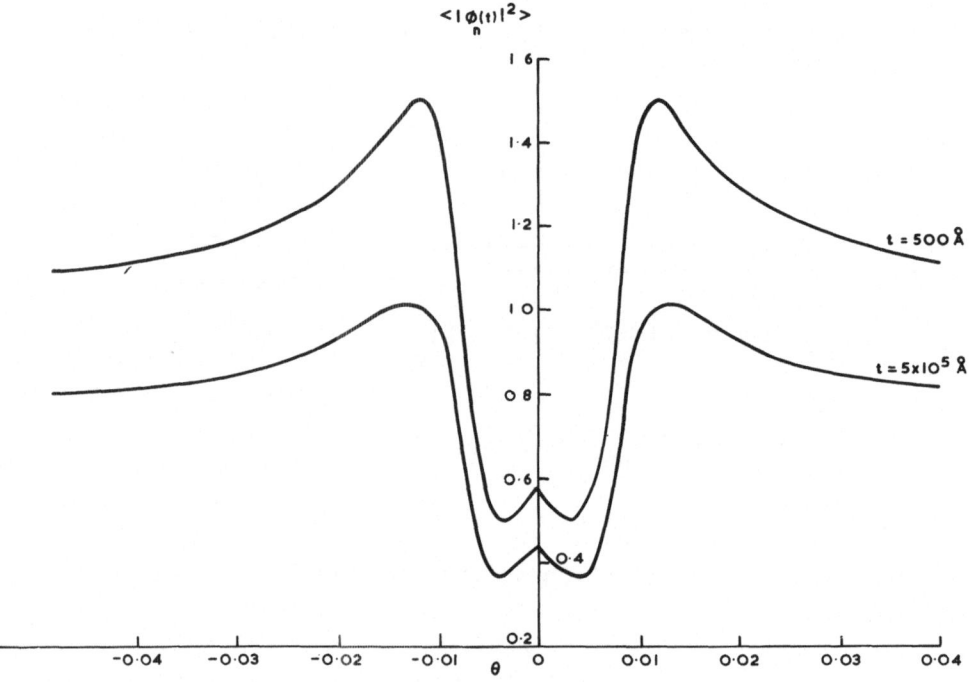

Fig. 1. Calculated emitted intensity distribution for positrons
using Moliere potential.

    As regards the comparison of the present results with experi-
mental observations of Uggerhoj et al. [6,7], we notice that
the present results are in qualitative agreement with experiment.
However, quantitatively, one notices that the experimental angular
widths are 3 to 4 times larger than the presented calculated ones.
To this end, we note that Uggerhoj observed the emitted intensity
along the [001] axis in copper where we know that four different
sets of planes intersect, and DeWames and Hall [4] have correctly
stressed that Uggerhoj's [6,7] results need not agree with those
obtained by calculating the diffraction due to a single plane.
Moreover, the angular resolution in the experiments of Uggerhoj
lies between 0.3° and 0.5° whereas the Bragg angle for 200 keV
electrons is itself θ_B = 0.0076 radians = 0.43°. Therefore one
cannot expect to observe the diffraction effects.
    We notice from equation (2) that $\text{Im}C_{hg}$ adds an imaginary part
to the potential Fourier component and hence gives rise to the
absorption of the corresponding diffracted waves. In particular
for single diffraction (two beam theory) $\text{Im}C_{ho}$ determined the
absorption of the two waves. In order to determine the relative
magnitude of the absorption we use potential (5) to calculate the
ratio $\text{Im } C_{ho}/V_h$ and get

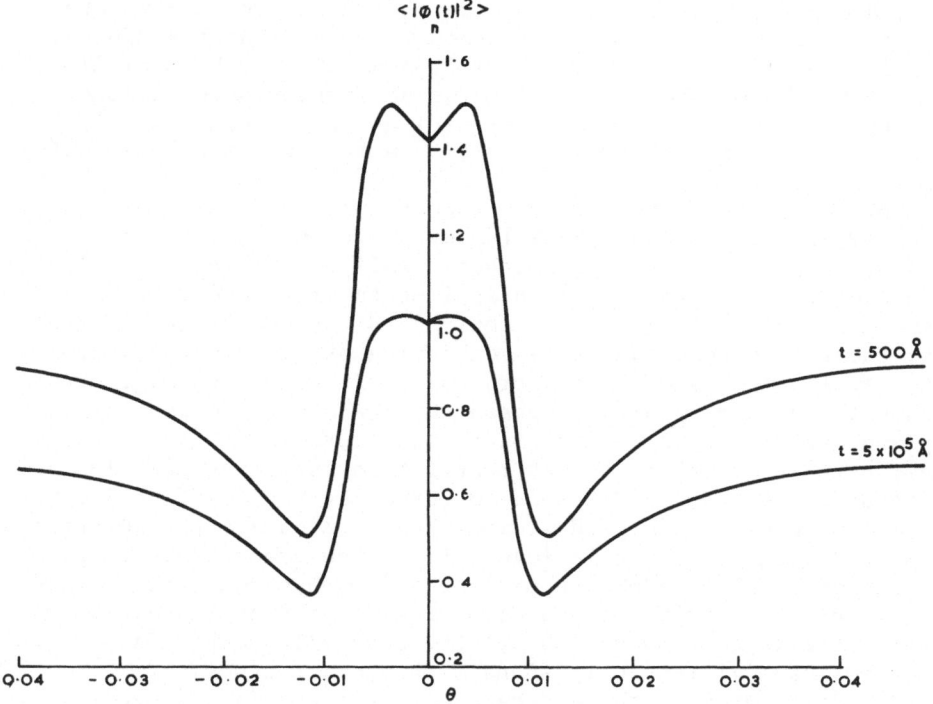

Fig. 2. Calculated emitted intensity distribution for electrons using Moliere potential.

$$\frac{\text{Im } C_{ho}}{V_h} = - (\frac{2m_o}{E_p})^{1/2} \frac{Z_1 Z_2 e^2}{4MC} [\sum_i \frac{\alpha_i^2}{(\Lambda_i^2 + K_h^2)} F_o(\Lambda_i, T) /$$

$$\sum_i \frac{\alpha_i}{(\Lambda_i^2 + K_h^2)}] \qquad (6)$$

where

$$F_o(\Lambda_i, T) = f_o + 2f_o (T/\theta_D) [1 - \exp(-\theta_D/T.)] - \Lambda_i \{ \tan^{-1} \frac{f_o}{\Lambda_i}$$

$$+ Ci(\frac{\Lambda_i \hbar c}{k_B T}) Sin(\frac{\Lambda_i \hbar c}{k_B T}) - Si(\frac{\Lambda_i \hbar c}{k_B T}) Cos(\frac{\Lambda_i \hbar c}{k_B T}) \}$$

with $Ci(x) = - \int_x^\infty \frac{\cos t}{t} dt, \quad si(x) = - \int_x^\infty \frac{Sin t}{t} dt$

and f_o and θ_D are the Debye frequency and the Debye temperature respectively. On numerical evaluation we found that for 200 keV electrons and positrons Im $C_{ho}/V_h = 1.33 \times 10^{-3}$ which is small compared to the absorption obtained using complex potentials of the optical model (Hashimoto et al. [8] and Hall and Hirsch [9]) although the Z_1 and Z_2 dependence is in agreement with these diffraction theories.

The deviation of the present results with those of Hall and Hirsch [9] and Hashimoto et al. [8] seems to be contained in the difference in the two approaches. Hall and Hirsch used kinematical diffraction theory which applies to small and thin crystallites, and dynamical theory is required for large crystals. Moreover, these workers assumed the scattering of Bloch waves into plane waves which is only an approximate situation in large crystals because there is no a priori reason why the scattered waves should be plane waves. We do not assume the scattered waves to be plane waves; neither do we impose any restriction regarding the size of the crystal. Another difference concerns the region of validity of the two results. While Hall and Hirsch calculate the adsorption in the Einstein model and include multiphonon processes, the present work includes only one phonon processes in Debye model for lattice vibrations and therefore the present results are applicable to only those particles that move close to the crystallographic directions and do not have violent collisions with lattice atoms.

The difference in the approach of Howie et al. [5] and the present calculations is that Howie et al. consider motion of the particle in the z direction (along the crystallographic direction) as independent from motion in the x-y plane while we start by considering (Pathak and Yussouff [3]) the Schrodinger equation for the whole system of crystal and particles and separation of particle equation automatically gives rise to absorption effects due to phonon scattering.

Conclusions

We see from the calculated emission patterns for positrons (Fig. 1) and electrons (Fig. 2) that the curves for t = 500Å are very much the same as would be obtained without attenuation (DeWames and Hall [4]). However, for the emission patterns due to emitters placed deep inside the crystal (t = 5 x 10^5Å), the phonon adsorption effects are clearly seen. In particular, we note that for positrons, the decrease in the normal value of $<|\phi_n(t)|^2>$ is about two times greater than the decrease in the minimum value. The decrease in the shoulders is still greater (about 3.5 times the decrease in the minimum value). For electrons, the situation is completely different. The decrease in normal yield is about half of the decrease in the maximum value. The decrease in shoulders is still smaller (about 1/3.5 times the decrease in the maximum value). Physically, this difference between the behaviour of electrons and positrons is due to the fact that electrons have an attractive interaction with the crystal ions and therefore, as t increases, they encounter more and more interference effects along principal planes and directions than along random directions so that the increase in their absorption along principal planes due to increase in t is more than that along random directions. On the other hand, the positrons are repelled by the ions in the principal planes hence as t increases, the increase in their absorption due to phonon excitation is larger in random directions than along principal planes and directions. Similar arguments apply for the shoulders. To see the details of changes in the shoulders with t, one may look at the situation in the framework of the two wave channeling picture rather more easily than in the particle picture. One notices that while at $\theta=0$ in Fig. 2, the large decrease between the two curves id due to the fact that the wave with nodes at the planes is mainly excited (Hashimoto et al. [8]; Howie [10]; Whelan [11]), the smaller decrease between the two curves in the shoulders can be ascribed to the fact that the main wave excited here is that with nodes between the planes, and which therefore channels easily. For positrons (Fig. 1), because of the change in the sign of the potential, the symmetries of the waves are interchanged (Whelan [11]) and hence the reverse behaviour at shoulders with respect to the dips is seen.

Acknowledgments

The auther is grateful to Dr. A. B. Lidiard for some comments and to Drs. J. H. Tait and J. S. Briggs for reading the original manuscript.

References

[1] H. Yoshioka, J. Phys. Soc. (Japan) 12, 618, 1957.
[2] H. Yoshioka and Y. Kainuma, J. Phys. Soc. (Japan) 17, Suppl. B-II, 134 (1962).
[3] A. P. Pathak and M. Yussouff, Phys. Rev. B 3, 3702 (1971).
[4] R. E. DeWames and W. F. Hall, Acta Cryst. A24, 206 (1968); Phys. Rev. 174, 342 (1968).
[5] A. Howie, M. S. Spring and P. N. Tomlinson, Atomic Collision Phenomena in Solids (North Holland-1970) p. 34.
[6] E. Uggerhoj, Phys. Lett. 22, 382 (1966).
[7] E. Uggerhoj and J. U. Andersen, Can. J. Phys. 46, 543 (1968).
[8] M. Hashimoto, A. Howie and M. J. Whelan, Phil. Mag. 5, 967 (1960); Proc. Roy. Soc. A269, 80 (1962).
[9] C. R. Hall and P. B. Hirsch, Proc. Roy. Soc. A286, 158 (1965).
[10] A. Howie, Phil. Mag. 14, 223 (1966).
[11] M. J. Whelan, Atomic Collision Phenomena in Solids (North Holland-1970) p. 3.

CHANNELING OF 25 MeV POSITRONS AND ELECTRONS

A. NEUFERT, U. SCHIEBEL, and G. CLAUSNITZER
Strahlenzentrum, Institut für Kernphysik
D-6300 Giessen, West Germany

ABSTRACT

Channeling effects of 25 MeV positrons and electrons in MgO-single-crystals were measured. The influence of string scattering on the transmission profiles is discussed as a function of crystal thickness.

Introduction

Detailed investigations of directional effects of electrons and positrons in the MeV-region showed that classical orbital pictures are applicable for these highly energetic light particles [1,2,3]. Especially positrons show effects, which are similar to the channeling of protons. For electrons, however, the relatively large angular momentum with respect to the atomic strings causes a strong reduction of the blocking effect [1] which can be observed for energies around 1 MeV [4].

The transmission profiles of 25 MeV electrons do not show a pronounced narrow dip for small incident angles Ψ_i with respect to the crystal axis, which would correspond to the positron channeling peak. They show a narrow central peak and a wide minimum, similar to the positron shoulders. The occurrence of the central maximum was interpreted due to rosette motion by Kumm et al. [1]. A main purpose of this contribution is a further support of our explanation, that the transmission profile is mainly influenced by "string scattering" [6].

Apparatus

Figure 1 shows the experimental set-up. A beam of 25 MeV electrons and positrons is obtained from the Giessen electron linear accelerator. The positron converter is located between the two

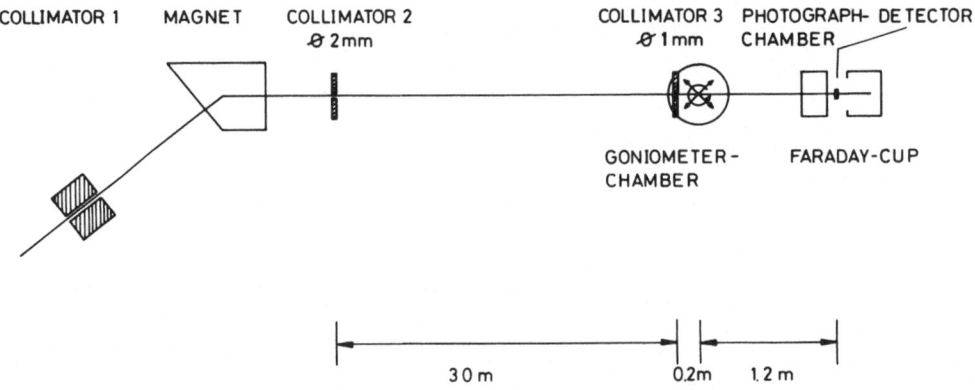

Fig. 1.   Schematic diagram of the experimental set-up.

accelerating sections; the beam transport system consists of a 50°
achromatic deflection and quadrupole focussing. Two aluminum
collimators at a distance of 3 meters lead to a beam divergence of
about 0.03° and an energy resolution $\Delta E/E < 10^{-3}$.

The single crystals were mounted on a three-axis goniometer;
the goniometer is provided with stepping motors, which allow an
orientation of the crystals within 0.01° in all directions. The
thickness of the MgO single crystals ranged from 22 to 170 microns;
the crystals are cut normal to the <100> direction.

The particles transmitted in forward direction were detected
with plastic scintillators of different diameter; thus the acceptance angle of the detectors θ could be varied from 0.1° to 0.65°.
The total beam current was monitored in a faraday cup following
the scintillator (typical beam currents 10^{-13} A for e^+, 10^{-12} A
for e^-).

The directional distributions of transmitted particles could
be recorded in a separate vacuum chamber on x-ray films.

Results and Discussion

Positrons

It was already mentioned that the positron directional effects
should be similar to the proton channeling data. In the relativistic expression for the critical angle, the energy E has to be replaced by 1/2 pv. For 25 MeV positrons this quantity has a value
of 12.8 MeV. Since no proton experiments in this energy region
were available, we compared the widths of our positron peaks with
data of Campisano et al. [7] for 1.5 MeV protons. Thereby it was
assumed that for protons the energy dependence of the widths is
proportional to $E^{-1/2}$. This led to a value of 0.2° for the full

width at half maximum of the channeling peaks, which is in good
agreement with our results (Fig. 2).

The influence of crystal thickness on the positron transmission
profiles produces a strong decrease of the channeling effect for
thicker crystals. Nevertheless the effect is still present for a
thickness of 170μ MgO.

The width of the peaks increases with crystal thickness. But
this dependence is presently disregarded, until further measurements
have been performed to clear the influence of other effects on the
peak widths, such as planar channeling and mosaic spread.

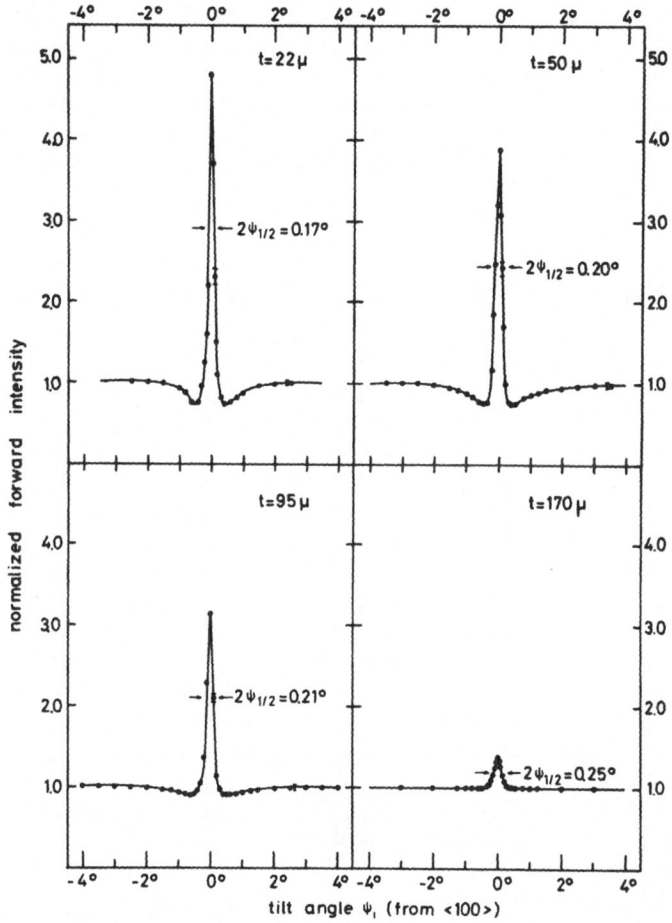

Fig. 2: Transmission profiles of 25 MeV positrons for MgO single
crystals of different thicknesses t and detector acceptance angles θ.
(θ = 0.1° for t = 22 μ; θ = 0.25° for t = 50 μ, 95 μ, and 170 μ).

Another interesting detail is the formation of shoulders in
the positron transmission profiles (Fig. 2, 3). Already Andersen
et al. [3,8] gave an explanation of this effect: for incident
angles $\Psi_i \lesssim \Psi_1$ there is enhanced scattering into directions corresponding to conservation of transverse energy. Thereby a fraction
of the transmitted particles is scattered into a solid angle greater
than the detector acceptance, which leads to a reduction of the
measured forward intensity in the shoulder region.

Transverse energy conservation produces a ring-shaped angular
distribution of the transmitted particles for small Ψ_i. For larger
angles of incidence the patterns have the shape of a circular arch
[9, 10, 6]. Although with increasing crystal thickness the transverse energy conservation is violated, those patterns are obtained
for 25 MeV positrons transmitted through 22μ MgO (Fig. 4). Of
course they are smeared out by multiple scattering. In addition
the photographic reproduction does not contain all details of the
original exposures. For thicker crystals the directional distributions become more and more symmetric, since multiple scattering
prevails the tendency of transverse energy conservation. Therefore

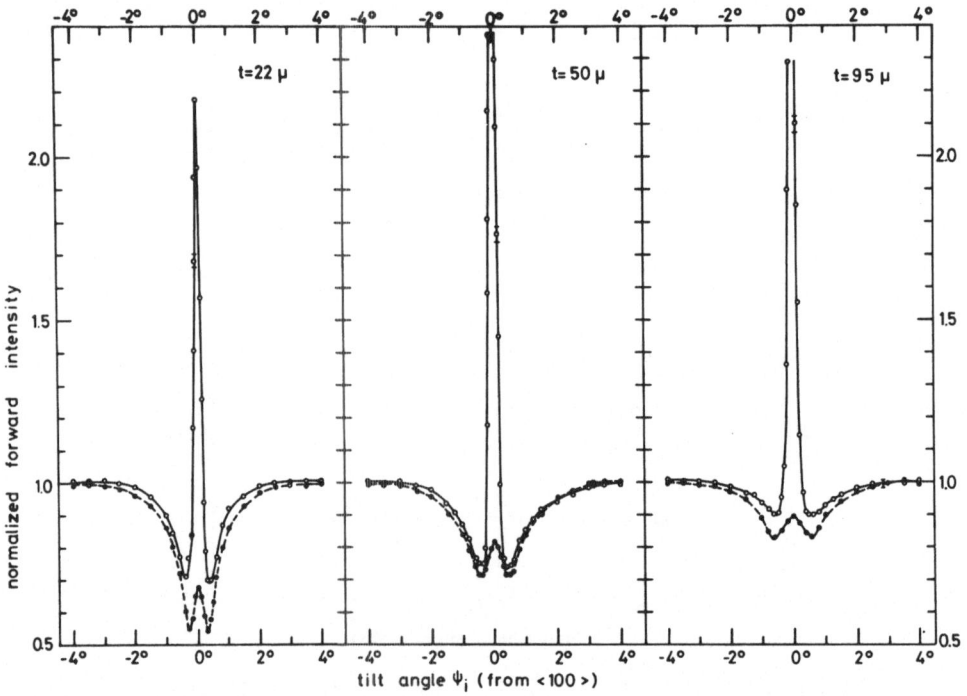

Fig. 3. Transmission profiles of 25 MeV positrons (full lines)
and electrons for MgO single crystals of different thicknesses
(t – 22, 50 and 170 μ with θ = 0.25°)

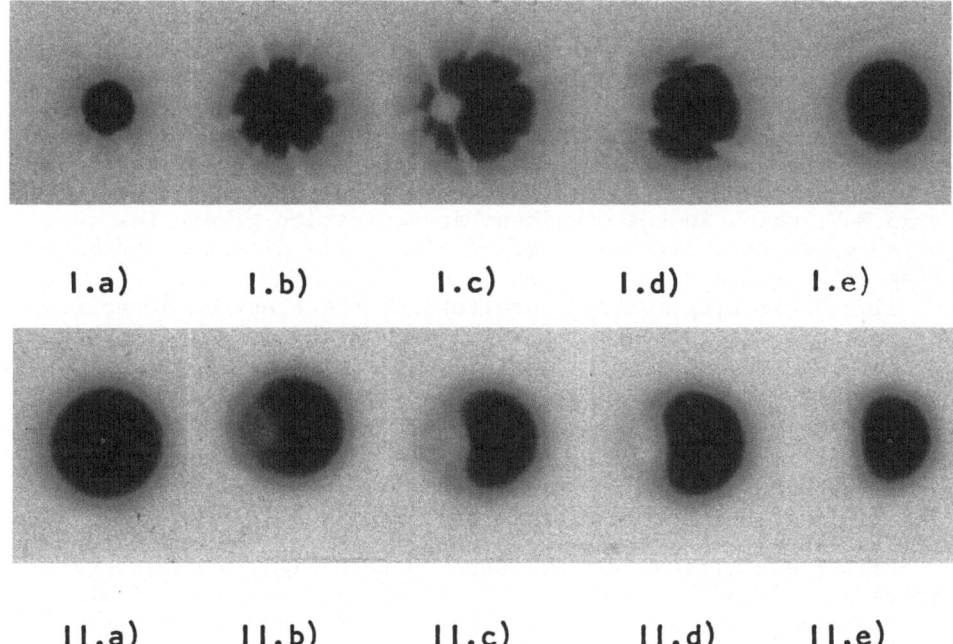

Fig. 4. Photographic recordings of directional distributions for
25 MeV positrons and electrons transmitted through 22μ thick MgO
at different incidence angles Ψ_i along <100>:
  I. positrons: a) Ψ_i = 0.0°   b) Ψ_i = 0.25°   c) Ψ_i = 0.5°
                d) Ψ_i = 1.0°   e) Ψ_i = 2.5°
 II. electrons: a) Ψ_i = 0.0°   b) Ψ_i = 0.4°   c) Ψ_i = 0.6°
                d) Ψ_i = 0.75° 3) Ψ_i = 1.0°

the shoulders in the positron transmission profiles are less pronounced with increasing crystal thickness (Fig. 2).

    Electrons

    Angular distributions of 25 MeV electrons lead to directional
patterns (Fig. 4) similar to those discussed in the previous section.
They also can be interpreted by transverse energy conservation,
which is due to correlated scattering with atoms of a single string,
the string-effect [6]. The analogy to the positron results is
further demonstrated on Fig. 3. The reduction of the transmitted
electron intensity in the region of the positron shoulders is
evidently caused by the same effect. The electron transmission
profiles exhibit a narrow central maximum, which was first observed
by Kreiner et al. [5]. The authors assumed bound particle (rosette)
motion along lattice rows as a possible explanation [1,5]. We
want to give a partly different interpretation. There are two

effects, which contribute to the total shape of the transmission profiles (Fig. 5):

i. For small values of Ψ_i there is an increased probability of hitting the nuclei in the string, which causes enhancement of wide angle Rutherford scattering. It is interesting, however, that this effect is much smaller for 25 MeV than for 1 MeV electrons [2,4]. This can be explained considering the influence of the large angular momentum of the electrons with respect to the strings for 25 MeV, which in the continuum approximation gives rise to a repulsive effective string potential for small distances from the string [1].

ii. Neglecting multiple scattering, the electron directional distributions are shaped as shown in Fig. 6 and described in the previous chapter. Since the correlated scattering is predominantly effective for small angles Ψ_i, the mean rotation angle $\bar{\delta}$ decreases with increasing Ψ_i. Simultaneously the radii of the patterns, which are proportional to Ψ_i, increase linearly. Therefore, the mean deflection angle, corresponding roughly to the size of the figures (Fig. 6.), shows a maximum for a certain Ψ_i. This explains the observed minimum in the transmission profiles at Ψ_{min} (Fig. 5).

Fig. 7 exhibits a distinct dependence of Ψ_{min} from crystal thickness. The half width of the central peak increases from a value of $0.3°$ for a 22μ to $0.6°$ for a 95μ crystal in the case of 25 MeV electrons. Other measurements [1] report $0.25°$ for a 8μ thick MgO crystal and 20 MeV. This result is in good agreement with our experimental data.

The thickness dependence of Ψ_{min} can be qualitatively understood by the increase of the mean rotation angle $\bar{\delta}$ for thicker crystals, as indicated on Fig. 6.

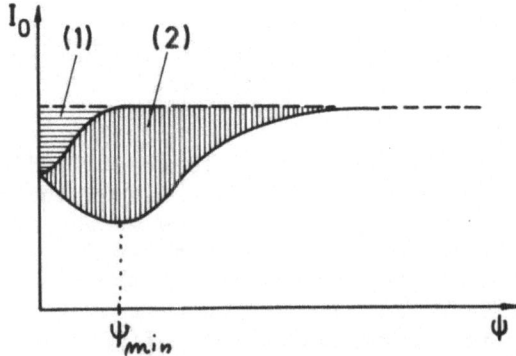

Fig. 5. Schematical drawing of an electron transmission profile. The reduction of the transmitted forward intensity for small angles Ψ_i is due to: (1) enhanced wide angle Rutherford scattering; (2) string scattering.

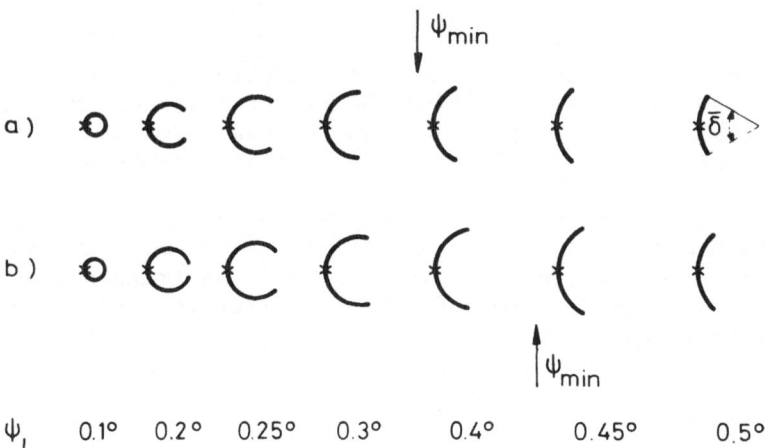

Fig. 6. Schematical drawing of directional distributions for high energy electrons transmitted through: (a) a thin and (b) a thicker single crystal for various incident angles Ψ_i. Multiple scattering has been neglected.

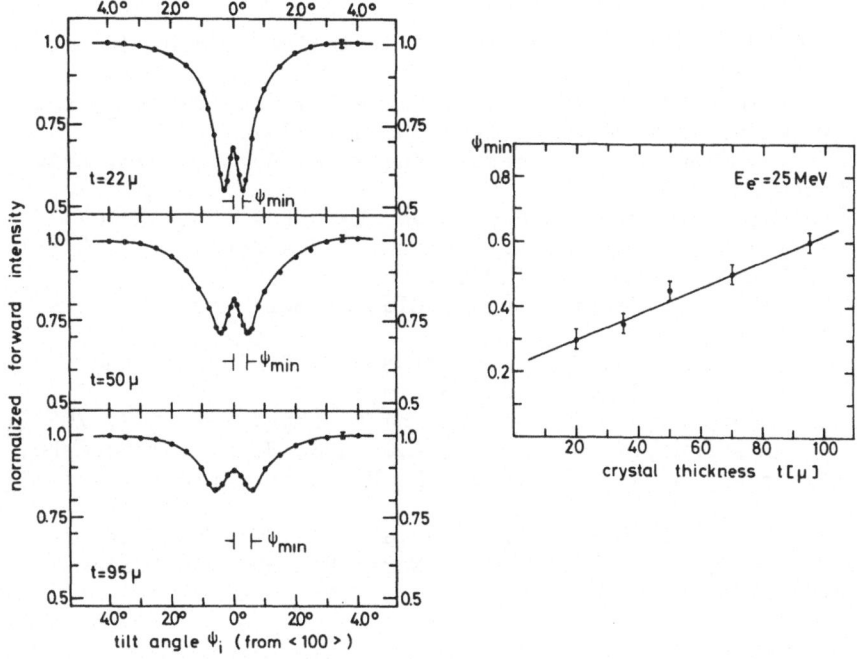

Fig. 7. Transmission profiles of 25 MeV electrons for MgO single crystals of different thickness t. (t = 22, 50 and 170µ with θ = 0.25°.) Half width of the center peak Ψ_{min} versus crystal thickness.

Conclusions

Electron and positron transmission profiles and directional distributions show a rather similar behaviour for incident angles in the shoulder region. In both cases the reduction of the transmitted intensity can be explained by the effect of string scattering.

In addition, the central maximum in the electron transmission profiles is understood by the same effect. Especially the thickness dependence of the angular width Ψ_{min} of the transmission minimum can be described in terms of string scattering, which supports our interpretation.

Acknowledgments

We want to thank our colleagues for their interest and help with the experiments. This research was supported by the Bundesministerium für Bildung und Wissenschaft, W—Germany.

References

[1] H. Kumm, F. Bell, R. Sizmann, H. J. Kreiner and D. Harder, Rad. Effects 12, 53 (1972).

[2] R. L. Walker, B. L. Berman, R. C. Der, T. M. Kavanagh, and J. M. Khan, Phys. Rev. Letters 25, 5 (1970).

[3] M. J. Pedersen, J. U. Andersen and W. M. Augustyniak, Rad. Effects 12, 47 (1972).

[4] E. Uggerhøj and F. Frandsen, Phys. Rev. B 2, 582 (1970).

[5] H. J. Kreiner, F. Bell, R. Sizmann, D. Harder, and W. Hüttl, Phys. Letters 33 A, 135 (1970).

[6] U. Schiebel, A. Neufert and G. Clausnitzer, Phys. Letters 42 A, 45 (1972).

[7] S. U. Campisano, G. Foti, F. Grasso, I. F. Quercia and E. Rimini, Rad. Effects 13, 23 (1972).

[8] J. U. Andersen and W. M. Augustyniak, Phys. Rev. B 3, 705 (1971).

[9] D. D. Armstrong, W. M. Gibson, A.Goland, J. A. Golovchenko, R. A. Levesque, R. L. Meek and H. E. Wegner, Rad. Effects 12, 143 (1972).

[10] K. Ecker, V. Lottner, S. F. Schmid and R. Sizmann, Atomic Collision Phenomena in Solids, North Holland, Amsterdam (1970) p. 455.

THE CHANNELING-BLOCKING EFFECT OF ENERGETIC ELECTRONS

F. FUJIMOTO, K. KOMAKI, H. FUJITA*, N. SUMITA*, Y. UCHIDA**,
K. KAMBE***, and G. LEHMPFUHL***
College of General Education
University of Tokyo, Tokyo, Japan

ABSTRACT

The blocking effect of electrons with high energy is observed by using the convergent-beam technique of electron diffraction and the electron microscopic images. The spot- and ring-like peaks inside the blocking dip are observed and are explained by the dynamical and classical theories.

Introduction

When an electron beam with high energy comes into a crystal parallel to a crystal axis, the intensity of the Rutherford scattering is stronger than that for the beam not parallel to the axis. This effect is known as electron channeling [1,2]. Fujimoto et al. [3] studied electron channeling, applying reciprocity, by use of photographic plates and investigated the variation of the intensity distribution around the axial direction with the energy of electrons and potential depth.

In the incident beam direction, an intensity minimum can be expected, since the incident beam is scattered away by strong Rutherford scattering. However, Kumm et al. [4] observed a peak at high energy (5 ∿ 20 MeV) and explained the peak by a rosette motion of electrons around the string. Similar effect is observed in the planar case and is known as the so-called Bloch-wave channeling" [5].

In the present work, we study the variation of the intensity distribution around the crystal axis and lattice plane with the electron energy and potential depth, utilizing the convergent beam technique of electron diffraction and the electron microscopic images made by using a large objective aperture. Further we interpret the observed effect by the dynamical theory of electron diffraction and the classical theory.

Experiment

The principle of experiment is explained in Fig. 1a, b and c.
Fig. 1b is the experimental arrangement of Kumm et al. [4]. Fig. 1a
is the convergent beam technique and the intensity distribution can
be observed on a photographic plate. Another method given in
Fig. 1c is the arrangement to measure the intensity distribution
by use of the electron microscopic image of a curved crystal. The
three methods are almost equivalent to one another.

The pictures observed by using the convergent beam with the
electron energy of 100 keV from a silicon crystal with the surface
normal to the <100> axis are shown in Fig. 2a and b. Fig. 2a,
called "Kossel pattern" [6], is qualitatively the same as the
Kikuchi pattern from a thick crystal observed around the incident
beam direction (deficit pattern). Fig. 2b was taken by use of an
aperture with the ring form and represents the intensity distribution of the Rutherford scattering and has just a reverse intensity
distribution against that of Fig. 2a. As the interaction parameter,
$\alpha_a = 0.57$, given in the previous paper [3] is small in the case of
Fig. 2, the pattern around the <100> axis is the "geometrical" one.

Fig. 3a is the Kossel pattern around the <100> axis of a gold
film at the incident energy 1 MeV, where $\alpha_a = 14.5$. This picture
shows a dark (reverse) "flower" pattern. We can see a deep dip
at the center, while there are eight peaks around the dip which are
identified as the {310} reflection. The opening angle of the overall dip is equal to two times of about 0.65 ψ_1, ψ_1 being Lindhard's
critical angle. This result is the same as that for the Rutherford
scattering reported in the previous work [3].

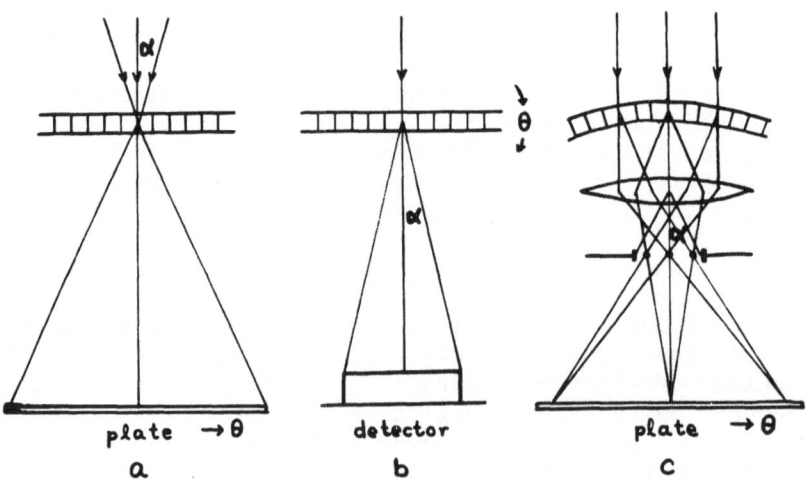

Fig. 1. Experimental arrangements for blocking observation.

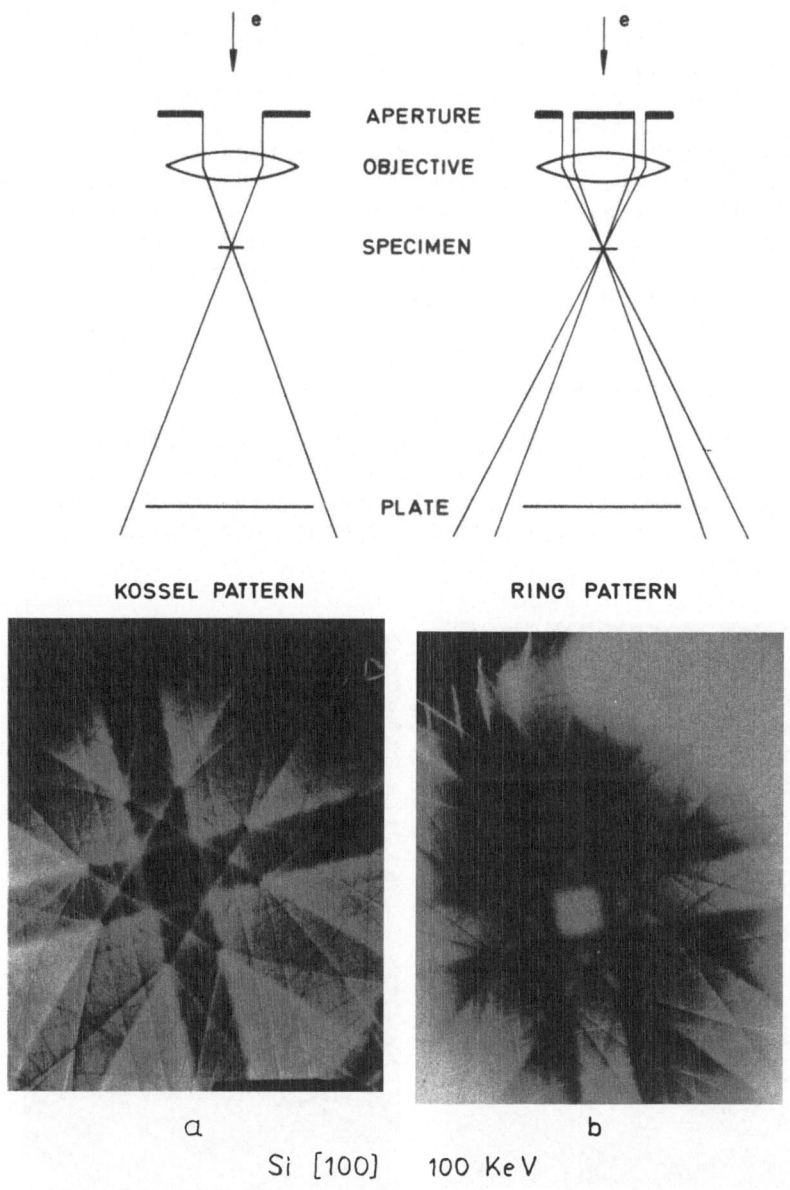

Fig. 2. (a) Kossel pattern and (b) Kikuchi pattern obtained from
a silicon crystal around the <100> axis with 100 keV electrons.
Thickness 2500 Å.

From the result of Fig. 2a and b we expect to observe the eight dark dips in the flower pattern observed in the Rutherford scattering. In fact, we can find these in Fig. 3b which is the flower pattern corresponding to Fig. 3a and was presented in the previous paper [3].

Fig. 4a and b are the electron microscope image of a tungsten thin film curved around the <111> axis which are taken at 1 and 2 MeV, respectively, with an objective aperture including thirteen Laue spots. The values of α_a are 28.0 and 46.7 for 1 and 2 MeV, respectively. In Fig. 4a at 1 MeV, a continuous ring around the center appears besides the six peaks. At 2 MeV as is seen in Fig. 4b, the intensity of this ring becomes strong, and the second ring outside the first one can also be observed.

Theoretical Interpretation and Discussions

Lehmpfuhl [7] has calculated the wave fields of electrons, which enter a MgO crystal nearly parallel to the <110> axis by means of the dynamical theory of 43 beams. In order to understand the peaks and rings observed in the present experiment, we extended

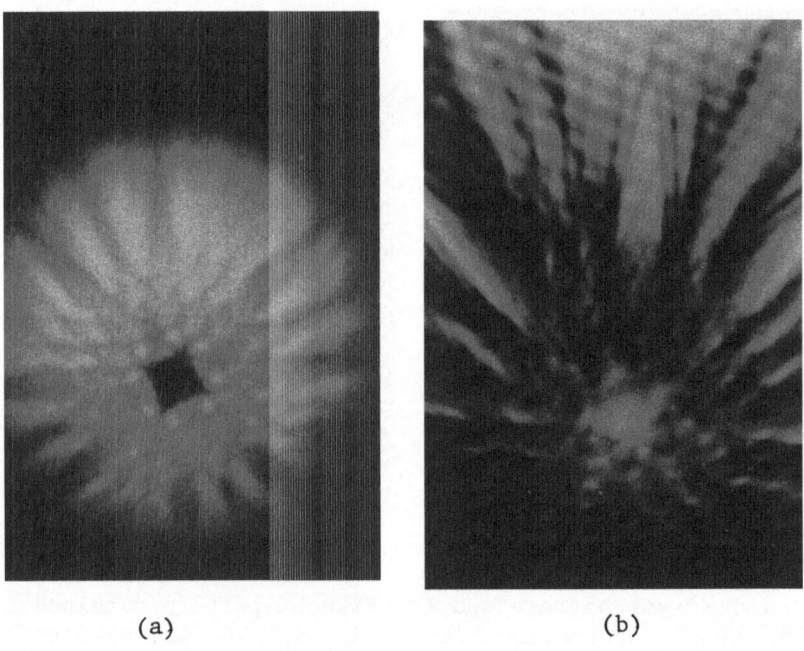

(a)   (b)

Fig. 3. (a) Kossel pattern and (b) Kikuchi pattern at a large scattering angle (ca. 6°) obtained from a gold crystal around the <100> axis with 1 MeV electrons.

his calculations to higher energies and studied the variation of the wave fields.

Fig. 5 and 6 give the electron densities of each wave field and the corresponding transversal energies, respecteively, at the incident energies 50 keV and 2 MeV. In Fig. 5 only the wave fields with high density around magnesium—atom strings are shown. At low energies, both the wave field maxima at the atom sites and that with maxima at the interstitial sites and zero at the atom sites are excited. As the energy increases, the transversal energy of each wave field comes down into the valley of atom potential and the electron densities are concentrated around the atom string. Besides, new types of wave fields are created. From the concepts of the band theory, we can identify 1s, 2s, 2p, 3d, as seen in Fig. 6. The wave fields 2p and 3d are just the quantized rosette motion and may correspond to the first and second rings in Fig. 4b. We find bright rings because the absorption effect for 2p and 3d wave fields is relatively small.

The ring around the center can be explained as follows: If the energy is not high enough, the transversal energy of the 2p wave field is not deep, and the string potential has no circular

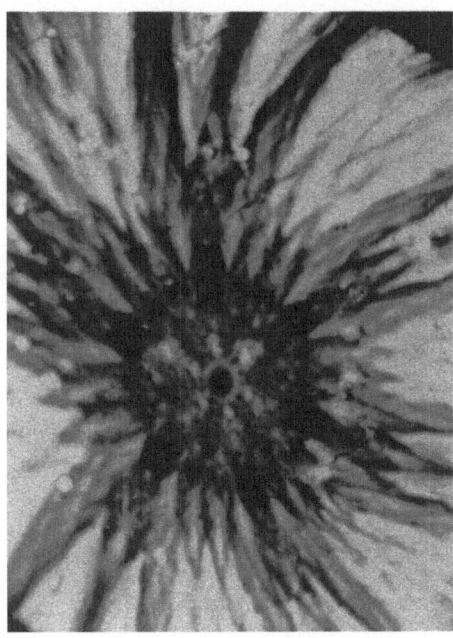 

(a) 1 MeV   (b) 2 MeV

Fig. 4. Electron microscopic images of a tungsten film curved around the <111> axis with a large objective aperture at (a) 1 MeV and (b) 2 MeV.

50 KeV   2 MeV

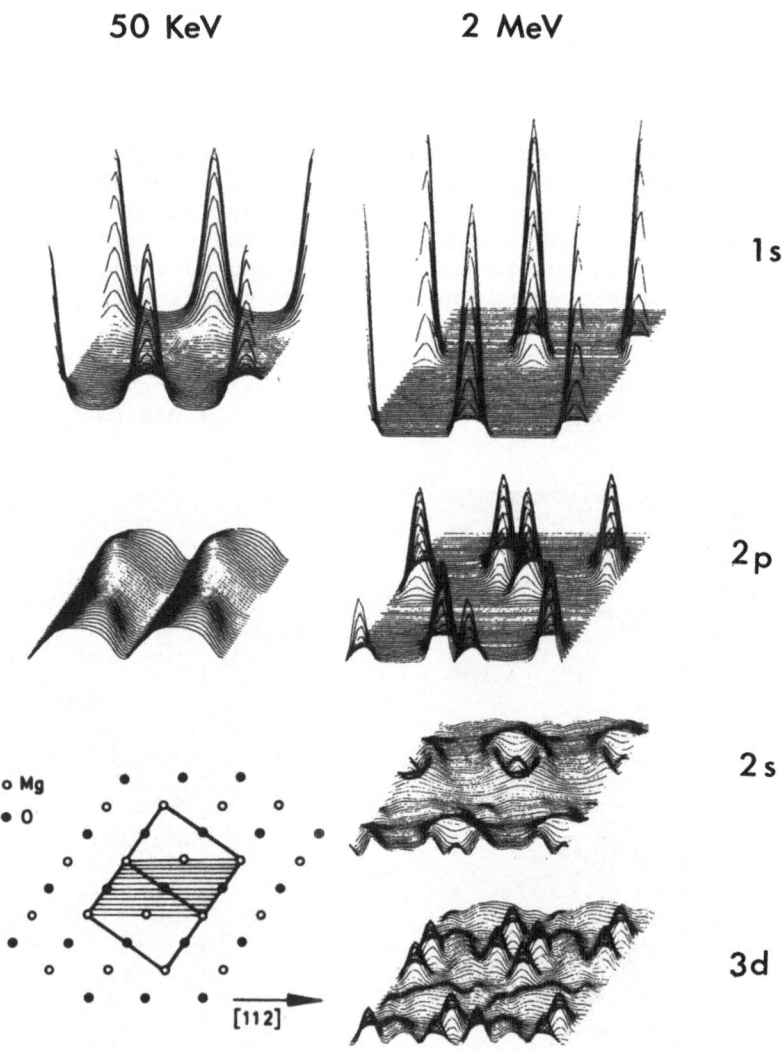

Fig. 5. Electron density of Bloch waves in a MgO crystal for
incidence near the <110> axis at 50 keV and 2 MeV.

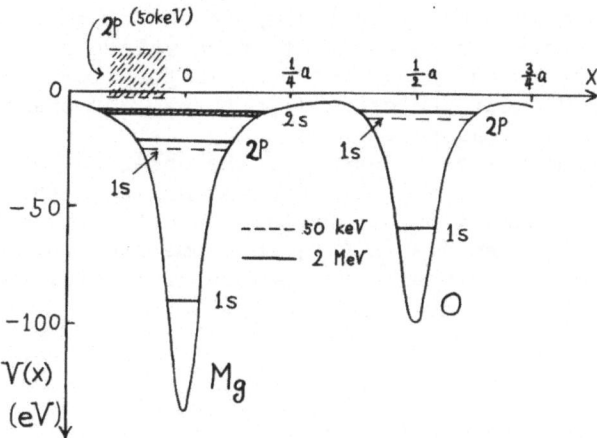

Fig. 6. Transversal energy levels of Bloch waves.

symmetry. Therefore, the wave field is excited only in the definite directions of the incident beam. At high energies, the
transversal energy is low and the string potential is circularly
symmetric. Then the excitation strength of the wave field is
independent of the incident beam direction.

The abnormal transmission in the axial case arises from the
asymmetric wave function such as 2p and 3d. On the other hand,
in the planar case a symmetric wave field is transmitted strongly
(so-called Bloch-wave channeling). This abnormal transmission
can be explained by the classical orbit as shown in Fig. 7.

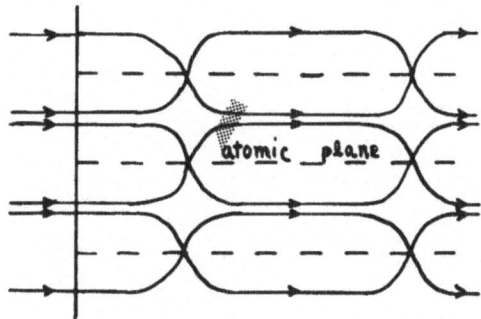

Fig. 7. Path of electrons in the "Bloch-wave channeling."

We discussed in this paper the spots and rings in the blocking dip of electrons for Au and W at large values of α_a, but we
have found the spots and rings not only in the flower patterns
but also in the deficit geometrical patterns, for example, for
Cu<111> [3]. If we measure the intensity distribution by the
conventional method using a detector, the observed curve depends
very much on the beam divergence, opening angle of the detector
and the azimuthal direction of crystal and detector. The difference among the measurements [2,4,8] may come from these situations.

More detailed accounts of the present work will be published
separately.

References

  *Department of Metallurgy, Osaka University, Osaka, Japan.
 **Japan Electron Optics Laboratory Co. Ltd., Tokyo, Japan.
***Fritz–Haber–Institut der Max–Planck–Gesellschaft, Berlin,
   Germany.
[1] E. Uggerhøj and J. U. Andersen, Can. J. Phys. 46, 543 (1968).
[2] E. Uggerhøj and F. Frandsen, Phys. Rev. B 2, 582 (1970).
[3] F. Fujimoto, S. Takagi, K. Komaki, H. Koike, and Y. Uchida,
    Rad. Effects 12, 153 (1972).
[4] K. Kumm, F. Bell, R. Sizmann, H. J. Kreiner, and D. Harder,
    Rad. Effects 12, 53 (1972).
[5] L. E. Thomas and C. J. Humphreys, Phys. Stat. Sol. (a) 3,
    599 (1970).
[6] P. Goodman, Acta Cryst. A 28, 92 (1972).
[7] G. Lehmpfuhl, Z. Naturforsch. 28a, 1 (1973).
[8] A. A. Vorobiev, A. Ya. Bobudaev, S. A. Vorobiev, and V. V.
    Caplin, Phys. Lett. 40A, 108 (1972).

COHERENT PHOTON EMISSION BY FAST PARTICLE EXCITED
ATOMS IN ELECTROMAGNETIC FIELD

M. I. RIAZANOV

Moscow Physical-Engineering Institute
Moscow, U.S.S.R.

A particle moving in a substance continuously excites the atoms with which it collides. It gives rise to an incoherent emission of photons by excited atoms (Cherenkov emission is an exception). A method which permits one to calculate coherent photon emission by excited atoms in electromagnetic field has been developed.

Let us consider the kinematics of photon emission with the frequency ω and the momentum $\vec{k}$ (h=c=1) by the atom excited due to interaction with the particle and the electromagnetic field. Let $\vec{p}$ be the initial momentum of the particle, E initial energy, $\vec{p}-\vec{q}$ and $E-\vec{q}\vec{v}$ the final momentum and final energy of the particle ($\vec{v} = \vec{p}/E$ is the particle velocity). Let $\vec{k}_1$ and ω_1 be the change of momentum and energy of electromagnetic field. The condition required for identification of the coherent photon emission by different atoms is the identification of initial and final states of an atom. Therefore, the law of energy and momentum conservation may be written as

$$\vec{q} + \vec{k}_1 - \vec{k} = 0$$

$$\vec{q}\vec{v} + \omega_1 - \omega = 0 \ .$$

$$(1)$$

The angle ϕ between the direction of the coherent photon emission and the particle velocity is defined by

$$\cos\theta = \frac{1}{v\sqrt{e}} (1 - \frac{\omega_1 - \vec{k}_1 \vec{v}}{\omega}) \ .$$

$$(2)$$

At $\omega_1 \to 0$, $\vec{k}_1 \to 0$ (2) assumes the well-known form for the Cherenkov emission [1].

It is convenient to choose the external electromagnetic field

$$\vec{E}_0(\vec{r},t) = \vec{E}_{01} \cos(\vec{\kappa}_1\vec{r}-\omega_1 t+\theta_1)$$

$$+ \vec{E}_{02} \cos(\vec{\kappa}_2\vec{r}-\omega_2 t+\theta_2) \, (\vec{E}_{01}||\vec{E}_{02}) \qquad (3)$$

resonant with respect to the energy levels E_3, E_2, E_1, namely:
$\omega_1 = E_2-E_1+\epsilon_1$; $\omega_2 = E_3-E_2+\epsilon_2$; $\epsilon_1<<\omega_1$; $\epsilon_2<<\omega_2$. The process is the
effect of both the field E_0 and the particle field

$$\vec{E}_p(\vec{r},t) = \int d^3q \vec{E}_p(\vec{q}) \ell^{i\vec{q}\vec{r}-i\vec{q}\vec{v}t}$$

$$\vec{E}_p(\vec{q}) = \frac{ie}{2\pi^2\epsilon} \quad \frac{\vec{v}(\vec{q}\vec{v})\epsilon-\vec{q}}{q^2-(\vec{q}\vec{v})^2\epsilon} \qquad (4)$$

ϵ is the dielectric constant of a substance. In the frequency
range $\gamma<<\omega-\omega_{30}<<\omega_{30}$ $(\omega_{30}\equiv E_3-E_0)$ (γ is the excited level energy
width) the dipole moment induced by the fields $\vec{E}_0$ and $\vec{E}_p$ in the
atom is expressed by $\vec{d}(t) = \int d^3\kappa \int d\omega \vec{d}(\vec{\kappa},\omega) \exp(i\vec{\kappa}\vec{R}_a-i\omega t)$, $\vec{R}_a$ is the
atom radius-vector;

$$\vec{d}(\vec{\kappa},\omega) = -\vec{d}_{03}(\vec{d}_{32}\vec{E}_{02})(\vec{d}_{31}\vec{E}_{01})(\vec{d}_{10}\vec{E}_p(\vec{\kappa}-\vec{\kappa}_1-\vec{\kappa}_2))$$

$$\cdot \exp(i\theta_1+i\theta_2)\delta(\omega-\omega_1-\omega_2-\vec{\kappa}\vec{v}+\vec{\kappa}_1\vec{v}+\vec{\kappa}_2\vec{v})$$

$$\cdot \{4(\omega-\omega_{30})(\omega-\omega_{30}-\epsilon_2)(\omega-\omega_{30}-\epsilon_1-\epsilon_2)-|\vec{d}_{21}\vec{E}_{01}|^2$$

$$\cdot (\omega-\omega_{30})-|\vec{d}_{32}\vec{E}_{02}|^2(\omega-\omega_{30}-\epsilon_1-\epsilon_2)\}^{-1} . \qquad (5)$$

Applying (5) one can get the energy coherently emitted by the
excited atoms into a solid angle $d\Omega$ within the frequency interval
$d\omega$ in unit time

$$d\epsilon = \omega^4 dw d\Omega \frac{2\pi^3}{q} n_0^2|v_{21}^0|^2|d_{03}|^2|d_{10}|^2[\vec{n}\vec{e}]^2$$

$$\cdot |[\vec{n}\vec{E}_p(\vec{\kappa}-\vec{\kappa}_1-\vec{\kappa}_2)]|^2\delta(\omega-\omega_1-\omega_2-\vec{\kappa}\vec{v}+\vec{\kappa}_1\vec{v}+\vec{\kappa}_2\vec{v})$$

$$\cdot|(\omega-\omega_{30})(\omega-\omega_{30}-\epsilon_1)(\omega-\omega_{30}-\epsilon_1-\epsilon_2)-|v_{21}^0|^2(\omega-\omega_{30})-|v_{32}^0|^2(\omega-\omega_{30}-\epsilon_1-\epsilon_2)|^{-2}$$

$$(6)$$

$$(\vec{E}_{01}\equiv\vec{e}E_{01}; \ \vec{E}_{02}\equiv\vec{e}E_{02})$$

$$|v_{21}^0|^2 = \frac{1}{12} |d_{21}E_{01}|^2, \quad |v_{32}^0|^2 = \frac{1}{12} |d_{32}E_{02}|^2$$

n_0 is the number of atoms per unit volume. It should be stressed that small values of the denominator in (6) may correspond to the smallness of $(\omega-\omega_{30})$, i.e., the frequency range, where the absorption is very noticeable as well as to the region where absorption is lacking. For example, at $\varepsilon_1 = \varepsilon_2 = 0$ and

$$\omega-\omega_{30} \simeq \pm\{|v_{12}^0|^2 + |v_{32}^0|^2\}^{1/2}$$

at

$$(\omega-\omega_{30}) \gg \gamma \;, \quad |v_{12}^0| \gg \gamma \quad \text{and} \quad |v_{32}^0| \gg \gamma \tag{7}$$

It is worthwhile to compare (6) with the Vavilov–Cherenkov radiation intensity in media with $\varepsilon-1 \sim 1$. The ratio is

$$(d\varepsilon/d\varepsilon_{V-Ch}) \sim (n_0 r_{at}^3)^2 (\omega_{at}/\Delta\omega) \tag{8}$$

where r_{at} is the atom size; ω_{at} is the frequency approximate to the atomic one, $\Delta\omega$ is the maximum from the values ε_1, ε_2, γ, $\omega-\omega_{30}$. The condition (7) requires that the external electromagnetic field be sufficiently strong. Therefore, the major obstacle faced by the experimentors is the proper choice of strong external electromagnetic field frequency that should accurately coincide with the frequencies of transitions in atoms. Nevertheless, it can be seen from (8) that the intensity of the process involved may greatly exceed that of the Vavilov–Cherenkov radiation. The possibility of using the particular effect to detect particles has been discussed elsewhere [2].

References

[1] J. V. Jelley, "Cherenkov Radiation and Its Applications,"
    Pergamon Press (1958).
[2] M. I. Riazanov, ZhETF Pis. Red, 15, 437 (1972).

EFFECT OF THE LEVELS OF THE TRANSVERSE MOTION OF ELECTRONS ON THE ELECTROMAGNETIC PROCESSES IN MONOCRYSTALS

N. P. KALASHNIKOV, E. A. KOPTELOV and M. I. RIAZANOV
The Moscow Physical-Engineering Institute
Moscow, U.S.S.R.

ABSTRACT

The orientational dependence of the elastic scattering of fast nonrelativistic electrons in monocrystals has been investigated when the energy of transverse motion $E_\perp = E\,\theta_0^2$ is smaller than the energy of the "bound" state of the atomic plane. The presence of the "bound" state of transverse motion in the effective attractive potential of the crystallographic planes leads to the radiation gamma-quantum when the electron transits into the discrete spectrum state. This effect is responsible for the production of the orientational maxima in the bremsstrahlung spectrum of nonrelativistic electrons.

The motion of the fast charged particle in monocrystal can be represented as motion in the total poential,

$$U(\vec{r}) = \sum_i U_o(\vec{r}-\vec{R}_i) = a^{-3}\sum_{\vec{n}} U_o(\frac{2\pi}{a}\,\vec{n})\,e^{i\frac{2\pi}{a}\,\vec{n}\vec{r}} \ . \tag{1}$$

The interaction of the fast charged particle with the lattice atoms is characterized by the small longitudinal momentum transfer

$$(\Delta p)_{||} \sim \frac{1}{pR_{eff}^2} << \frac{1}{a} \tag{2}$$

where $R_{eff} = \{me^2\}^{-1/3}\}^{-1}$ is the Thomas-Fermi radius of the atom, "a" is a lattice constant and $\hbar = c = 1$. The longitudinal momentum transfer becomes infinitesimally small when the incident momentum goes to the infinity. In this case (2) the atoms of monocrystal, which are at the length $\ell \sim (\Delta p_{||})^{-1}$, scatter coherently. It follows that the average crystallographic plane potential [1] gives the adequate description of the scattering problem, when the entrance angle is small enough. Thus the total potential (1) is divided into two parts $U = \overline{U}_{plane} + W(\vec{r})$, where

$$U_{plane} = \sum_{\vec{n}=(n_x,o,o)} U_o(\frac{2\pi}{a}\vec{n}) \exp(i\frac{2\pi}{a}\vec{n}\vec{r})\frac{1}{a^3}$$

$$= \sum_s \frac{1}{a_y a_z} \int dy \int dz\ U_o(\ (x-X_s)^2+y^2+z^2)$$

$$= \sum_s V_o(x-X_s) \ .$$ (3)

The potential $U_o(r)$ is a potential of a single atom

$$U_o(\vec{r}) = -\frac{Ze^2}{r} e^{-\kappa r}$$

and the second part of the total potential $W(\vec{r})$ takes into account the deviation of the exact potential from the average plane potential, i.e., the thermal vibrations and the discreteness of the lattice potential. The processes determined by the potential $W(\vec{r})$ will be considered by means of the perturbation theory.

Let us investigate the motion of the fast nonrelativistic electrons in the average crystallographic plane potential. The motion of the negative charged particle is characterized by the presence of the "bound" state of the transverse motion.

It is of great interest to consider the case when the negative particle has only one energy level of the discrete spectrum in the potential (3). The energy of this level is [2]

$$\epsilon = -\frac{\lambda^2}{2m}$$ (4)

where $\lambda = \frac{4\pi Z^{2/3}}{\kappa a^2}$. (5)

The only level of the discrete spectrum occurs in the relatively small potential well providing that

$$\frac{4\pi Z^{2/3}}{(\kappa a)^2} \ll 1 \ .$$ (6)

The wave function of the only discrete state takes the form [2]

$$\phi_\lambda(x) = \begin{cases} \sqrt{\lambda}\ \exp(-\lambda|x|) & \kappa^{-1} \leq x \leq \frac{a}{2} \\ \sqrt{\lambda} & |x| < \kappa^{-1} \end{cases} . \tag{7}$$

The presence of the only energy level does not permit the classical treatment of the transverse motion of the electron in monocrystals. It should be noted that the scattering problem in the potential of the crystallographic planes (3) reduces to the scattering problem in the field of the single crystallographic plane when the splitting of the discrete energy level due to the effect of the neighoring crystallographic planes is negligibly small, i.e.,

$$\left|\frac{\Delta\varepsilon}{\varepsilon}\right| \simeq \frac{2\kappa}{\lambda} \exp(-\lambda a) \ll 1 . \tag{8}$$

It follows that

$$\exp\left(-\frac{4\pi Z^{2/3}}{\kappa a}\right) \ll \frac{4\ Z^{2/3}}{(\ a)^2} \ll 1 . \tag{9}$$

The total wave function $\psi_{\vec{p}}(x,y)$, which describes the electron motion in the average crystallographic plane potential and has the boundary condition $\psi_{\vec{p}}(x,y=0) = e^{i\ p\theta_0 x}$, is written as

$$\psi_p(x,y\geq o) = e^{ipy(1 - \frac{\theta_o^2}{2})} \left\{ e^{ip\theta_o x} - \frac{1}{2(\lambda^2+p^2\theta^2)} \right.$$

$$\cdot\ [e^{ip\theta_o|x|}(\lambda^2-i\lambda p\theta_o)(1-\Phi(\sqrt{\frac{px^2}{i2y}} - \sqrt{\frac{p0_o^2 y}{i2}}))]$$

$$\left. + e^{-ip\theta_o|x|}(\lambda^2+i\lambda p\theta_o)\ [1-\Phi(\sqrt{\frac{px^2}{i2y}} + \sqrt{\frac{p\theta_o^2 y}{i2}})]\right\}$$

$$+ e^{i(p+\frac{\lambda^2}{2p})y-\lambda|x|}\frac{\lambda^2}{\lambda^2+p^2\theta_o^2}\ [1+\Phi(\sqrt{i\frac{\lambda^2 y}{2p}} - \sqrt{\frac{px^2}{i2y}})],$$

$$\tag{10}$$

where $\Phi(z) = \frac{2}{\sqrt{\pi}}\int_o^z dt\ e^{-t^2}$ is the error integral.

It follows from the formula (10) that the legth of the essential reconstruction of the wave function is

$$L_o \gtrsim 2p/\lambda^2$$

which increases with the incident momentum.

Let us consider the scattering of the fast nonrelativistic electron through the large angle when the angle between the incident electron momentum and crystallographic plane is small. The transition matrix element is determined by the potential $W(\vec{r})$ and is written as

$$<\vec{P}_f|M|\vec{P}_i> = \sum_\alpha <\vec{P}_f|M_o|\vec{P}_i>e^{-i\vec{q}\vec{R}a} \{1 + \frac{i\lambda p\theta_o}{\lambda^2+p^2\theta_o^2} \theta (\frac{p\theta_o^2 Y_a}{12})$$

$$+ \frac{\lambda^2}{\lambda^2+p^2\theta_o^2} e^{i\frac{\lambda^2+p^2\theta_o^2}{2p}} Y_a \theta(i\frac{\lambda^2 Y_a}{2p}) . \qquad (11)$$

The differential scattering cross section is determined by the squared modulus of the expression (11). In the case $\theta_o = 0$ we have

$$\frac{d\delta}{d\Omega} = \frac{d\delta_o}{d\Omega} N_{tot} \{1+ \frac{2}{\eta 2} [\cos\eta^2(C(\eta)-S(\eta)) + \sin\eta^2(C(\eta)+S(\eta))$$

$$- \frac{2}{\sqrt{\pi}} \frac{2}{\eta} + 2 \int_0^1 d\tau [C^2(\eta\tau^{1/2})+S^2(\eta\tau^{1/2})]\} , \qquad (12)$$

where $\frac{d\delta_o}{d\Omega}$ is the differential scattering cross section by the single atom, N_{tot} is the total number of the lattice atoms, $C(\eta)$ and $S(\eta)$ are the Fresnel functions, $\eta^2 = \frac{\lambda^2 L}{2p}$, and L is the thickness of the monocrystal along the incident motion direction. When $\eta<<1$, the differential scattering cross section closely resembles the usual Rutherford scattering cross section. But in the opposite case $\eta>>1$, from (12) we obtain:

$$\frac{d\delta}{d\Omega} \simeq \frac{d\delta_o}{d\Omega} N_{tot} \cdot 2 . \qquad (13)$$

This increase of the differential scattering cross section is due to the essential reconstruction of the electron wave function in the "channeling" regime.

When the entrance angle is not equal to zero, instead of (13) we obtain

$$\frac{d\delta'}{d\Omega} = \frac{d\delta_o}{d\Omega} \ N_{tot} \ (1 + \frac{\lambda^2}{\lambda^2 + p^2 \theta_o^2}) \ , \quad \eta >> 1 \ . \tag{14}$$

These theoretical results for the plane "channeling" qualitatively agree with the available experimental data.

Let us investigate the bremsstrahlung of the fast nonrelativistic electrons in a single crystal when the incident electron momentum is nearly parallel to the crystallographic plane YOZ. The energy-momentum conservation law permits the radiation gamma-quantum with the momentum "$\vec{k}$" only if the longitudinal momentum transfer is larger than

$$(\Delta p_{||})_{min} = |\vec{p}_i| - |\vec{p}_f| - |\vec{k}| \ . \tag{15}$$

If the incident and final states of electron are free then

$$(\Delta p_{||})_{min} = \omega(v^{-1}-1) \ (\hbar=c=1), \tag{16}$$

where ω is the frequency of the radiated gamma-quantum. The reciprocal value of the longitudinal momentum transfer determines the effective length of the bremsstrahlung. Therefore it follows from (16) that for nonrelativistic particles the effective length of the bremsstrahlung is smaller than the lattice constant, and in this case the crystal structure has no influence on the bremsstrahlung spectrum.

However, the "channeling" effect changes the situation essentially. The presence of the "bound" state of transverse motion in the effective attractive potential of the crystallographic plane leads to the gamma-quantum radiation, even when the longitudinal momentum transfer is equal to zero. The effective range of the bremsstrahlung is the monocrystal thickness L. Now we consider the bremsstrahlung process, when the initial state of an electron is a plane wave and the initial state of an electron is a plane wave and the final state is the state of the discrete spectrum of the transverse motion. In this case the matrix element of the bremsstrahlung takes the form [5]

$$M_{i \rightarrow f}^{(1)} = \frac{ie}{2\omega \ m} \int \psi_f^* \ \vec{e} \ \vec{p} \ e^{i\vec{k}\vec{r}-\omega t} \psi_i \ d^3 r dt, \tag{17}$$

where ψ_i and ψ_f are the electron wave functions in the initial and final states. By substituting the explicit expressions for ψ_i and ψ_f (10) and performing the integration one can obtain

$$M_{i \to f}^{(1)} = \frac{ie}{\sqrt{2\omega}\, m}\, (\vec{e}\vec{p}_o)(2\pi)^3 \delta(E_i - E_f - \omega)\delta(p_{i_x} - p_{f_x} - k_x)\delta(p_{fy} - k_y)$$

$$\cdot\, 2\alpha^{3/2}\, \frac{\cos\frac{k_z}{2\kappa} + \frac{k_z}{k_z}\sin\frac{k_z}{2\kappa}}{\alpha^2 + k_z^2} \cdot \tag{18}$$

Therefore, the differential bremsstrahlung cross section is written as

$$d\delta = 2\, e^2(1 - \cos^2\theta) L_x L_y\, \sin\Theta\, d\theta\, d\phi\, \omega d\omega\, \left(\frac{mU_o}{\kappa}\right)^3$$

$$\cdot \left[\frac{\cos\frac{\omega\sin\theta\,\cos\phi}{2\kappa} + \frac{Mu_o}{\omega\kappa\sin\theta\,\cos\phi}\sin\frac{\omega\sin\theta\,\cos\phi}{2\kappa}}{\left(\frac{mU_o}{\kappa}\right)^2 + \omega^2\sin^2\theta\,\cos^2\phi} \right]^2 \cdot \tag{19}$$

In the particular case corresponding with the experimental conditions [6], $\sin\theta\cos\phi = 1$, the differential bremsstrahlung is simplified

$$\frac{d\delta}{d\omega} \sim \omega\, \frac{\left(\cos\frac{\omega}{2\kappa} + \frac{\alpha}{\kappa}\sin\frac{\omega}{2\kappa}\right)^2}{\left(1 + \frac{\omega^2}{\alpha^2}\right)^2} \tag{20}$$

where α is determined by (5).

It follows that the presence of the "bound" state of the transverse motion in the effective attractive potential of the crystallographic plane leads to the production of the orientational maxima in the bremsstrahlung spectrum of nonrelativistic electrons.

References

[1] J. Lindhard, Mat-Fys. Medd. Dan. Vid. Selsk. 34, No. 14 (1965).
[2] L. D. Landau and E. M. Lifshitz, Quantum Mechanics, Addison-Wesley (1965).
[3] E. Uggerhoj, F. Frandsen, Phys. Rev. 2B, 582 (1970).
[4] Yu. S. Korobochko, E. L. Berezovskii, B. D. Grachev, S. S. Kozlovskii, and V. I. Mineev, Soviet Physics, Solid State Physics 14, 3316, (1972)

[5] A. I. Akhiezer and V. B. Berestetskii, Kvantovaya Elektrodinamika
    (Quantum Electrodynamics), Moscow (1969).
[6] Yu. S. Korobochko, V. E. Kosmach, B. I. Mineev, ZHETP <u>48</u> (5),
    1248 (1965).

SECTION VII
SURFACE SCATTERING

MEDIUM-ENERGY ION SCATTERING BY CRYSTAL SURFACES

V. A. MOLCHANOV
Institute of Nuclear Physics
Moscow State University
Moscow 117234, U.S.S.R.

ABSTRACT

The results of the experimental studies of medium-energy ion scattering by crystalline surfaces published of recent years are discussed. Main attention is paid to such peculiarities which resulted from "near-surface" character of the ion scattering process.

SOME DIRECTIONAL EFFECTS IN FORWARD ION SCATTERING BY CRYSTAL SURFACES

E. S. MASHKOVA and V. A. MOLCHANOV
Institute of Nuclear Physics
Moscow State University
Moscow 117234, U.S.S.R.

ABSTRACT

The results of the experimental study of 10-30 keV argon ion scattering by (110) copper face both in primary ions incident plane and in space are presented. The inlet and outlet blocking effects as well as energy loss behavior were examined. The obtained data are discussed in the framework of isolated string.

ON THE SCATTERING OF LOW ENERGY H^+ AND He^+ IONS FROM A (001) COPPER SURFACE

H. H. W. FEIJEN, L. K. VERHEY, E. P. Th. M. SUURMEIJER
and A. L. BOERS
*Technical Physics Laboratory, State University
Groningen, The Netherlands*

ABSTRACT

Energy spectra and angle dependent yield distributions of 3-9 keV H^+ and He^+ ions scattered from a (001) face of a copper single crystal were measured. Spectra recorded for scattering planes making small angles with a <110> direction on the surface are influenced by the focusing action of the surface semi-channels in this direction. Computer simulations were carried out to understand the main regularities of this focusing effect.

Introduction

Under certain conditions the scattering of ions on monocrystalline surfaces will be seriously influenced by the presence of so-called surface "semi-channels," formed by the atomic rows of the first and second crystal layers. Theoretical predictions of focusing effects resulting from these channel-structures [1,2,3] were confirmed experimentally by Mashkova et al. [4] and Yurasova et al. [5]. In these experimental investigations, (001) copper surfaces were bombarded with Ar^+ ions, being rather heavy projectiles and consequently giving rise to large critical angles for channeling. In our experiments H^+ and He^+ ions have been applied because their critical angles are smaller and therefore the characteristic features of semi-channel scattering may show up more clearly [6].

Experimental

    The experimental set-up has been described in detail else-
where [7,8]. Some experimental data of the apparatus are:
    - primary ion energy: 2-10 keV
    - primary ion current density $\sim 5.10^{-9}$ A/cm^2;
    - residual gas pressure in collision chamber $\sim 5.10^{-10}$ Torr;
    - the target is a Cu (001) single crystal;
    - energy selection by means of a .75% electrostatic analyzer;
    - ion detection by means of a specially designed Daly conver-
       ter [8].
In all measurements the total scattering angle θ was fixed at 30°.
Detection occurred in the plane of incidence; i.e., the azimuthal
scattering angle was always zero, the glancing angle ψ could be
varied between 0° and 30° and the azimuthal angle of target rotation ϕ between -5° and 55°. $\phi=0$° coincides with a <100> direction
on the surface (see Fig. 1).

Results and Discussion

    Energy distributions and ϕ-dependence

    It is well-known that the amount of reflection of energetic
(keV) protons from the bulk of a monocrystalline target depends
upon the orientation of the crystal with respect to the incoming
ion beam. Thus, in our case we may expect that the energy distributions of reflected particles will vary with both ϕ and ψ (θ fixed
at 30°).
    Figure 2a shows the influence of ϕ on the energy distributions
of 3 keV protons (the incident beam actually consists of 6 keV H_2^+

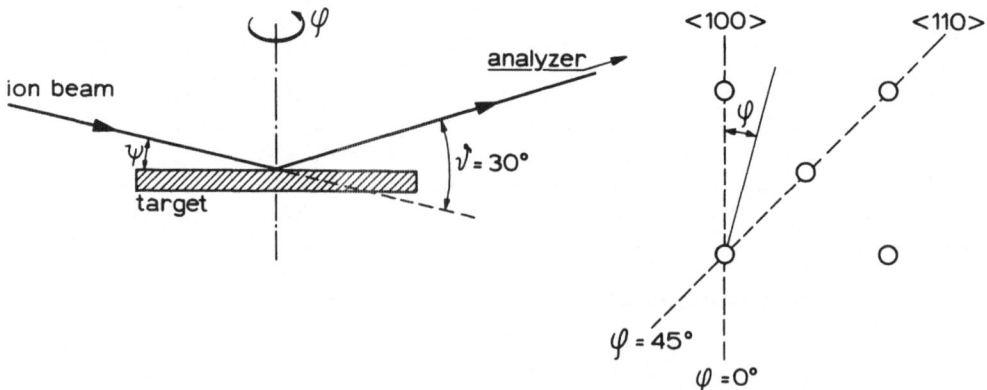

Fig. 1. Schematic diagram of the target arrangement. ψ - incidence angle; ϕ - azimuthal angle of target rotation; θ - total
scattering angle.

ions) scattered through 30° on a (001) copper crystal. Spectra
are presented for ϕ = 0°, 45° and 41°. For ϕ = 0° (incidence plane
parallel to a <100> direction on the surface) only a narrow peak is
observed corresponding to reflection from atoms in the outermost
surface layer, since under these conditions the yield from below
the surface is seriously reduced by planar channeling. For ϕ = 45°
the distribution becomes higher and broader and the maximum moves
to lower reflected energies [6]. Thus, while under these circum-
stances the spectrum would be expected to be influenced by planar
channeling too, the contribution of the second and lower layers to
the reflection apparently dominates the reflection from the outermost layer.

In order to get a better understanding of the atomic interaction processes leading to these interesting observations around
ϕ = 45°, the ϕ-dependence of the reflected yield has been measured
as a function of the reflected ion energy. Figure 2b shows the
result for three different reflected energies (a small tilt of the
target causes the difference in heights between the left and right
peak in this and the following figures). The peaked structure of
these distributions and their strong dependence on the reflected
ion energy indicates that we are dealing with a kind of focusing

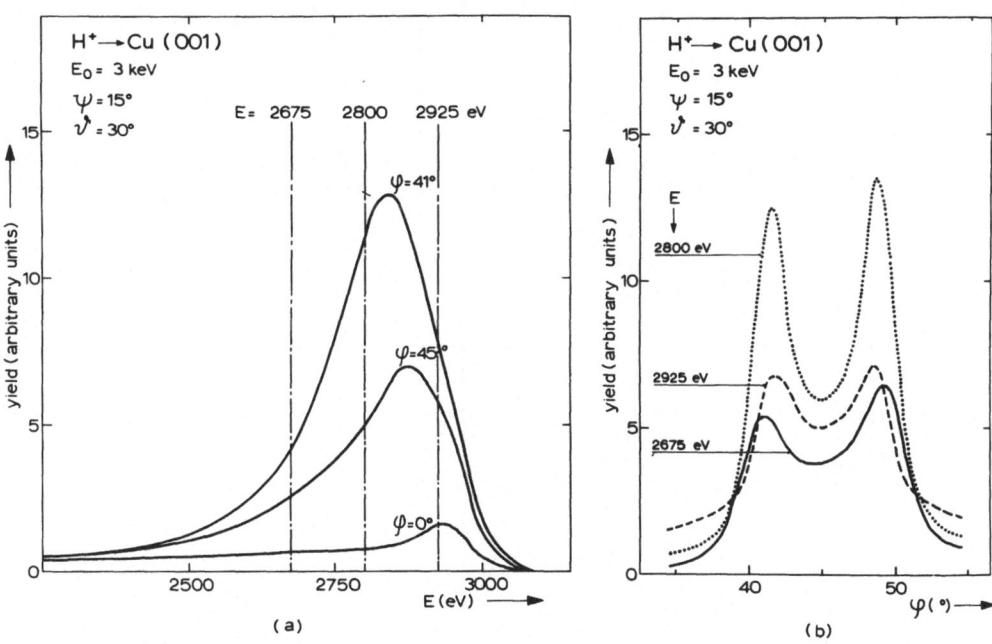

Fig. 2. a) Energy distributions of 3 keV H⁺ ions incident on a
Cu(001) surface for different angles ϕ (see Figure 1). b) Reflected
yields of H⁺ ions scattered through an angle θ = 30° as a function
of ϕ for different reflected ion energies E.

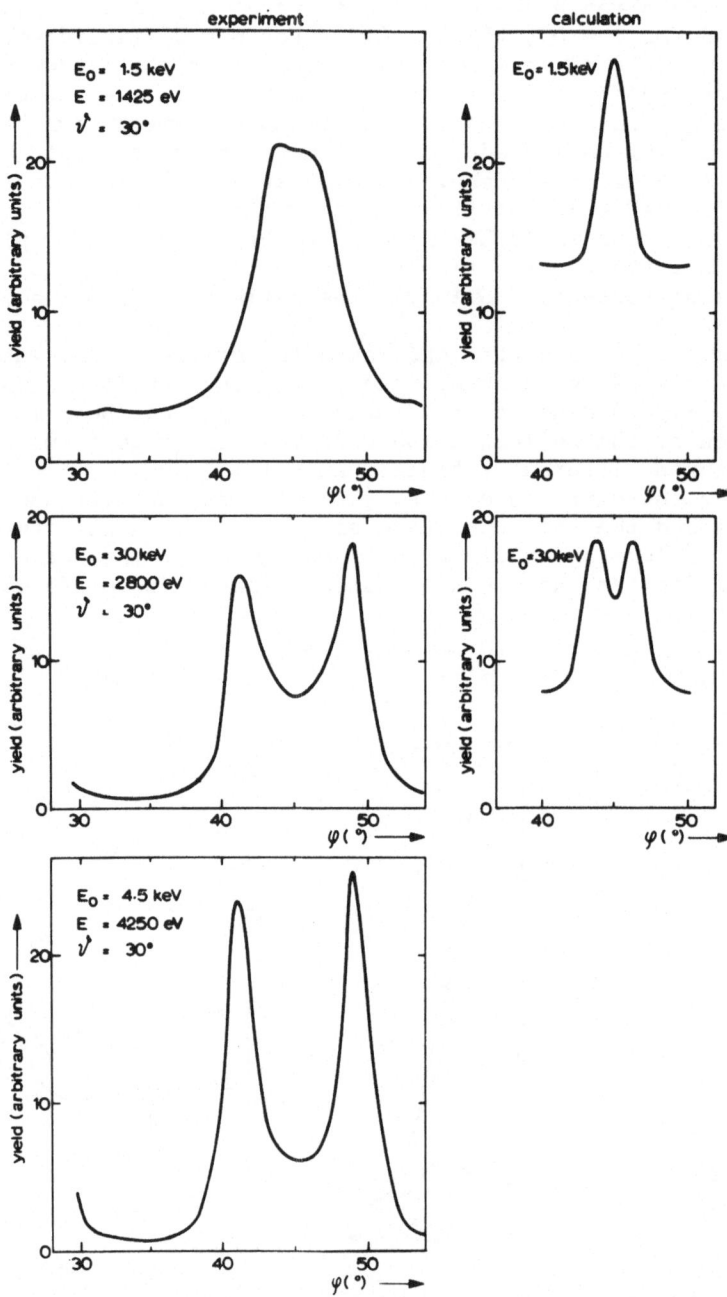

Fig. 3. Left: reflected yields of H$^+$ ions from Cu(001) at ψ=15°
as a function of ϕ for different primary energies Eo. Right: calculated total reflected yield (see text) versus ϕ for different
primary energies Eo.

mechanism having to do with the presence of the so-called surface
semi-channels in the <110> directions of a (001) copper surface.
In the next sections these ideas will be discussed in detail on
the basis of more systematic investigations, both numerical and
experimental.

Focusing effects related to surface semi-channels

The presence of semi-channels on the surface of a monocrystalline target may give rise to different focusing effects. a) The
parallel incident ion flux is partly transformed into a wedge-
shaped one due to deflection on the atomic chains of the first
layer in directions perpendicularly to them. Under certain conditions one or both edges of this wedge-shaped beam will be focused
on the atomic chains of the second layer ("wedge-focusing") and
the reflected yield from the second layer will increase. b) For
small incidence angle ψ focusing also occurs in the direction of
the atomic chains forming the edges of the semi-channels (chain-
effect). It should be noted that the so-called surface semi-
channeling only occurs when both focusing conditions are fulfilled.
As concerned to the observations made in Figure 2 we tend to
believe that they are due to the so-called "wedge-focusing" effect.
Further verification of this statement has been made by numerical
and experimental investigations of the energy dependence of the
effect.
The experimental results presented at the left side of Figure
3 concern ϕ-dependence, similarly to those presented in Figure 2b,
for different incident ion energies. The reflected ion energy
corresponds to the energy for which the observed peaked distribution
is most pronounced. For 4.5 keV incident energy the distribution
shows two large peaks with a deep minimum in between. For 3 keV the
separation between the peaks is less and for 1.5 keV only one peak
remains. Convincing evidence for the occurrence of a focusing mechanism follows from the fact that the yield in the peak is several
times the "random" yield [6] (for 1.5 keV H⁺ about 5 times).
Computer simulations of the scattering process have been carried out in order to interpret the experimental observations. The
results are presented at the right side of Figure 3. As can be
noticed there is a qualitative agreement. The fact that the experimental distributions are broader than the calculated ones may be
due to the choice of the interatomic potential function. The
background is too high because in the calculated curves all reflected particles, irrespective of their charge, energy and outgoing
angle, are included. A more precise examination of the computer
calculations indeed sustains the above assumption that the observed
focusing effect is due to the presence of <110> semi-channels for
$\phi = 45°$. For 3 keV H⁺ ions the reflected yield in the peaks mainly
originates from primaries with impact points between the <110>
atomic chains of the first and second layer. Under influence of
the chains in the first layer, forming the so-called "walls" of

the semi-channels, these particles are focused on the second layer
chains in between them (semi-channel base). The result of this
focusing is a high reflected yield from the second layer. For
1.5 keV H^+ ions the deflection by the surface chains is so large
that focusing on the second layer occurs for ϕ-values close to
$\phi = 45°$. For ϕ-values further away from 45° the second layer gradually disappears in the shadow of the first layer.

When the described model is correct the focusing effect should
also be ψ-dependent. Therefore in Figure 4 the reflected yield of
4.5 keV H^+ ions is plotted against ϕ for different incidence angles
ψ. As appears from this figure, for decreasing ψ-values the two
peaks shift together and for $\psi = 9°$ only one peak remains.

In Figure 5 the angle between the peaks has been plotted as a
function of ψ for three different reflected energies (curves B).
The explanation for these observations is as follows. Starting
from $\phi = 45°$ and rotating the target, the reflected yield will increase and pass a maximum when the edge of the shadow of one of
the <110> semi-channel walls passes through its base [9]. When
further rotating the target, the yield will again decrease, because

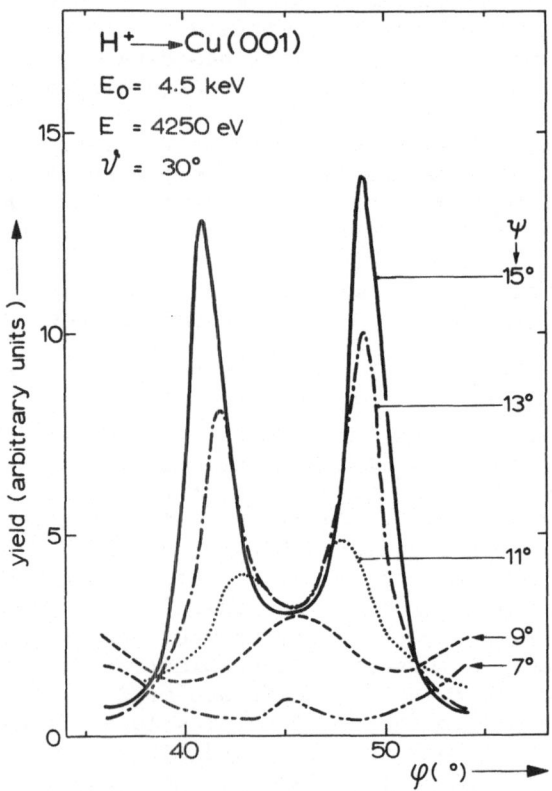

Fig. 4. Reflected yields of 4.5 keV H^+ ions as a function of ϕ
for different angles of incidence ψ.

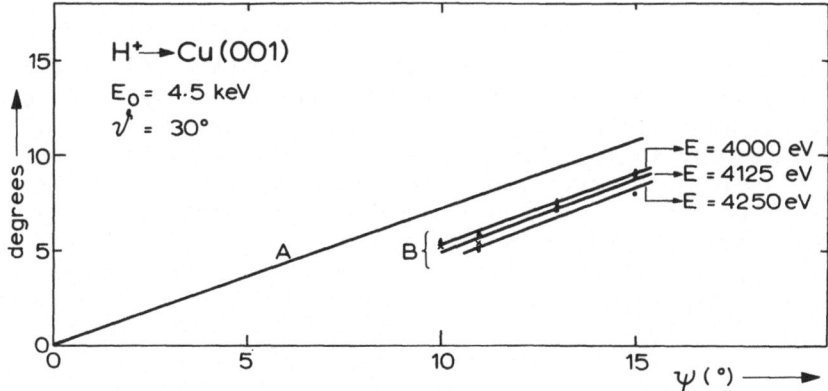

A = angle between $\varphi_{(110)}$ and $\varphi_{(111)}$
B = angle between peaks for different reflected energies

Fig. 5. Curve A: calculated angle between $\phi = 45°$ and $\phi(111)$ as a function of ψ for 4.5 keV H⁺ incident on Cu(001). (For definition of $\phi(111)$ see text.) Curves B: angular distance between the two peaks in the yield curves of Figure 4.

the second layer gradually shifts into the shadow of the first layer, and reach a minimum value for that position of the incidence plane $\phi(111)$ for which the incident ion beam is parallel to one of the (111) planes forming the walls of the semi-channel. Since the angle between $\phi = 45°$ and both possible $\phi(111)$ values on either side of $\phi = 45°$ will decrease with decreasing ψ (curve A in Figure 5) the yield maxima consequently will shift together and finally coincide. Returning to Figure 4 it can be seen that the height of the peaks decreases as ψ decreases. This can be caused by two effects. 1) For ϕ in the neighborhood of 45° the spatial distribution of the reflected particles will not only be influenced by the above described "wedge-focusing" of the incident beam. In addition to this there is another effect, namely the focusing of the particles <u>reflected</u> from the second layer by the two adjoining chains of the first layer; i.e., the semi-channel walls. Since for $\psi = 15°$ the target position is symmetric with respect to the incoming and outgoing direction ($\theta = 30°$), in case of protons (energy loss in atomic collisions negligibly small) both effects will be superimposed and a relatively high yield is expected. For lower ψ-values the influence of the walls on the outgoing particles, scattered from the second layer in the incidence plane, decreases. 2) When decreasing ψ the scattering along the <110> chains of the first and second layer is confined to angles around specular directions (chain effect). Thus because in our case the scattering angle θ is fixed at 30° the reflected yield will decrease.

Chain effect

In order to be able to distinguish between the influence of both effects mentioned above on the yield curves in Figure 4, similar measurements have been performed with He$^+$ ions. It is well-known that the ion reflection of low energy ($\lesssim$10 keV) He$^+$ ions, because of their high neutralization probability at a metal surface, mainly consists of ions scattered on the outermost surface layer (10). Thus by proper choice of the reflected ion energy (namely in the peak of the reflected He$^+$ energy distribution) it should be possible to obtain unique information from the reflection of the atomic chains in the first layer, and consequently to eliminate the influence of the "wedge-focusing" effect.

The experimental results for 6 keV He$^+$, presented in Figure 6, are in agreement with this assumption; the peaked structure, as observed in the H$^+$ yield curves, appears to be absent. For $\psi \gtrsim 13°$ no structure at all is observed in the yield curves. When decreasing ψ to values $\lesssim 11°$ a valley appears around $\phi = 45°$ as a result of the chain effect in the <110> direction (critical angle for 6 keV He in 110 direction $\sim$8.5°). The absence of a peak at $\phi = 45°$ in the valley of the yield curves of He$^+$ indicates that in case of H$^+$ this peak originates uniquely from reflection on the

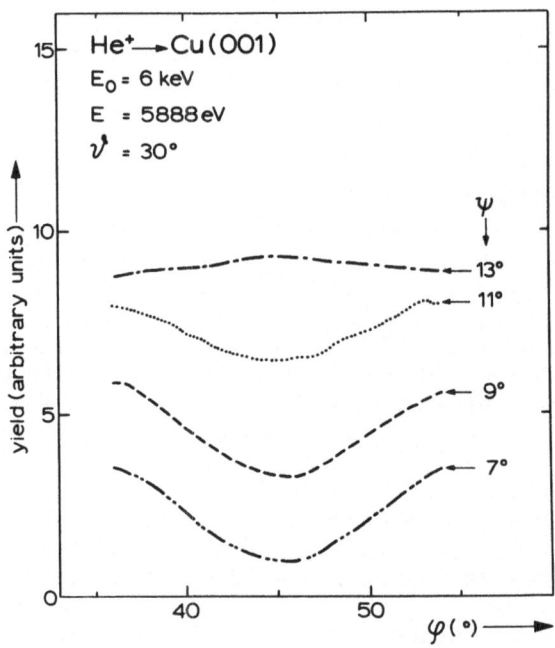

Fig. 6. Reflected yields of 6 keV He$^+$ ions as a function of ϕ for different angles of incidence ψ.

base of the semi-channel. Of course, this reflection from the second layer will as well be influenced by the chain effect. Apparently, however, the "wedge-focusing" effect is so strong that a small peak remains at ϕ = 45°.

Concluding Remarks

We may conclude that the experimentally observed anisotropy effects together with the computer simulations give a fairly complete picture of the focusing mechanisms, related to the presence of surface semi-channels. In addition, under some conditions it appears to be possible to obtain supplementary information on the reflection of H^+ from the first and second layer by comparison with similar He^+ measurements.

Acknowledgment

The authors are indebted to Mr. B. Poepjes and Mr. W. Spijkervet for their assistance in the numerical work presented in this paper.

This work was performed as part of the research program of the "Stichting voor Fundamenteel Onderzoek der Materie" (F.O.M.) with financial support from the "Nederlandse Organisatie voor Zuiver Wetenschappelijk Onderzoek" (Z.W.O.).

References

[1] V. E. Yurasova, V. I. Shulga, and D. S. Karpusov, Proc. 8th Intern. Conf. Phen. in Ionized Gases, Wien (1967), p. 50.
[2] V. M. Kivilis, E. S. Parilis, and N. Yu Turaev, Sov. Ph. Dokladay 15, 587 (1970).
[3] S. D. Marchenko, E. S. Parilis, and N. Yu Turaev., Proc. 10th Intern. Conf. Phen. in Ionized Gases, Oxford (1971), p. 79.
[4] E. S. Mashkova and V. A. Molchanov, Rad. Eff. 13, 183 (1972).
[5] V. E. Yurasova, I. G. Bunin, V. I. Shulga, and B. M. Mamaev, Rad. Eff. 12, 175 (1972).
[6] H. H. W. Feijen, L. K. Verhey, and A. L. Boers, Phys. Lett. 45A, No. 1, 31 (1973).
[7] E. P. Th. M. Suurmeijer and A. L. Boers, Journ. Phys. E: Sci. Instr. 4, 660 (1971).
[8] E. P. Th. M. Suurmeijer, Thesis, Groningen, The Netherlands (1973).
[9] S. A. Drentje, Thesis, Groningen, The Netherlands (1968).
[10] D. P. Smith, Surf. Sci. 25, 171 (1971).

INFLUENCE OF THERMAL LATTICE VIBRATIONS
ON MULTIPLE ION SCATTERING

L. K. VERHEY and A. L. BOERS
Technical Physics Laboratory
State University Groningen, The Netherlands

ABSTRACT

The influence of the target temperature on the
multiple scattering of 6 keV Ar^+ ions on a (100)-face
of a copper single crystal has been investigated.
Scattered Ar^+ energy distributions have been measured
as a function of the target temperature. Parameters
in the experiment were the ion incidence angle ($\psi < 15°$)
and the azimuthal angle of target rotation.
Comparison of the experimentally observed temperature dependencies with computer simulations allow to
estimate the influence of the thermal lattice vibrations
on the ion scattering characteristics.

Introduction

The backscattering of noble gas ions with primary energies in
the keV range can be used as a tool in structure analysis of crystal surfaces [1,2,3]. The method for this analysis is essentially
based on the multiple scattering effect. The energy and the spatial
distribution of the reflected ions are determined by the structure
of the surface, i.e., the relative position of the atoms. Interpretation of the measurements is possible by comparing these with
computer simulations. For these simulations the fundamental processes leading to multiple scattering and other processes which
may disturb correct interpretations have to be known.

Qualitatively, multiple scattering is quite well understood.
Experimentally it has been observed by several authors [3,4,5,6]
and the main regularities have been explained theoretically by
the simple chain model [7] and by calculations in which the whole
surface is taken into account [8,12,13]. However, a more accurate
examination of experiments and theory shows still large differences

which until now prevents multiple scattering to be a useful tool
in studies of most types of surface structures [3].

In general, the calculated energies for single and double
collisions, i.e., collision sequences with one or with two close
collisions and several weaker ones, are too high. In addition
calculations predict a narrower spatial distribution as found
experimentally.

Parilis et al.[9,14] explained these discrepancies by taking
into account thermal vibrations. In his model he supposed the
chain to be in a mode which corresponds to a complete zig-zag
configuration, i.e., only the vibrational mode with the highest
frequency is taken into account.

Figure 1 shows the calculated relationship between the relative energy after scattering E/E_0 and the angle of incidence ψ
together with some of our experimental results. The calculations
are carried out for a straight chain (a) and for a vibrating chain
according to the model of Parilis (b). It appears that the vibrating chain model is in better agreement than the straight chain
model, though it fails to give a complete description of the experimental results.

The aim of this paper is to give a further insight into the
influence of thermal vibrations on the basis of both experimental
and calculational studies.

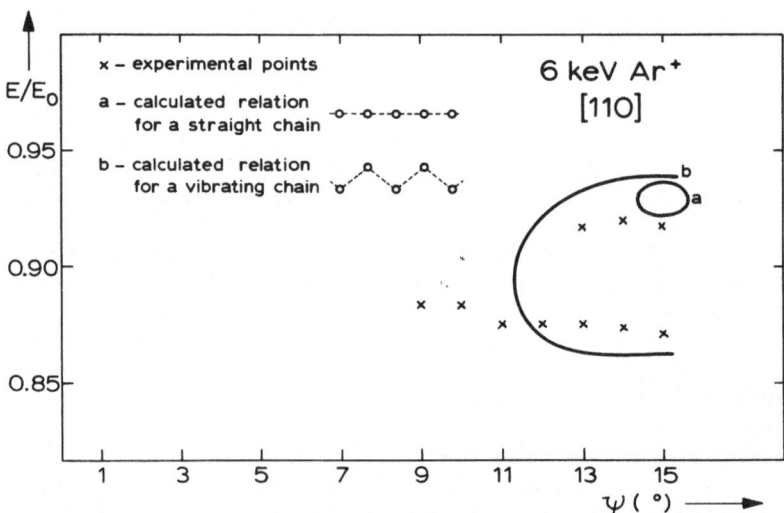

Fig. 1. Calculated and measured relationship between the relative
energy E/E_0 and the angle of incidence ψ for 6 keV Ar^+ ions reflected from a (100) Cu surface. The scattering angle θ is 30°, the
scattering plane is a (110) plane. The vibrational amplitude A in
curve b is .2 Å

Experimental

The basic experimental set-up used for our ion scattering experiments has been described elsewhere [10,11]. Therefore only some relevant characteristics and changes of the apparatus will be mentioned here. A scheme of the system is shown in Figure 2. The primary beam section consists of an ion source of the Nier type (energy spread 1%) and a magnetic mass analyser ($\frac{M}{\Delta M}$ = 160). The secondary section consists of a 85° electrostatic energy analyser and a Daly detector [11]. The energy resolution of the energy analyser amounts to .75%; its angular resolution is ± .5°.

Both primary and secondary sections are fixed to the collision chamber. Thus only scattering over a fixed angle can be measured, which in this work amounts to 30°. Target rotation is possible around two axes. The angle of incidence ψ and the azimuthal angle of target rotation are both adjustable to within .2°. Temperature control of the target is possible between 200 and 900 K.

The residual gas pressure in the collision chamber is below 10^{-9} Torr. In this work only 6 keV Ar$^+$ ions have been used; the

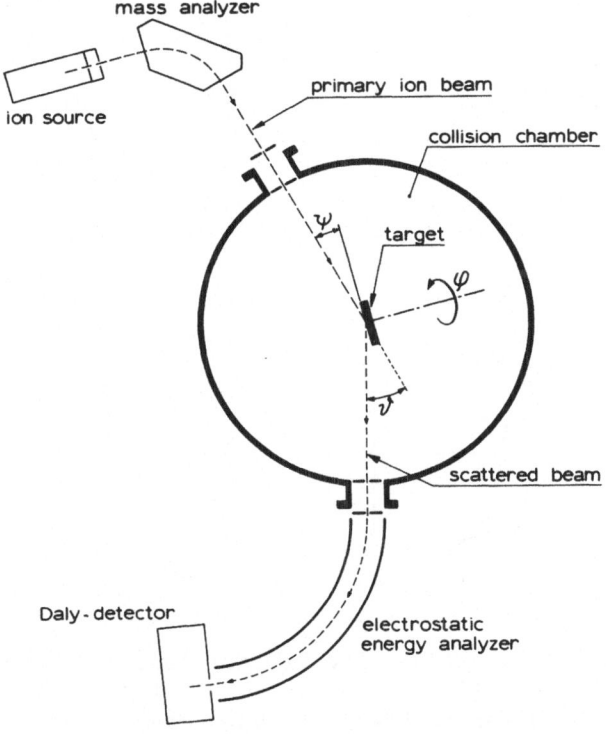

Fig. 2. Scheme of the experimental set-up. θ = scattering angle, ψ = angle of incidence relative to the surface, ϕ = angle of incidence relative to the crystallographic direction. The scattering plane is normal to the surface.

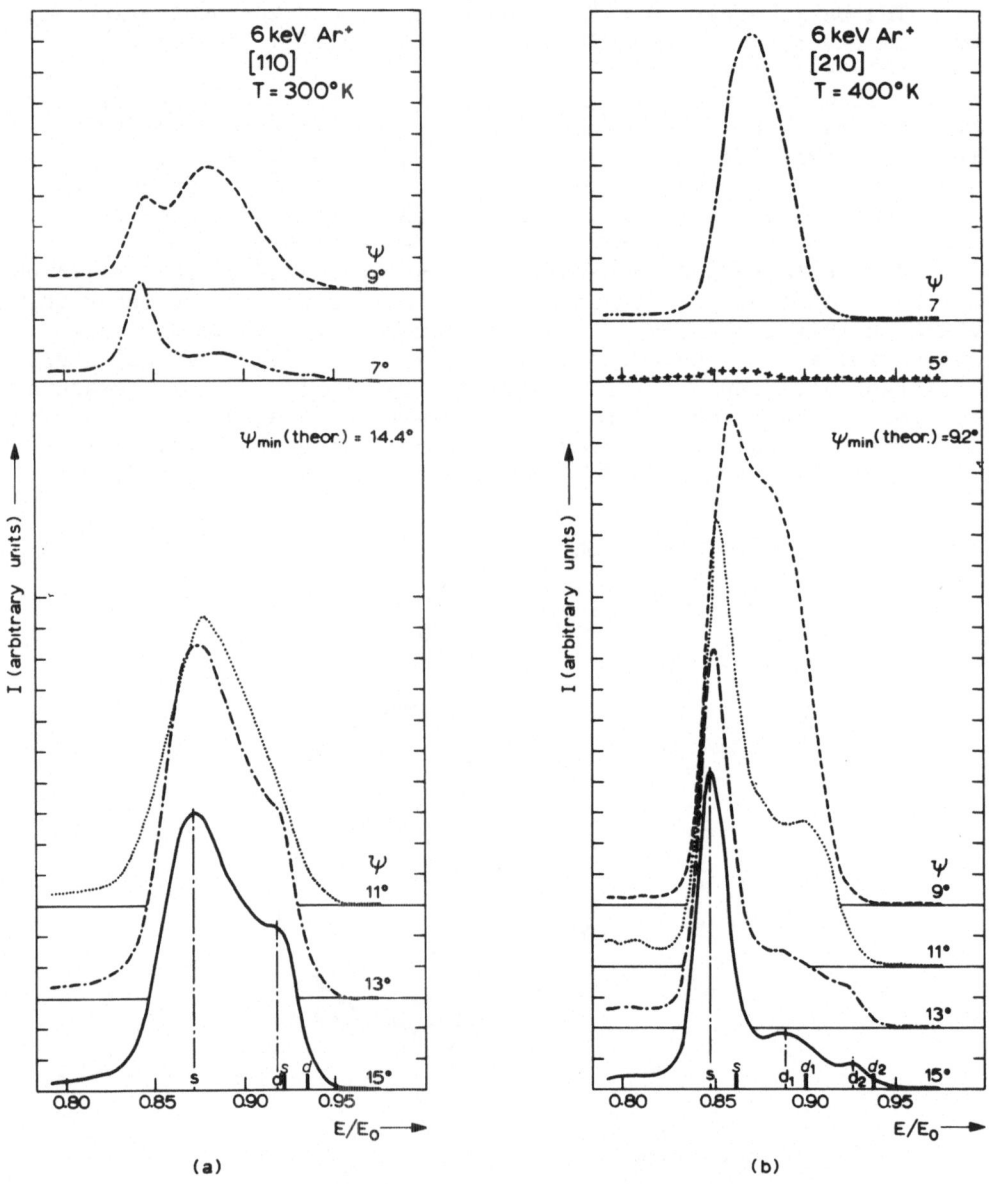

Fig. 3. Energy distribution of 6 keV Ar+ ions reflected from a (100) Cu surface for different angles of incidence ψ. The scattering angle θ is 30°, the scattering planes are a (110) (3a) and a (210) (3b) plane. The experimental and calculated energies for single (s) and double (d, d1, d2) collisions are indicated for ψ = 15°. The intensities are not corrected for the geometry.

primary ion current was 6–8 10^{-9} A. Except for temperatures ≥ 800 K
there was no evidence of impurities on the target. At high target
temperatures the pressure in the collision chamber increased to
values above 10^{-9} Torr. In addition a high background in the
reflection spectra could be observed under these conditions.

Experimental Results

    Influence of the angle of incidence and the scattering plane

    A detailed study of the influence of the angle of incidence ψ
on the reflection spectra has been conducted. Four different directions on the surface have been investigated, namely the [100],
[310], [210] and [110] directions (the projection of incident and
scattered beam both coincide with the direction). The general
features of the ψ dependence can be seen in Figure 3 together with
some calculational results. In the calculations a Born–Mayer
potential $V = A \exp(-b \cdot r)$ is used with A = 10000 eV and b = 4.5 $\overset{\circ}{A}^{-1}$.
For the [110] and [100] directions the chain model is used, while
for the [310] and [210] directions the whole surface is taken into
account [12].
    In Figure 3a where the ions have been scattered in a (110)
plane both the single (s) and the double (d) collision peaks could
be seen. However, the peak energies at $\psi = 15°$ greatly differ from
the calculated energies. Decreasing the angle of incidence leads
to a disappearance of the double collision peak, while the high
energy flank of the single collision peak broadens. Between $\psi = 9°$
and $\psi = 7°$ the single collision peak disappears also, and for smaller
angles only structure peaks are observed [1]. Once more comparison
with calculations shows a large deviation; according to the calculations the minimum angle of incidence ψ_{min} for which scattering
over 30° from this chain is possible is 14.4° (see also Fig. 1).
If zig–zag collisions are taken into account this limit is somewhat
smaller; however, the influence of these collisions on the spectra
is not yet clear.
    Figure 3b shows reflection spectra for the [210] direction.
Calculations predict three peaks (strictly speaking, four; the two
single collision peaks having the same energy) which can be found
in the spectrum at $\psi = 15°$ at a somewhat lower energy.
    A decrease of the angle of incidence leads to an increase of
the intensity of the high energy double collision peak (d2), while
the low energy peak (d1) can no longer be distinguished. Between
11° and 7° the double and single collision peaks show a shift
towards each other as theory predicts. Below $\psi = 7°$ the peak
decreases very fast and at 5° it has disappeared. Theoretically
this should occur for $\psi \approx 9.2°$.
    From these results, and the measurements in the [100] and [310]
directions, it appears that for directions with small atom distances

large deviations between theory and experiment occur, while for directions with larger atom distances these deviations are smaller [14].

In this paper we have confined ourselves to only one of the mentioned deviations, namely the way in which the low energy ("single collision") peak disappears with decreasing ψ. In Figure 4 the intensity of this peak is plotted against ψ. The main features which may be noticed are 1) the shift of the curves to smaller ψ values when going from small atom distances ([110]) to large atom distances ([210]), and 2) the dependence of the slope of the small angle flanks on the atom distance. The shift of the curves can be explained qualitatively by calculations with a straight chain [14]. The dependence of the slope on the atom distance suggests an influence of thermal vibrations. At T =0, i.e., a straight chain, the small angle flank should be almost vertical [13]. At temperatures above zero a distribution of vibrational amplitudes will be present. Relative to the atom distance this distribution will be narrower for directions with large atom separations than for directions with smaller ones. That is why the behaviour in these directions shows more resemblance to the theoretical T = 0 case.

Temperature dependence

A better way to examine the influence of thermal vibrations is to look at the temperature dependence of the curves of Figure 4.

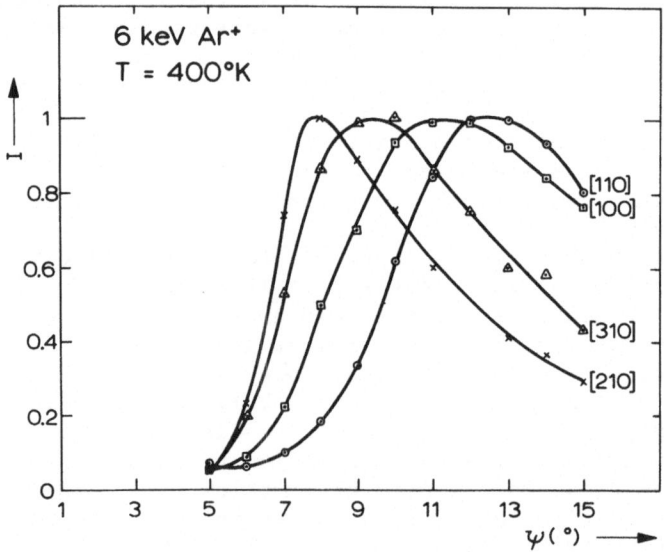

Fig. 4. Intensity of the single collision peak versus the angle of incidence ψ for the [110], [100], [310] and [210] direction. The target temperature is 400° K.

That is why we have repeated the measurements at several other temperatures in the range between 200 and 800 K. Figure 5 shows the results for the [110] and the [210] directions. For increasing T the curves shift to smaller angles of incidence for both directions. However, in the case of scattering in the (110) plane not only the small angle flank shows a shift, but also the maximum of the curve, while in the case of the [210] direction only a shift of the flank is observed. Probably, a further rise of the temperature will show a different behaviour. At 800 K the maximum of the curve does show a shift; however, the vacuum conditions during this measurement were rather bad. Therefore, more definite conclusions cannot be made.

The smaller shift of the curve in the case of the [210] direction stems from the larger atom distance D in this direction. For large D the relative amplitude A/D grows more slowly when rising the temperature. For all directions a decrease in temperature leads to a narrowing of the curve, which may be explained by the narrowing of the distribution of the amplitudes.

The intensity curves of the [210] direction suggest that the T = 0 limit is reached and the thermal vibrations only affect the flank. Comparison of this limit with the calculations show only a difference at ca. 1°. In the case of the [110] it is obvious that the limit is not yet reached. Even at the lowest temperature the curve is still rather broad and the maximum, the "experimental limit," still shifts.

In Figure 6 the temperature behaviour of the single scattering peak is given for all investigated directions. As a measure for the limiting angle of incidence, ψ_{min}, the angle ψ_h has been taken for which the reflected intensity has fallen to half of its maximum value. In Figure 6 ψ_h is plotted as a function of the square root of the temperature. The energy of the vibrating atoms will be

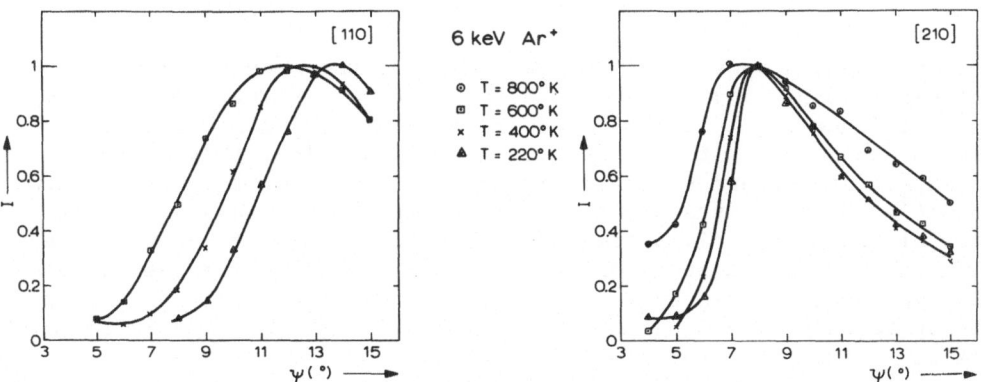

Fig. 5. Temperature dependencies of the single collision intensity – angle of incidence relationship for the [110] and [210] directions.

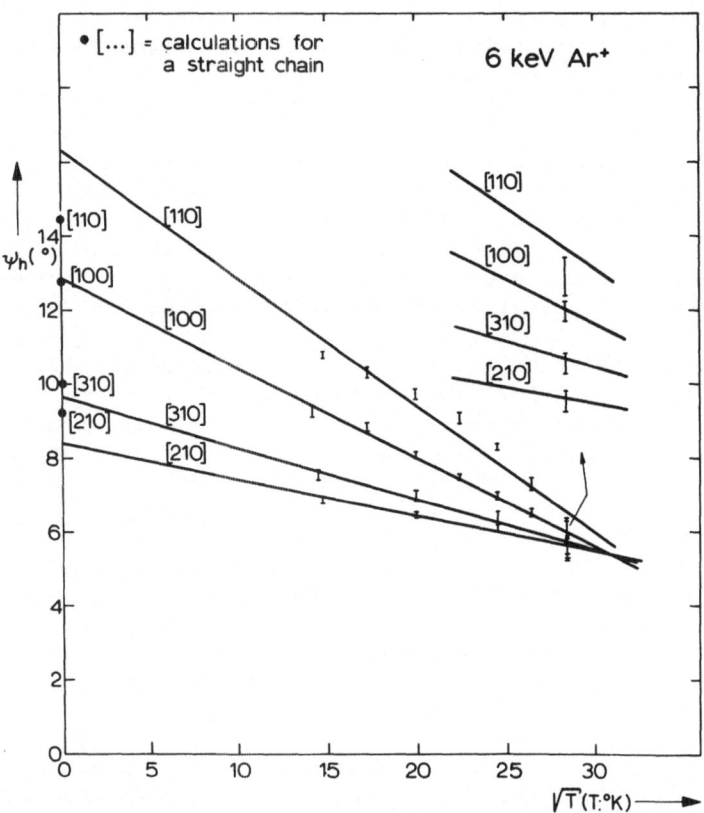

Fig. 6. Relationship between the angle of incidence ψ_h at which
the intensity of the single collision peak has fallen to half of
its maximum and the square root of the temperature. Calculations
for a straight chain (0°K) are indicated (●).

approximately proportional to the square of the mean amplitude of
the vibrations. Thus, Figure 6 actually shows the relation between
the incidence angle limit ψ_h and the mean vibrational amplitude.
The points do not fit very well to a straight line; they suggest
a temperature dependence which is somewaht stronger than proportional
to $T^{1/2}$. Measurements at still lower temperatures should provide
more clearness on this point. Unfortunately, this is not possible
with the present manipulator. As a rough approximation the curves
in Figure 6 have been extrapolated to T = 0. The resulting values
for ψ_h found at T = 0 agree reasonably well with the calculated
limits, except for the [110] direction.

Calculations on Vibration Models

Vibration models

Until now only Parilis et al. [9] have published calculations
on thermal vibrations. In his model the chains were thought to
vibrate in waves with a wave length of 2D. Calculations with this
model predict in our case, $\theta = 30°$, a minimum angle of incidence at
about 8.3° as curve 1 of Figure 7 shows. The occurrence of this
limit for ψ_{min} is not surprising, in as much as the waves, with
increasing amplitude will finally transform the lattice in a new
one in which the atom distance has become twice as large, leading
to a 2D-periodicity with its own specific limit. This feature in
fact does appear for all periodic thermal motions. Experimentally,
however, for this no evidence could be found. That is why we have
performed calculations with a different model. In this model all
atoms are presumed to be at rest except one vibrating atom. Curve 2
of Figure 7 shows the dpendence of the limiting angle ψ_{min} on the
vibrational amplitude A for this model. ψ_{min} is now almost proportional to the amplitude at least in the ψ-interval between 4 and 12°.
(For large amplitudes the curve will deviate more from a straight
line and will approach zero when the amplitude goes to infinity.)

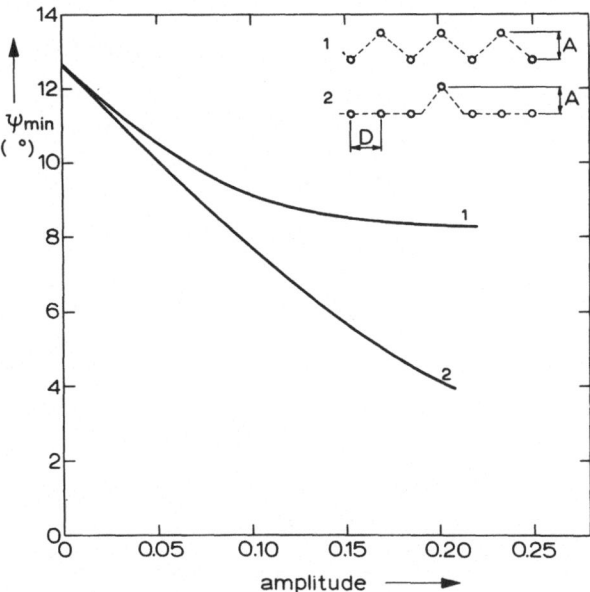

Fig. 7. Calculated relationship between ψ_{min} and the vibrational
amplitude A for a vibrating chain (1) and for a chain with one
vibrating atom (2). ψ_{min} is the minimum angle of incidence for which,
at an amplitude A, scattering over 30° is possible. A is measured
in lattice distances (3.16 Å for Cu).

The experimentally observed relationship between ψ_h and the mean amplitude $\overline{A}$ has been found to be approximately linear too. A proportionality can be explained by the "one vibrating atom" model, but not by a wave model. However, a stronger dependence cannot be explained.

Comparison with the experiment

For a direct comparison with the experiment Figure 7 cannot be used. This is because in practice we have to do with a distribution of amplitudes rather than with only one specific amplitude as assumed in Figure 7. The mean value of this distribution will be characteristic for the target temperature. Calculations which take this distribution into account are in progress.

In Figure 8 preliminary results are shown for the [110] direction. In these calculations the amplitude distribution $F(A) = A \exp(-K\,A^2)$ has been taken ($K = \dfrac{\text{constant}}{kT}$), being the function which up to now gives the best agreement with the experiment. In the figure the calculated ψ_h, i.e., the angle of incidence for which the intensity is half as large as the maximum, is plotted against the mean amplitude. Experimental points are plotted also.

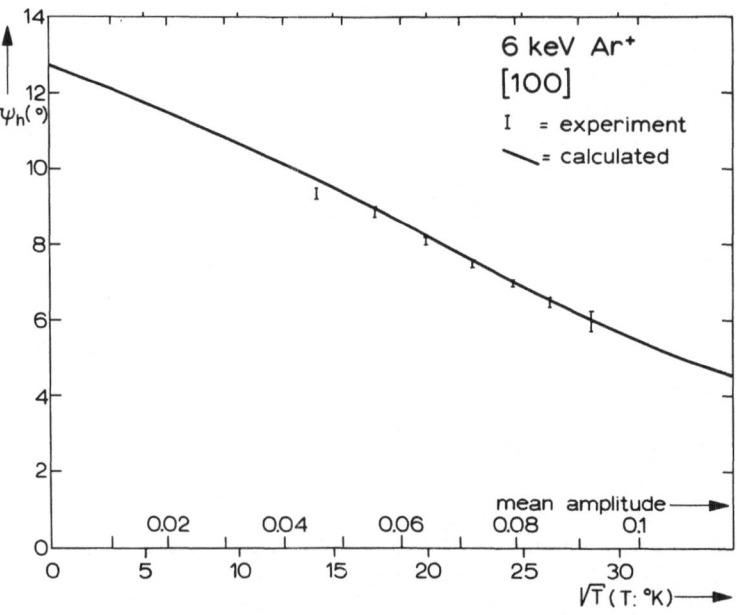

Fig. 8. Calculated relationship between ψ_h (see text) and the mean vibrational amplitude $\overline{A}$ together with experimental results. The temperature and the amplitude scale are fitted to each other. $\overline{A}$ is measured in lattice distances (3.16 Å for Cu).

The temperature scale has been transformed into an amplitude scale in such a way that the experimental and theoretical curves have equal slopes. This leads to a mean amplitude of

$$\overline{A} = .031 \sqrt{T}$$

with T measured in degrees Kelvin and $\overline{A}$ in units of lattice spacing D (for copper 3.61 Å). Thus at room temperature we find $\overline{A}$ = .055 ≃ .2 Å, which is in good agreement with other estimations.

Conclusions

Multiple scattering of keV ions can effectively be used for studying thermal vibrations of the surface atoms of a crystal lattice. Correlated vibrational modes seem to play a minor role in the vibrations of the surface atoms. Further research on this subject has to show if either this conclusion is correct, or our results must be explained by our experimental conditions (small angle ion scattering).

Acknowledgments

We wish to thank B. Poepjes and W. Spijkervet for their assistance with the measurements and the calculations.
This work is part of the research program of the Stichting voor Fundamenteel Onderzoek der Materie (FOM) and was made possible by financial support from the Nederlandse Organisatie voor Zuiver Wetenschappelijk Onderzoek (ZWO).

References

[1] S. H. A. Begemann, A. L. Boers, Surface Sci. 30, 134 (1972).
[2] H. H. Brongersma, P. M. Mul, Surface Sci. 35, 393 (1973).
[3] W. Heiland, H. G. Schäffler, E. Taglauer, Surface Sci. 35, 381 (1973).
[4] I. N. Evdokimov, E. S. Mashkova, V. A. Molchanov, Sov. Phys. Doklady 14, 467 (1969).
[5] E. S. Mashkova, V. A. Molchanov, E. S. Parilis, N. Yu. Turaev, Phys. Lett. 18, 7 (1965).
[6] W. F. v.d. Weg, D. J. Bierman, Physica 35, 406 (1968).
[7] V. M. Kivilis, E. S. Parilis, N. Yu. Turaev, Sov. Phys. Doklady 12, 328 (1967).
[8] V. E. Yurasova, V. I. Shulga, D. S. Karpuzov, Proc. 8th Intern. Conf. Phen. in Ionized Gases, Wien, 50 (1967).
[9] E. S. Parilis, N. Yu. Turaev, V. M. Kivilis, Proc. 8th Intern. Conf. Phen. in Ionized Gases, Wien, 47 (1967)

[10] E. P. Th. M. Suurmeijer, A. L. Boers, J. Phys. E: Sci. Instrum.
     $\underline{4}$, 663 (1971).
[11] E. P. Th. M. Suurmeijer, Thesis (1973).
[12] A. van Veen, J. Haak, Phys. Lett. $\underline{40A}$, $\underline{5}$, 378 (1972).
[13] V. E. Yurasova, V. I. Shulga, D. S. Karpuzov, Can. J. Phys.
     $\underline{46}$, 759 (1968)
[14] V. M. Kivilis, E. S. Parilis, N. Yu. Turaev, Sov. Phys. Doklady
     $\underline{15}$, 587 (1970).

X-RAY PRODUCTION AND ENERGY LOSS IN LOW-ANGLE ION SCATTERING AT SOLID SURFACES

B. W. FARMERY, G. S. HARBINSON, A. D. MARWICK,*
H. PABST and M. W. THOMPSON
School of Mathematical and Physical Sciences
University of Sussex, Brighton, BN1 9QH
Sussex, England

Introduction

In previous publications [1,2,3] we have reported experiments in which protons and helium ions with energies between 0.1 and 1 MeV have been scattered at grazing incidence from a flat target surface of either amorphous tungsten oxide [2], or a tungsten crystal [1,3].

The angular distribution of scattered ions in the amorphous case has an approximately gaussian form truncated by the surface plane. From a crystal surface, when the angle of incidence is greater than the critical channeling angle, a blocking pattern is seen. When the angle of incidence is less than the critical angle, the angular distribution shows a strong peak in the specular reflection position where ions conserve transverse energy with respect to the surface plane [1].

The energy spectrum of the scattered ions has the form of a skewed peak with a tail on its low-energy side. For an amorphous surface the most probable energy loss increases with scattering angle for a fixed angle of incidence. These observations were shown to be consistent with a theory based on multiple scattering from randomly arranged centers [2]. In the crystal case the spectral peak is sharper and the most probable energy loss decreases with scattering angle for a fixed angle of incidence. When the detector is set in the angular position corresponding to conserved transverse energy with respect to a crystallographic row, the intensity near the peak of the energy spectrum shows maxima for positions at which rows in sub-surface layers can scatter trajectories up through the windows formed by spaces between the surface layer of rows [3].

Observations

In our most recent work, to be reported at this conference, we have studied the production of x rays by protons in the range 100 to 275 keV from both tungsten (M shell) and oxygen (K shell) and simultaneously observed the energy spectra of the scattered protons. In the case of amorphous tungsten oxide the x-ray spectrum shows both oxygen-K and tungsten-M peaks, and each increases with angle of incidence ϕ, for small angles, roughly like $\phi^{0.6}$. The dependence of the x-ray yield with ion energy E_1 is like E_1^4, as is usual for thick-target yields generated by protons in this energy range at normal incidence. The dependence on ϕ is interpreted as being due to an increasing reflection probability as ϕ decreases, thus removing a fraction of the protons which would have slowed down in the target and generated x rays. A quantitative theory has been developed along these lines based on the model of multiple-scattering in a three-dimensional random array of centers used above [2] to calculate the energy spectrum of scattered ions. A good prediction of the observed behaviour is obtained and it is possible then to deduce the cross sections for x ray production by protons from the oxygen K-shell and the tungsten M-shell. The former extends the range of previously known data [4]; the latter confirms earlier observations [5].

With tungsten crystal targets only the tungsten M-shell peak was studied in detail. The yield was first measured in a series of runs with the angle of incidence ϕ held constant while the crystal was rotated about its surface normal, through an angle designated θ. Crystals either had a {100} or a {110} surface plane. In all cases minima were observed in yield versus θ whenever the plane of incidence contained a simple crystallographic row. Repeating for several values of ϕ gave a family of curves showing a progressive increase in yield with increasing ϕ, but at a much faster rate than the $\phi^{0.6}$ behaviour seen in the amorphous case.

Another set of yield measurements were made with incidence close to atomic row directions and with ϕ and θ adjusted together to give a constant value of ψ, the angle to the row axis and thus a constant value of the transverse energy. For a particular row a graph of yield versus ϕ then showed a set of yield values for each value of ψ which increased with ϕ. In most cases this was the most pronounced effect, subsidiary to which was a tendency for larger ψ to give larger yields. Large differences, by as much as a factor of five, were seen between different crystallographic row directions. Proton energy had a very marked effect.

The data were plotted in a different way on a polar diagram with the radius vector along the projected direction of incidence in the transverse plane of the crystal and its length indicating the x-ray yield for constant ψ. These diagrams show a family of curves with a slow expansion as ψ increases. For trajectories near the <100> row which approach the surface normally in transverse projection, a minimum is seen for large ψ which turns into a maximum at small ψ. It is thus clear that both the projected direction of the trajectory in the transverse plane, and its transverse energy, affect the yield.

Measurements of the energy spectrum of scattered protons have extended the range of our previous observations. They again show that the low-energy tail of the spectrum increases markedly as the plane of incidence moves towards a row axis and that the most probable energy energy loss generally decreases as the scattering angle increases for fixed ϕ. But these measurements showed an initial increase followed by a decrease, when the plane of incidence contained the <100> axis, yet a monotonic decrease 3° away from <100>. The scattering angle and ϕ have also been varied so that the detector is set at the position of specular reflection (i.e., conserved transverse energy). The most probable energy loss then decreases with ϕ at first, increasing later as ϕ approaches the critical angle for channeling.

Theory

A theory has been developed [6] which give the yield of any process that depends on the distance of a trajectory from an atomic row. It solves an equation of motion in the transverse plane for an encounter with a single row, or plane, which incorporates the continuum potential $U(r)$ and the excitation function $S(r)$ (yield per unit path length at distance r). The calculations give yield versus ψ for several values of impact parameter and for different forms of $S(r)$.

The yield per encounter is very roughly proportional to ψ^P, for either planes or rows. The exponent p is determined by the forms of $U(r)$ and $S(r)$ but always p > -1. This means that in normal transmission channeling experiment the total yield per transmitted ion always increases with ψ, but in a surface reflection experiment it could either increase or decrease. For an inner shell process, in which $S(r)$ falls off more rapidly than $U(r)$, we find p > 1; for outer shell processes, where $S(r)$ falls off more slowly, p can be negative. The theory has been worked out both for exponentially varying functions $U(r)$ and $S(r)$ and for power-law functions and the qualitative results are the same.

The surface scattering problem has been considered as a multiple collision problem in the transverse plane in which the projected trajectory diffuses, or is scattered, by one row after another. Both the energy transverse to the surface, $\phi^2 E_1$ and the energy transverse to the rows $\psi^2 E_1$ are shown to be of prime importance in explaining the x-ray data. Multiple scattering in two dimensions by a random array of centers was taken as a rough model for motion in the transverse plane and a quantitative theory developed. This model appears to be consistant with the results of a computer simulation. From this the energy spectrum of scattered ions can be deduced and can be shown to have a flatter tail, as observed, as one approaches a row. This is due to trajectories which have turned through angles, in projection, greater than 2π and which this theory shows to be of great importance. The theory also shows why ϕ, which determines the energy transverse to the surface, is an important parameter in determining the x-ray yield by its effect on the reflection probability.

References

*Metallurgy Division, A.E.R.E., Harwell, Didcot, Berkshire, England.

[1] B. W. Farmery, A. D. Marwick, M. W. Thompson, "Atomic Coll. Phen. in Solids," North Holland, 1970, p. 589.

[2] A. D. Marwick, M. W. Thompson, B. W. Farmery and G. S. Harbinson, Rad. Effects 10, 49 (1971).

[3] A. D. Marwick, M. W. Thompson, B. W. Farmery and G. S. Harbinson, Rad. Effects 15, 195 (1972).

[4] R. Hart, F. Reuter, H. Smith and J. Khan, Phys. Rev. 179, 4 (1969).

[5] P. Needham. B. Sartwell, Phys. Rev. A2, 1686 (1970).

[6] M. W. Thompson and H. Pabst, Rad. Effects, to published (1974).

SURFACE SCATTERING OF LOW ENERGY IONS

W. HEILAND, H. G. SCHÄFFLER and E. TAGLAUER

Max-Planck-Institut für Plasmaphysik, EURATOM Association
8046 Garching, Germany

ABSTRACT

Angular and energy distributions of Ne^+ ions
(200 eV to 2 keV) backscattered from Ni(110) and
(111) surfaces were measured and compared with
numerical calculations. In the energy distributions peaks due to multiple and single scattering
events can be identified. Multiple scattering
occurs between a minimum and a maximum scattering
angle. Experimentally, both angles depend on the
lattice parameters, the angle of incidence, the
energy and the ion-atom combination. These effects
are qualitatively understood in the atomic chain
model of multiple scattering. The main discrepancies between model and experiment are slight shifts
of the measured energies towards lower values and
deviations of the maximum scattering angle towards
higher values. The angular deviation can be account-
ed for by fitting the screening length in the Molière
approximation of the Thomas-Fermi potential, which
was used for the calculations. The screening length
thereby obtained is about 30% below the theoretical
value of Lindhard. A comparison with other potentials from the literature indicates that the potential determined here is within theoretical expec-
tations.

Introduction

The energy spectra of low energy noble gas ions backscattered
from a solid surface usually show intensity peaks which depend on
the masses of the surface atoms [1]. Under certain experimental
conditions multiple scattering can also be observed [2], in which
case the energy of the peaks also depends on the geometric

conditions of the ion beam and surface atoms and on the interaction potentials. Therefore, the investigation of multiple scattering provides, in principle, the possibility of determining potential parameters for the interaction of ions with surface atoms.

The present paper describes an attempt to estimate the Ne^+-Ni interaction potential in the energy range 200 eV - 2 keV. For this purpose the experimental results are compared with numerical calculations for a model situation. The limitations of the model will be discussed as well as the problems arising from the influence of the neutralization of the ions at the surface.

Experimental Observations

The experiments are carried out in a UHV scattering chamber described elsewhere [2]. The beam of Ne^+ ions impinges on a Ni(111) or (110) surface. The target crystal can be rotated to provide different angles of incidence ψ between the ion beam and surface and azimuthal angles ϕ for scattering along different crystallographic directions. The energy of the positively charged scattered particles is analyzed with a cylindrical 127° analyzer. The angular resolution for the scattering angle θ as determined by the collimation of the incident beam and the acceptance of the analyzer is $\Delta\theta = 5°$.

Typical spectra are shown in Fig. 1 [3]. There is a high intensity peak whose energy is mostly close to the binary collision energy and a second peak at higher energies. Their energetic separation becomes very small at the ends θ_{min} and θ_{max} of a certain angular region, so that they can no longer be resolved. Also the intensity drops outside this region, which varies for different crystal directions. The dependence of the spectra on the angle of incidence, ψ, and on the crystal orientation (i.e., the lattice constant in the scattering plane) is in qualitative agreement with the model of ion scattering from atomic surface chains. However, the absolute calculated energies of the simple model do not agree with the experiments, the experimental energy values being mostly below the theoretical ones [2]. One of the results of these model calculations is that two energies of backscattered ions should be observed in a certain range of scattering angles, and that there should be no scattered intensity outside this region owing to a shadowing effect. Figure 2 demonstrates again the experimental observations.

If the energies of the scattered ions are plotted as a function of the scattering angle, the region where two peaks are resolved can be seen. The low angle θ_{min} of the "loop" is better defined experimentally than the high angle limit θ_{max}, owing to lower intensities at larger scattering angles. The intensity distribution shows that the backscattered intensity is mostly confined to the same angular region, the maximum of the intensity being near the minimum angle θ_{min}.

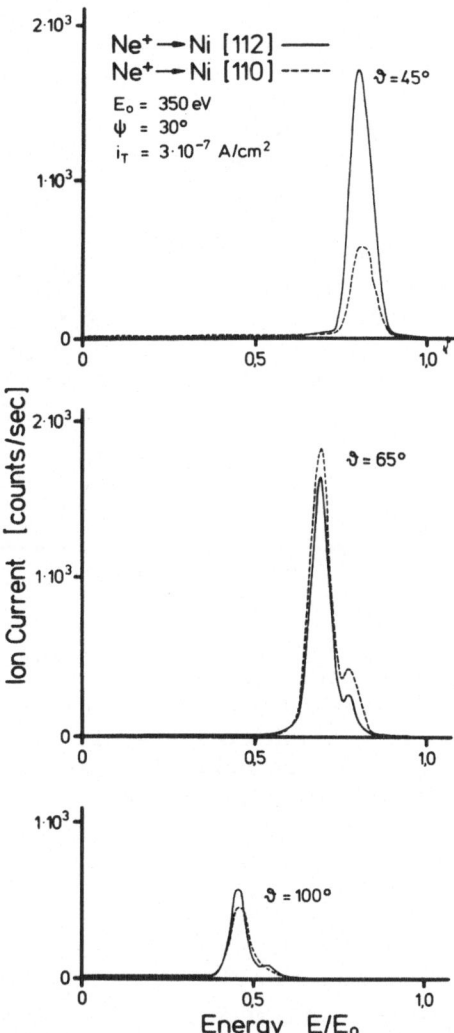

Fig. 1. Energy spectra for Ne[+] ions backscattered from a (111) Ni surface along two different crystallographic directions. θ = laboratory scattering angle, ψ = impact angle relative to the surface.

Figure 3 shows the intensity variation with the azimuthal angle φ. We find strong oscillations of the intensity near θ_{min} and θ_{max}. The minimum intensity thereby corresponds to scattering along the [110] direction in the surface, the maximum to the [112] direction. If we plot the intensity ratio of these minima and maxima as a function of the scattering angle θ, we get the curves of Fig. 4. Comparing the scattered ion intensities at constant angle of incidence and scattering angle may give a more significant

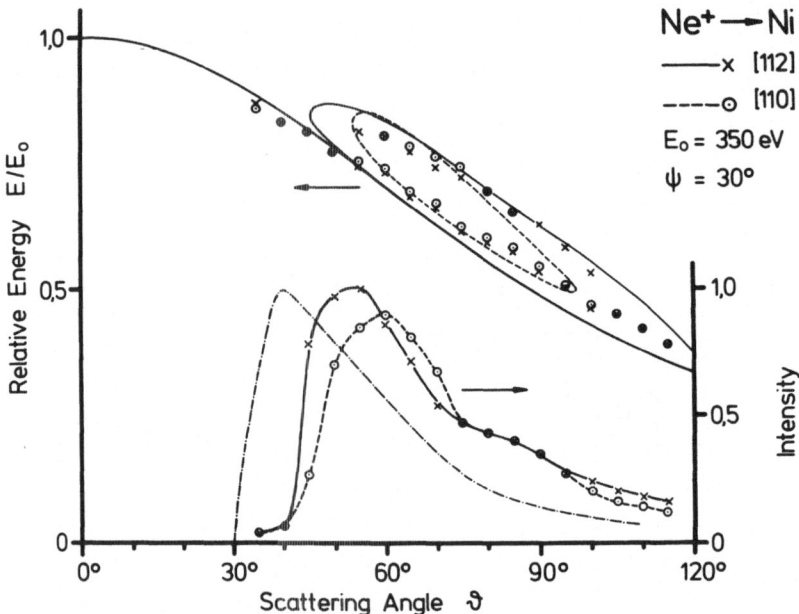

Fig. 2. Energy and intensity of the peaks of Fig. 1 as a function
of the scattering angle θ. The curves in the energy distributions
are calculated with a_{exp} (see text). The dash-dotted line in the
intensity distribution is a calculated example using the functions
$d\sigma/d\Omega$ and (1-P) from Fig. 6. The solid line starting at $E/E_0 = 1$
is the energy for a single binary collision.

intensity value because the influence of the scattering cross section
and neutralization probability is reduced, as will be discussed
later. For primary energies between 200 and 400 eV at about
$\theta \lesssim 55°$ the intensity for the [110] direction falls considerably
below the value for the [112] direction, whereas θ_{max} obviously
depends on the energy. It has to be discussed to what extent
these experimental results are in agreement with the model calculations for scattering from a linear chain of surface atoms and to
what extent the same results could be explained by mere cross section and neutralization considerations. Before doing so we can
see that the linear model gets some support from the data in Fig. 5.
Here the position of the intensity peaks is plotted for scattering
along the [110] direction from two different crystal faces. The
data for both cases, the (110) and the (111) planes, is in fairly
good agreement.

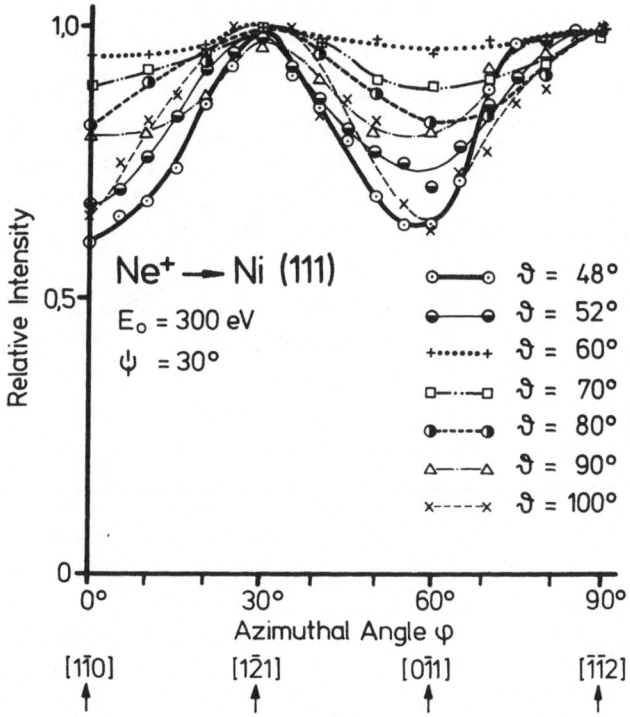

Fig. 3. Intensity of the main peaks in the energy spectrum for scattering along different crystal directions. Parameter is the scattering angle θ. The curves are normalized to 1 at θ = 90°.

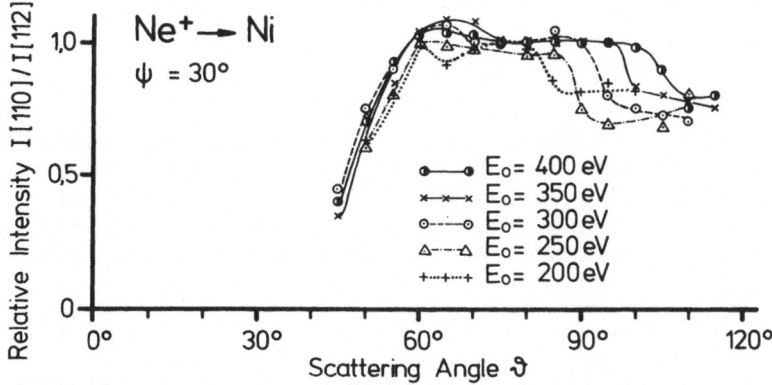

Fig. 4. Intensity ratios of the main energy peaks for scattering along [110] and [112] directions. Parameter is the primary energy E_0.

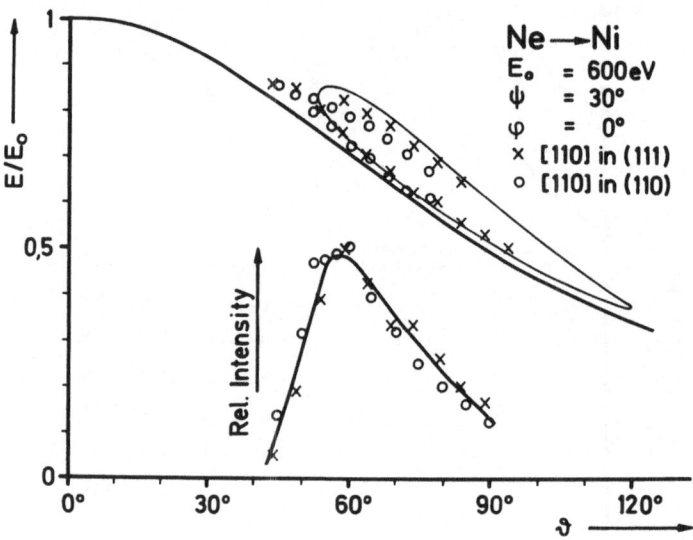

Fig. 5. Energy and intensity distributions of the peaks in the energy spectra for the [110] crystal direction in the (111) and (110) nickel surfaces. The solid line starting at $E/E_0 = 1$ is the energy for a single binary collision. The loop is calculated from the atomic chain model using a_{exp} for the interaction potential (see text).

Discussion

The numerical calculations are carried out with a screened Coulomb interaction potential

$$V(r) = \frac{e^2 Z_1 Z_2}{r} \phi \left(\frac{r}{a}\right) \qquad (1)$$

The approximation of Molière [4] was chosen for the Thomas–Fermi screening function ϕ. The screening length a can be used to fit the numerical results to the data. For the scattering of an ion from a chain of surface atoms the impact parameters p_i and p_{i+1} of two subsequent collisions are connected by a relation like

$$p_{i+1} = p_i \cos \theta'_i + x_i \sin \theta'_i + d \sin (\theta_{i-1} + \theta'_i) \qquad (2)$$

where d is the lattice constant and $\theta_{i+1} = \theta_i + \theta'_i$ the total scattering angle after i+1 collisions. The motion of the center of mass for each collision is taken into account by the factor x_i. As a result we obtain the typical loops in the $E - \theta$ plot which have also been calculated for other situations by several authors [4,5,6]. The results for E_0 = 350 eV, ψ = 30° and d = 2.49Å and 4.32Å are plotted in Fig. 2. If the simplified model situation were true, the ions would only be scattered in the angular region described by the loops, giving two energy values for one scattering angle. A relative maximum in the scattered intensity is to be expected for scattering angles where the two branches merge.

Besides these geometric effects the scattering cross section and the neutralization probability have to be taken into account. Figure 6 shows the calculated differential scattering cross section $d\sigma/d\Omega$ for Ne-Ni at primary energies of 500 to 2000 eV using a Born-Mayer potential. The cross section decreases strongly with increasing scattering angle θ.

Since only those particles which leave the surface as ions are measured in the present work, for all intensity considerations the neutralization probability P is an important factor. According to the models of Gobas and Lamb [8] and Hagstrum [9] the neutralization probability for an ion backscattered from the surface is

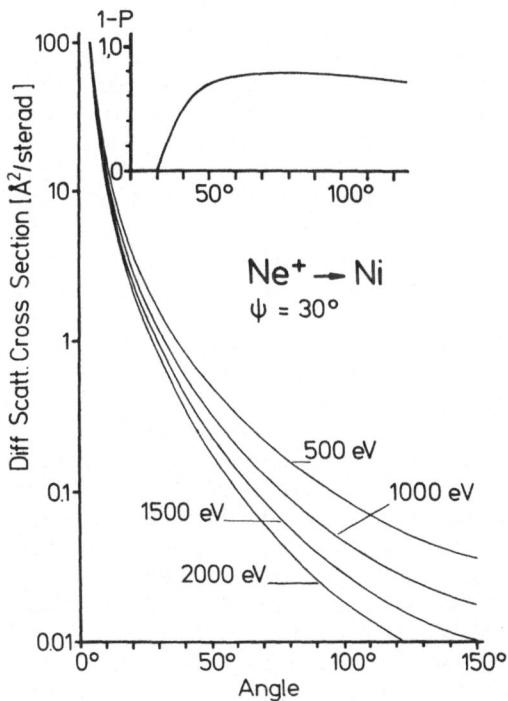

Fig. 6. Calculated differential scattering cross sections $d\sigma/d\Omega(\theta)$ using a Born-Mayer potential and Abrahamson's [14] constants. Inset on top: example for the neutralization function (1-P), calculated with v_0 = 0.9x10^{6}cm/sec. $d\sigma/d\Omega$ and 1-P are given as functions of the laboratory scattering angle θ.

$$P = 1 - e^{-v_o/v_\perp} \tag{3}$$

where $v_\perp$ is the velocity perpendicular to the surface and v_o is a
constant which depends on the electronic configurations of the ion
and the surface.

In the inset of Fig. 6 1-P is plotted as a function of θ for
$\psi = 30°$ and an arbitrarily chosen $v_o = 0.9 \times 10^6$ cm/sec. The leading
edge of the calculated curve is very sensitive to this parameter v_o.

The intensity distribution of the backscattered ions is

$$I(\theta) \sim \frac{d\sigma}{d\Omega} \cdot (1-P). \tag{4}$$

This can give a qualitative explanation for the intensity distribution of Fig. 2. The increase at small scattering angles can be due
to the neutralization factor (1-P), and is not necessarily determined by the shadowing effect of the surface chains; the decrease
at larger angles is determined by the cross section $d\sigma/d\Omega$. Since
the parameter v_o in equation (3) is not well known, a quantitative
comparison is not possible. In the case of He^+ the increase in
the intensity distribution starts at smaller angles than for Ne^+
[10]. For He^+ ions of the same energy multiple scattering is less
pronounced and the shift in the intensity distribution could be due
to shadowing. But again it has to be stated that by choosing a
proper v_o this intensity distribution could be matched. It has
to be concluded that the low angle part will always be influenced
by the neutralization probability.

However, the value of θ_{min} does not depend very much on the
interaction potential, as can be seen from both experiment and
calculation. The high angle limit is more sensitive to the interaction potential. It is due to the shadowing effect of those atoms
on the surface which are in front of the atom that causes the main
deflection of the incident particle. The intensity generally decreases for higher scattering angles but the neutralization effect
does not cause a steep intensity slope at the high angle limit
θ_{max} (if 1-P is matched to the slope near θ_{min} it varies only slowly
in the range of θ_{max}). This should also be valid if one takes into
account that the neutralization may be generally higher if the ion
trajectory runs along a more densely packed surface chain, namely
the [110] direction compared to the [112] direction. To evaluate
this angle more clearly, the ratio of $I_{[110]}/I_{[112]}$ has been plotted
in Fig. 4 and a steplike decrease is found at a certain scattering
angle, depending on the primary energy. This is also the scattering angle where the two peaks in the energy spectrum coincide to a
small single peak. But, plots like Fig. 3 or 4 are more suitable
for finding a definite value for θ_{max}.

The fact that the intensity does not drop to zero above θ_{max}
must of course be explained by scattering processes which do not
occur along perfect chains, i.e., either scattering from steps

and surface defects [11] or trajectories which leave the scattering
plane but are within the angular resolution width $\Delta\theta$.

Interaction Potential

Figure 7 shows the calculated limiting angles θ_{min} and θ_{max} as
a function of the value of the screening length a of the interaction
potential for Ne–Ni and different primary energies. As was mentioned
above, θ_{min} is not very sensitive to the potential parameter. The
$a(\theta_{max})$ curves, however, would allow an evaluation to find the parameter a with a reasonable experimental error for θ of about $\pm 5°$.
        If we accept the arguments given in the discussion and take
the steps in the $I_{[110]}/I_{[112]}$ plots for a number of energies as
the positions of θ_{max}, then we get the experimental points inserted

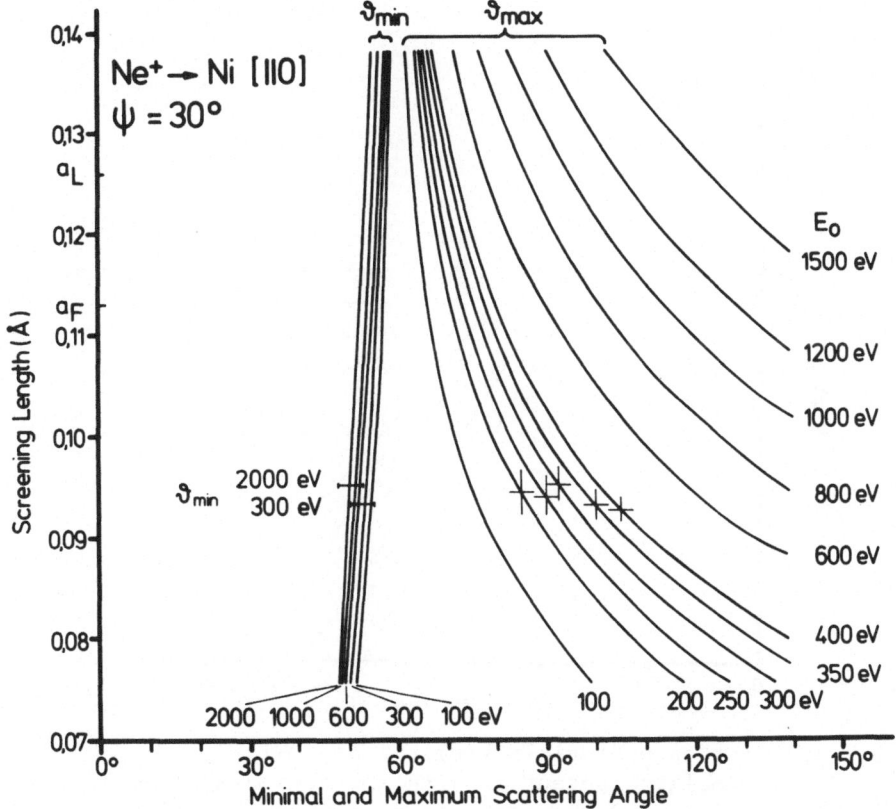

Fig. 7. Screening length as a function of the limiting scattering
angles θ_{min} and θ_{max}. The curves are calculated with a Moliere [4]
potential for different primary ion energies E_o. ⊢──┤ and + are
experimental values.

in Fig. 7. It is seen that this procedure yields a screening parameter of a_{exp} = 0.095 Å for the primary energies from 200 to 400 eV.
It also confirms the observation that we cannot find θ_{max} for energies E_0>600 eV owing to our maximum experimental scattering angle
of 120°.

    The loops in Fig. 2 are calculated with the screening parameter
deduced from Fig. 6. The theoretical values for a according to
Lindhard [12], a_L, and Firsov [13], a_F, are also indicated in Fig.
7. Both are higher than the value found here, a_F by about 15%,
a_L by 30%.

    The function V(r) for a_{exp} is plotted in Fig. 8 and compared
with a number of other potentials. The repulsive interaction is
weaker than for the other screened Coulomb potentials. Therefore,
the blocking effect for surface scattering is also smaller than
expected from calculations using these potentials. Compared with
Born–Mayer potentials, that determined here lies between the curves

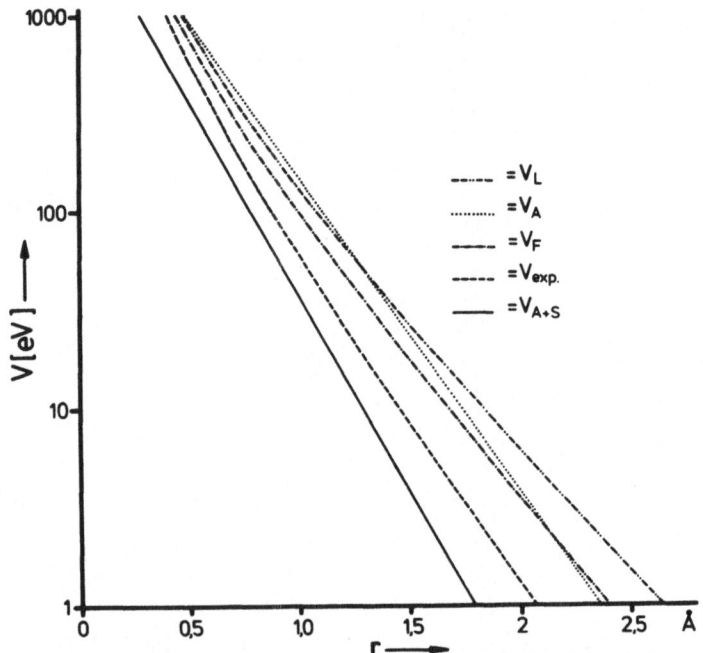

Fig. 8. Interaction potentials V(r) for Ne^+ –Ni.
    V_{exp} – Molière potential with the screening length
                a_{exp} = 0.095 determined from the data in Fig. 7.
    V_L and V_F – Molière potentials, using Lindhard's [12] and
                Firsov's [13] screening lengths, a_L and a_F.
    V_A and V_{A+S} – Born–Mayer potentials using the constants of
                Abrahamson [14] or Andersen and Sigmund [15].

using Abrahamson's [14] and Andersen and Sigmund's [15] constants. From the scattering of Ni^+ from Ne gas at higher energies Foster et al. [16] determined the interaction potential between 1 Å and 1.5 Å interatomic distance. Their values agree well with our results.

Acknowledgments

The technical assistance of Franz Schuster is gratefully acknowledged.

References

[1] D. P. Smith, Surf. Sci. 25, 171 (1971).
[2] E. Taglauer and W. Heiland, Surf. Sci. 33, 27 (1972).
[3] θ in the text is equivalent to ϑ in figures; φ in the text is equivalent to φ in the figures.
[4] G. Molière, Z. f. Naturforschg. 2a, 133 (1947).
[5] E. S. Parilis, Proc. 7th Int. Conf. on Phenomena in Ionized Gases, Belgrade 1965, p. 129.
[6] V. E. Yurasova, V. I. Shulga, and D. S. Karpuzov, Can. J. Phys. 46, 159 (1968).
[7] D. G. Armour, G. Carter and A. G. Smith, Rad. Eff. 3, 175 (1970).
[8] A. Cobas and W. E. Lamb, Phys. Rev. 65, 327 (1944).
[9] H. D. Hagstrum, Phys. Rev. 96, 336 (1954).
[10] W. Heiland and E. Taglauer, DPG Mtlg. VI, 7, 465 (1972) unpublished.
[11] W. Heiland and E. Taglauer, paper pres. at the Int. Conf. Ion Surf. Interaction-Sputtering and Rel. Phen., Garching 1972, Rad. Eff. 19, 1 (1973).
[12] J. Lindhard, V. Nielsen and M. Scharff, Mat. Fys. Medd. Dan. Vid. Selsk. 36, No. 10 (1968).
[13] O. B. Firsov, Sov. Phys. JETP 36, 1076 (1959).
[14] A. A. Abrahamson, Phys. Rev. 178, 76 (1969).
[15] H. H. Andersen and P. Sigmund, Risö Report No. 103 (1965).
[16] C. Foster, I. H. Wilson and M. W. Thompson, J. Phys. B 5, 1332, (1972).

HYBRID COMPUTATIONAL STUDIES OF ELASTIC SCATTERING OF ATOMS

J. PEARCE, R. E. CROSBIE, D. G. ARMOUR, and G. CARTER
*Department of Electrical Engineering, University of Salford
Salford M5 4WT, Lancashire, England*

ABSTRACT

Initially an analogue computer and subsequently a hybrid computer were used to simulate the trajectories of colliding atoms acting under a central Born–Mayer interatomic potential. Studies of the binary elastic interaction between a single Ar atom and an Au atom and the multiple collision between an Ar atom and two neighbouring Au atoms were undertaken and the relationships between energy retained by the scattered atom and scattering angle were deduced for various incidence conditions of initial energy, impact parameter and Born–Mayer constants. The results are shown to agree well with digital simulation studies and several new features of the interaction process are explored.

Introduction

The central problem in quantitative prediction of scattering parameters in the elastic collision of atomic particles is exact evaluation of the equation for the scattering angle

$$\phi = \pi - 2p \int_{R_m}^{\infty} \frac{dr}{r^2 (1 - \frac{V(r)}{E_R} - p^2/r^2)^{1/2}}$$

which is analytically insoluble when the interatomic potential $V(r)$ adopts any except the simplest inverse power form and even then for only restricted exponents of the power potential. For this reason computer evaluations of the integral for realistic interatomic potentials, or close analytic approximations to the scattering angle equations for realistic potentials, are usually adopted.

Computer evaluations are usually performed using a digital technique. When the primary atom collides consecutively or simultaneously with more than one target atom the problem of analytic solution of the trajectory equations becomes even more acute and computer simulation becomes almost mandatory. Digital simulation techniques have been developed for numerous atomic collision problems including range calculations in random and crystalline media [1,2]; channeled particle trajectories [3,4,5]; surface backscattering of low energy particles [6,7,8]; radiation damage and defect production [9,10,11]; and sputtering processes [12].

As in the simple binary collision case the interatomic potential is a critical parameter and is usually a variable in the simulations so that its importance to the resulting collision predictions can be assessed. Other variable parameters in simulations will include the primary particle energy and the impact geometry, and in order to obtain statistical information it is always necessary to carry out the calculations over a wide range of variation of these parameters. This, of course, renders the simulation technique expensive of computer time.

The analogue computer is an attractive alternative to the digital computer, since although of poorer accuracy it allows direct, manual access to parameter variation (an important consideration if it is required to rapidly attempt to match a simulation with experimental data). Additionally, the analogue computer has generally a more rapid run and display time which renders it valuable and economic for the accumulation of statistical data. In association with a digital computer, to render the system hybrid, benefits of both techniques can be derived.

In the present communication we present analogue computation of binary atom scattering of an Ar atom from a single Au target atom and hybrid computations of the scattering of an Ar atom from a pair of nearest neighbor Au atoms in the <110> direction of a [100] surface. The major intention of the study, at this stage, was not to deduce necessarily novel information of the scattering process, although as will be seen, some new information was accumulated, but to compare the results of the simulation with earlier digital evaluations and validate the method. This will be shown to have been achieved.

Computational Procedure and Results

The essential difference between analogue and digital computers lies in the manner in which the dependent variables of a problem are handled. In analogue machines the dependent variables are represented by continuously varying voltages whereas in digital computers all variables appear in discrete form. On an analogue computer a particular problem is represented by interconnected computing elements such as integrators, multipliers and function generators which operate in a parallel manner. It is this parallel

operation which allows rapid analogue solution of the large sets of
differential equations describing the simultaneous interaction between colliding atoms. On a digital computer such problems have
to be solved by iterative numerical methods, which for large systems can be very time consuming, even on the fastest digital
machines.

A simple binary collision may be reduced to an equivalent
single body central problem [13] and in the preliminary studies
the equations of motion of such a model representing an Ar ion-Au
atom interaction, with a Born-Mayer interatomic potential of the
form $V(r) = Ae^{-r/a}$ (where A and a for the heteronuclear collision
were derived from Abrahamson's homonuclear values), were solved on
an analogue computer. This method differs from the usual digital
approach which is to solve the scattering angle equation by means
of a numerical technique. Results obtained took the form of ion
trajectory plots and values of scattering angle for different impact parameters and ion energy. Fig. 1 shows the deflection angle/
impact parameter relationships for ion energies of 10eV, 50eV,
100eV, 1keV and 10keV and for comparison values calculated using
a digital technique by Smith and Carter [7,14]. The values computed by analogue computer are within 1° of the digital values and
this indicates the order of accuracy of the analogue approach.

When the interaction between an ion and n atoms (where n > 1)
is considered, the methods of classical mechanics show that the
following set of differential equations must be solved to trace
the motion of the ion:

$$M_0 \begin{vmatrix} \overset{..}{X}_0 \\ \overset{..}{Y}_0 \\ \overset{..}{Z}_0 \end{vmatrix} - \begin{vmatrix} X_0 - X_j \\ Y_0 - Y_j \\ Z_0 - Z_j \end{vmatrix} \cdot f(R_j) = 0 \quad \text{for } j = 1, 2 \ldots n$$

where M_0 is the ion mass, X_0, Y_0, and Z_0 the ion coordinates, X_j,
Y_j, and Z_j the jth atom coordinates and $f(r_j) = -\dfrac{1}{R_j} \cdot \dfrac{dv(r_j)}{dr_j}$
where $V(r_j)$ is the potential function. Using the largest analogue
computers currently available it would be possible to consider a
system of some ten to twenty atoms, because only a small analogue
computer was available for this work a cell sharing technique was
used in which an analogue cell representing a portion of the whole
problem is multiplexed.

The technique was implemented on a hybrid computer comprising
a sixty amplifier Solartron HS7/3D analogue machine and a PDP8/L
with 8K of core. The principle of the method was to set up the
analogue computer to represent the interaction between an ion and
a single atom. For a short time step the analogue cell solves for
the coordinates of a particular atom, while the remaining atom
coordinates are approximated by second order hold circuits. At

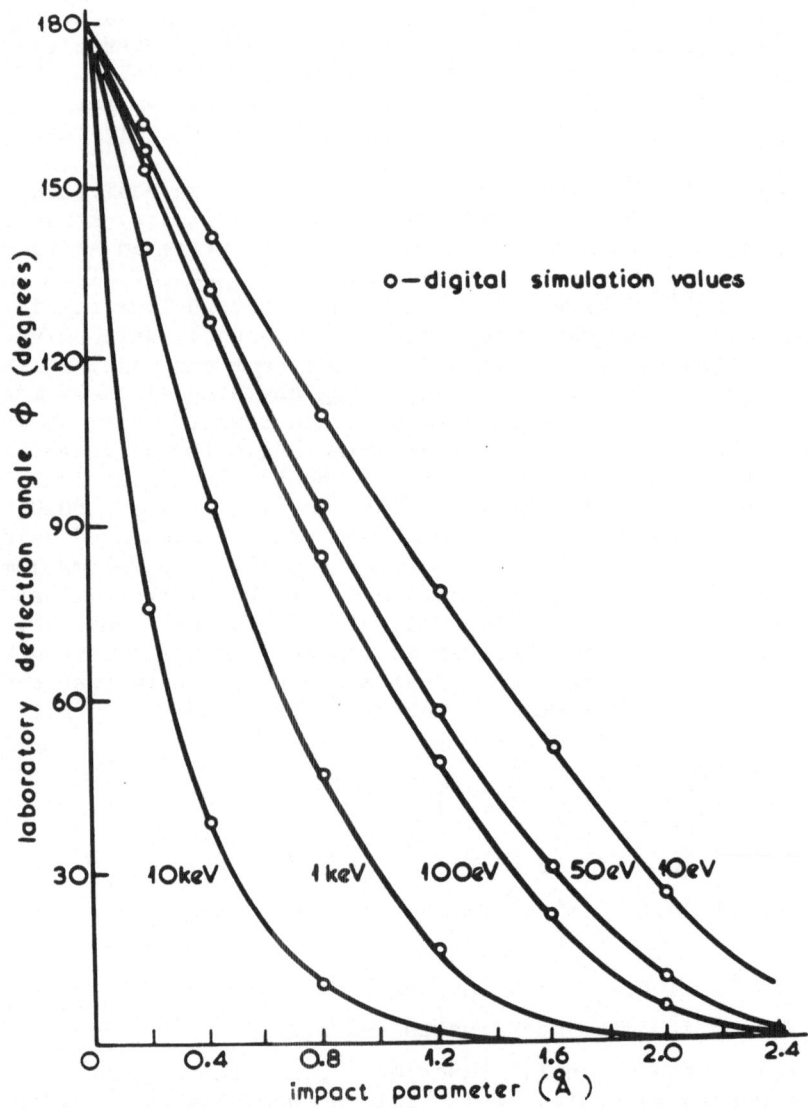

Fig. 1.   Deflection of an Ar ion by a single Au atom.

the end of the time step the atom coordinates are stored in the
digital computer, the analogue cell is set up to represent the
next ion-atom interaction and the same time step is repeated. This
process is continued until all the atoms have been treated and the
whole cycle is then repeated for subsequent time steps. In addition
to controlling the analogue computer and storing coordinate values,

the digital computer is responsible for determining initial conditions for given ion energy and impact parameter, scaling the analogue
cell, generating the potential function and calculating the final
deflection angle. Digital simulations of the technique have been
successfully performed·for a four atom case, although the hybrid
model has to-date only been used to simulate the relatively simple
Ar ion—two Au atom arrangement depicted in Fig. 2.

The primary aim of these simulations was to study the trajectories of the incident ions as the ion energy, impact parameter
relative to the central axis of the target atom pair and the constants of the Born-Mayer potential were varied. From the data
recorded it was then possible to determine the fractional energy
retained by the incident particle as a function of scattering angle,
data already reported using digital techniques and potentially by
experiment. Trajectories were determined as a function of impact
parameter to the central axis (<100> channel axis) and for the
Born-Mayer constants A = 42.38 keV, a = 0.219 Å, for the Ar atom
energies between 1 eV and 1 keV, and Figs. 3a to 3d display the
form of these trajectories for incident energies of 1 eV, 10 eV,
100 eV, and 1 keV. From the final Ar atom energy at the end of the
collision and the trajectories it was then possible to plot the
appropriate fractional energy retention E/E_0 as a function of
scattering angle as shown in Fig. 4. Pryde, Smith and Carter [14]

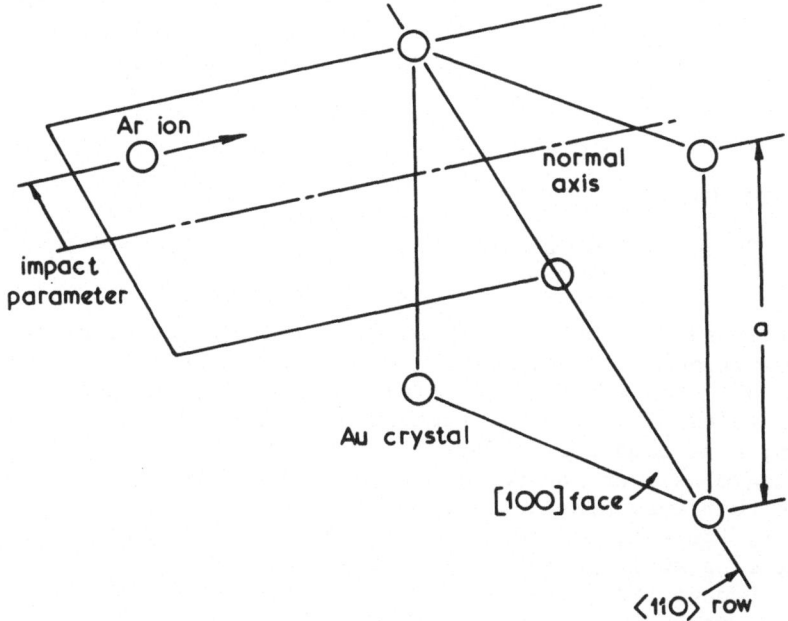

Fig. 2. Geometrical configuration for Ar-Au simulation.

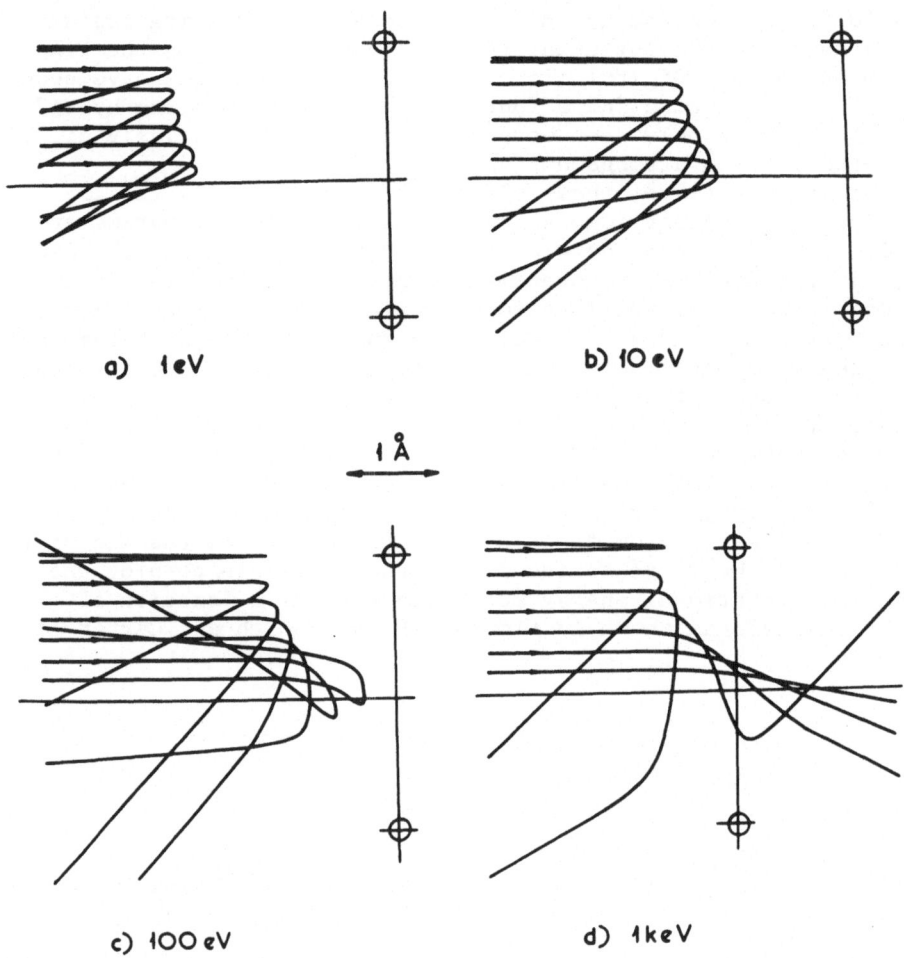

a) 1eV   b) 10 eV

1 Å

c) 100 eV   d) 1 keV

Fig. 3. Trajectories of an Ar ion interacting with two Au atoms.

have performed somewhat similar simulations corresponding to the
above conditions but employing a consecutive rather than a simultaneous collision model and digital computation, and the present
data was found to be closely similar to their deductions.   In
addition, some completely digital simulations of the simultaneous
scattering model were performed in the present work and it was
observed that agreement between the two computational methods was
excellent, for a given energy retention the two techniques differ
in scattering angle generally by considerably less than 1°, quite
accurate enough for comparison with low energy ion scattering experimental work.

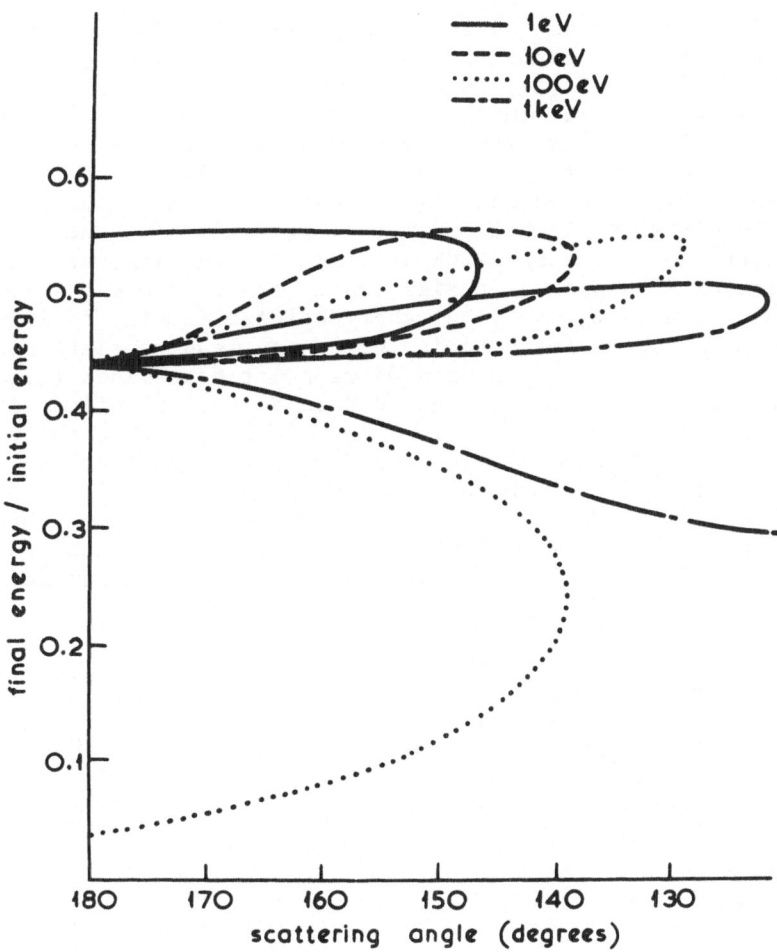

Fig. 4. Final energy/initial energy scattering angle relationship
for 1 eV to 1 keV ions.

    Several features of Fig. 4 are worthy of comment. First, at
low primary energies, the E/E_O vs ϕ curve is an unclosed loop where
the energy retention for d = 180° scattering can possess two different values. As primary energy increases the loop closes at
ϕ = 180° and then develops a second unclosed loop so that for all
scattering angles except ϕ = 180°, four values of E/E_O are possible.
As energy is further increased the unclosed loop becomes a single
valued function for E/E_O so that only one, two or three values of
E/E_O are determinable for different ϕ values. At the higher energies, the small scattering angles correspond to penetration of
the target atom pair where only a single value of energy retention

may be ensured and this penetration occurs, of course, for primary
particles incident closest to the atom pair axis. The feature of
closed loops and a second branch to the curves has been previously
observed by Pryde et al.[14], Begemann et al. [15], and
Heiland and Taglauer [16] in digital simulations, but the quadri-
valued nature of E/E_0 for certain ϕ values has not previously been
inferred. The origin of this behaviour is readily traced by study
of individual trajectories and Fig. 5 shows four different trajectories (with different impact parameter) which lead to the same
overall scattering angle but with different energy losses; the
three trajectories depicted in Figs. 5a to 5c are those previously
recorded by Pryde et al. [14] but that depicted in Fig. 5d is new.
Although the number of trajectories which can lead to scattering
of the forms shown in Figs. 5c and 5d were not determined (i.e.,
sufficient statistics were not recorded to adduce effective differential scattering cross sections) one could expect on the basis
of Pryde et al.'s work that these would be small, so that experimental observation, although possible, would be difficult.

At low energies where single closed or unclosed loops were
deduced, the higher energy retention always corresponds to a suc-
cession or simultaneity of two relatively large impact parameter

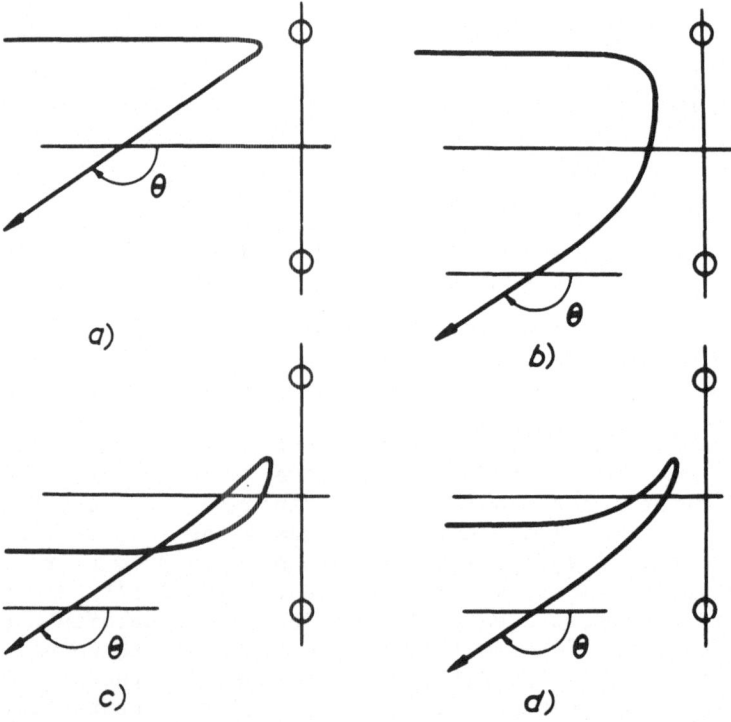

Fig. 5. Four types of collision leading to the same deflection angle.

collisions with each of the target atom pair as shown in Fig. 4b,
whereas the lower energy retention corresponds to a closer collision with one target atom and a very weak second collision. At
$\phi = 180°$ these collisions are respectively the result of an axially
directed atom, with simultaneous equal interaction with both target
atoms and a head on collision with one target atom. At very low
incident particle energies (1 eV) the loop is unclosed at $\phi = 180°$
and the energy retained by the primary particle incident along the
pair axis is greater than that retained in a head on impact. This
suggests that, at low impact energies, the atom pair commences to
act as an effectively heavier mass single atom. At low particle
energies the minimum distance of approach of the particle to either
target atom becomes large and at large distances the addition of
the two spherical potentials of the target atoms commences to adopt
more of a planar geometry. Thus the incident particle commences
to see the atom pair as a pseudo-planar surface of mass increasing
towards infinity as ion energy is reduced further. This feature
of very low energy scattering has been reported from digital
simulations by Smith and Carter [17] where the increasing retained
energy for axial and paraxial collisions was revealed. It must
be pointed out that, at such low energies, the model is highly
artifical since certainly the Born-Mayer repulsive potential is
most unlikely to be applicable and attractive terms in the potential
must be used, whilst thermal atomic vibrations would assume increasing importance in dictating the statistics of the energy transfer
processes also.

   In order to study the effect of varying the Born-Mayer potential
function constants, sets of trajectories were obtained for constants
of A x 10, A/10 and for a x 1.1 and a x 0.9. In Fig. 6 the fractional energy retention (E/E_0)-scattering angle relationships are
presented for initial ion energies of 100 eV with A x 10 and A/10.
As indicated in the figure, the 100 eV - A x 10 curve is identical
to that obtained for 10 eV ions using a constant A and similarly the
100 eV-A/10 curve corresponds to the 1 keV-A case. This shows that
multiplying A by a constant, K, has the same effect on the ion
trajectory as dividing its initial energy by K and this may be
seen to be the expected result by considering the Lagrangian of the
system,

$$L = T_o + T_1 + T_2 - A \exp(-r_1/a) - A \exp(-r_2/a)$$

where T_0, T_1 and T_2 are the ion and atom relative kinetic energies.
Multiplying the potential energy terms by K will give the same
spatial solution to the Lagrangian equations as dividing the kinetic
energy terms by K, although in the latter case the particle velocities will be reduced by $\sqrt{K}$. That the computer results agree with
theory shows the computer technique to be valid.

   The ion trajectories are more sensitive to variations in the
Born-Mayer constant a and since a occurs within the exponential,

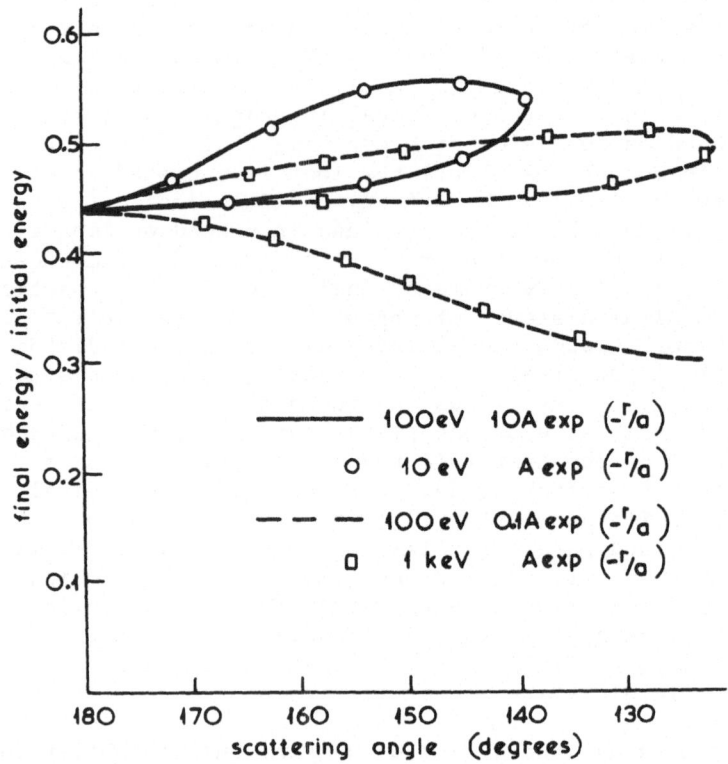

Fig. 6. Effect of varying Born–Mayer constant 'A' by a factor of 10.

there is no scaling with energy. A comparison is made in Fig. 7 between the E/E_0-scattering angle relationships for 100 eV ion using the constant values a x 1.1, a x 0.9 and a. Increasing a strengthens the potential and has a similar effect to reducing the initial ion energy. It is noted that a 10% change in a has nearly the same effect as a x 10 change in A.

The final topic of study was the deduction of the energy transfer from axially directed Ar atoms to each of the Au atoms as a function of the initial Ar atom energy. Similar digital computer simulations for Kr atoms incident upon W atoms have been performed by Smith and Carter [17] and theoretical predictions have been made in an approximation method called the Distant Collision Approximation (DCA) by Sigmund and Andersen [18]. The results of the present simulations are presented in Fig. 8, where the total energy transferred to both Au atoms (equal to that retained by the incident atom) is plotted as a function of the incident atom energy. This curve displays the peaked structure observed also by Smith

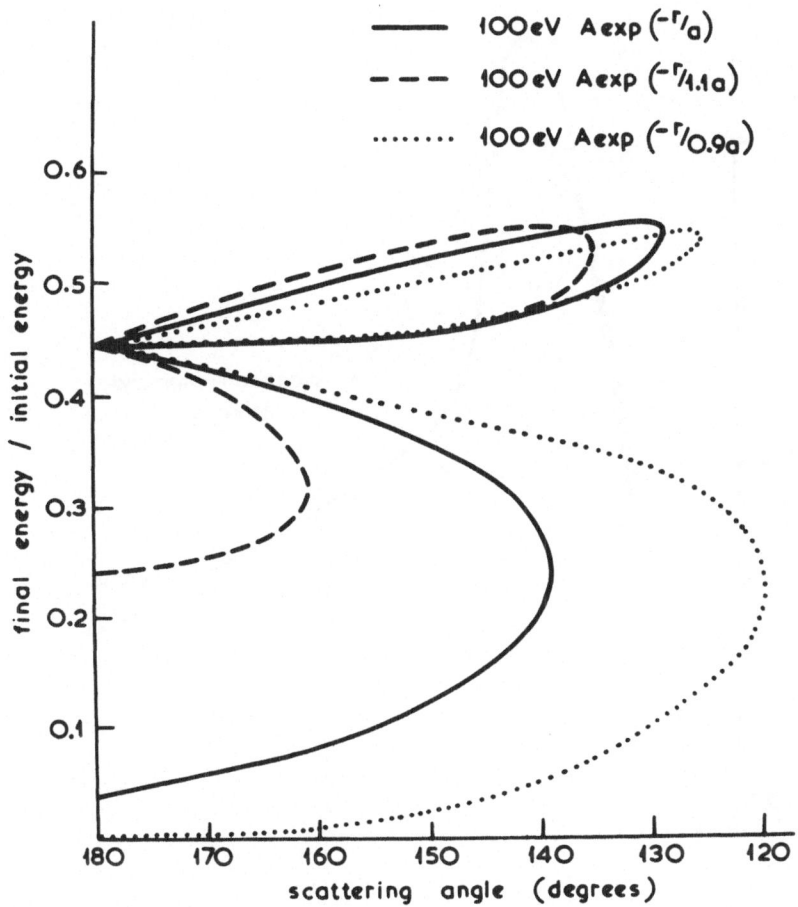

Fig. 7. Effect of varying Born–Mayer constant 'a' by 10%.

and Carter, although direct comparison is not possible since the interacting atoms and the potentials are different. Direct comparison with the DCA is possible, however, by adopting the appropriate number and type of particles in the Sigmund–Andersen model and the resulting theoretical curve is also displayed in Fig. 8. As in the case of the digital simulations by Smith and Carter and their comparison with the DCA, the simulations lead to a peak in the energy transfer at the same particle energy as theory predicts but the energy loss is always slightly smaller than the approximate predictions until an initial atom energy in excess of the energy where maximum transfer occurs. This comparable behaviour with the digital simulation again gives confidence in the reliability of the present simulations and further attestation to the rather good approximation to axial collisions afforded by the DCA.

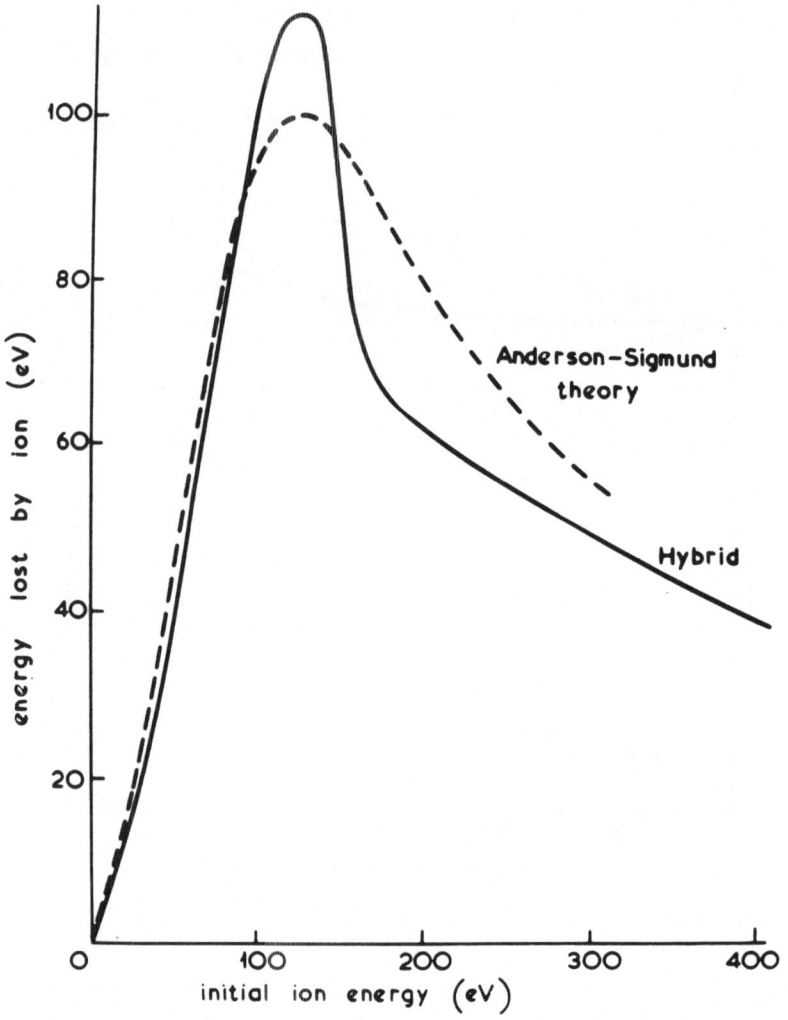

Fig. 8. Graph of energy lost by an Ar ion on a trajectory midway between two gold atoms against initial ion energy.

Conclusions

It has been demonstrated that analogue and hybrid computational methods are well suited to the simulation of, in the present case, simple atomic collision events. Trajectories and scattered energy/angle relationships for binary atomic collisions and simultaneous collisions between an incident atom and two target atoms are predictable with sufficient accuracy to ensure validity and comparability with experimental accuracy, particularly in the low energy atomic collision regime. A positive advantage of analogue and hybrid techniques is the ease of access to parameter variation, an important requirement if variations of interatomic potential constants, initial energies and impact parameters must be readily and rapidly effected. Run times on the analogue and hybrid machines for the type of problem explored here are typically seconds as contrasted to minutes of digital computation. This highlights the further advantage of the present techniques in that they are admirably suited to statistical operations, a fundamental require-ment of backscattering, atom reflection and forward scattering, atom penetration and channeling studies.

Although the data presented here are not entirely novel, they do confirm the observations of previous investigators using digital simulation methods, particularly with respect to the form of scat-tered energy/angle relationships for a simultaneous interaction between a projectile atom and two initially stationary atoms. Additional features of the energy transfer at low primary particle energies for axially directed atoms were found and a further type of plural collision observed.

It should be stated that such methods are quite readily adapted to more physically realistic collision problems by the addition of further components to the hybrid system. Thus, it would be possible to study, with the most powerful analogue compu-ter presently available in the United Kingdom, the simultaneous interaction of about twenty atoms. It is also possible without any major difficulty to include parameters such as thermal atomic motion and structural defects in a lattice, a time (or approach distance) dependent ion neutralization process and inelastic energy processes along a trajectory and these are developments which we are currently pursuing.

References

[1] O. S. Oen, D. K. Holmes and M. T. Robinson, J. Appl. Phys. 34, 302 (1963).
[2] M. T. Robinson, and O. S. Oen, Phys. Rev. 132, 2385 (1963).
[3] D. V. Morgan and D. Van Vliet, Can. J. Phys. 46, 503 (1968).
[4] D. Van Vliet, Rad. Effects 10, 137 (1971).
[5] D. V. Morgan and D. Van Vliet, Rad. Effects 12, 203 (1972).
[6] V. E. Yurosova, V. I. Shulga, and D. S. Karpuzov, Can. J. Phys. 49, 759 (1967).
[7] A. G. Smith and G. Carter, J. Phys. D2, 972 (1969).
[8] V. E. Yurosova, I. G. Bunin, V. I. Shulga and B. M. Mamaev, Rad. Effects 12, 175 (1972).
[9] M. Yoshida, J. Phys. Soc., Japan, 16, 44 (1961).
[10] J. B. Gibson, A. N. Goland, M. Milgram and G. H. Vineyard, Phys. Rev. 120, 1229 (1960).
[11] J. R. Beeler and D. G. Besco, J. Appl. Phys. 34, 2873 (1963).

[12] W. L. Gay and D. E. Harrison, Phys. Rev. <u>135</u>, A1780 (1964).

[13] H. Goldstein, "Classical Mechanics," Addison-Wesley Publishing Co., Inc., New York, 1950, Chapter 3.

[14] M. Pryde, A. G. Smith, and G. Carter, "Atomic Collision Phenomena in Solids," Eds. P. D. Townsend, D. W Palmer and M. W. Thompson, North-Holland Publishing Co. Ltd., Amsterdam, 1969, p. 573.

[15] S. H. A. Begemann, Surface Science <u>30</u>, 371 (1972).

[16] E. Taglauer and W. Heiland, Surface Science <u>33</u>, 27 (1972).

[17] A. G. Smith and G. Carter, Rad. Effects <u>12</u>, 63 (1972).

[18] H. H. Andersen and P. Sigmund, Mat. Fys. Medd. Kgl. Danske Videnskab Selskab <u>34</u>, 15 (1966).

SPUTTERING OF CONDENSED GASES BY PROTON BOMBARDMENT

S. K. ERENTS and G. M. McCRACKEN
UKAEA Culham Laboratory
Abingdon, Berkshire, England

Introduction

The desorption of gases from cryogenic surfaces by ion bombardment is of interest in certain types of plasma physics experiment, using either cryopumps or superconducting magnets, and in superconducting accelerators. In previous papers the results of thermal desorption [1] and of ion and electron bombardment of solid hydrogen and deuterium [2] were described. Yields in the range 10^4 - 10^5 atoms/ion were obtained for layers of condensed gas less than 1000 Å thick under proton bombardment and these values were explained in terms of a thermal spike in the copper substrate below the condensed gas layer. It was considered of interest to see how the yields of other gases compared with hydrogen, particularly gases with higher heats of sublimation. In this paper we present results of the desorption of heavier gases He, N_2, CO and Ar by 5 keV protons and also the effect on gas yields of making "sandwiches" of condensed gas consisting of H_2, N_2 and Ar in various thicknesses.

Experiment

The details of the experimental technique have been described previously [2]. The gas is condensed on a liquid helium cooled surface whose temperature is measured with a germanium resistance thermometer, Fig. 1. The target chamber is pumped continuously and the gas desorption rate is measured dynamically with a quadrupole mass spectrometer. The vacuum time constant is approximately 0.05 secs. Gas is introduced either through a palladium-silver leak (for hydrogen isotopes) or through a sintered Si C leak of low conductance. Gas is condensed on the surface for a known time at ambient pressure, from which the thickness of the layer can be deduced assuming a sticking coefficient. The ion beam is then turned on and the yield of gas from the target is measured as a function of time. The experiment is normally started by cleaning the surface with a high current density ion beam ($\sim$200 μA cm^{-2}).

Results and Discussion

   Argon, nitrogen and carbon monoxide

   A variety of gases were studied in order to compare them with
the previous results for hydrogen and deuterium. Measurements of
the yield as a function of time for a number of initial layer
thicknesses are shown in Fig. 2. Both nitrogen and argon show
similar behaviour to that found previously for hydrogen and deuterium with the difference that the yields are about a factor of 10^3
lower. The layer thickness is quoted in the directly measured
units of torr-sec exposure. The sticking coefficient of argon has
been measured experimentally and found to be greater than 0.95 [3]
at 4.2 K over the range of coverages of interest. The layer thickness in molecules cm^{-2} can thus be readily calculated. It has been
assumed that N_2 and CO will have similarly high sticking coefficients. For an initial layer thickness greater than 10^{-5} torr sec
it is seen that the yield increases with time as the gas layer is
eroded. At a certain thickness there is a maximum yield and

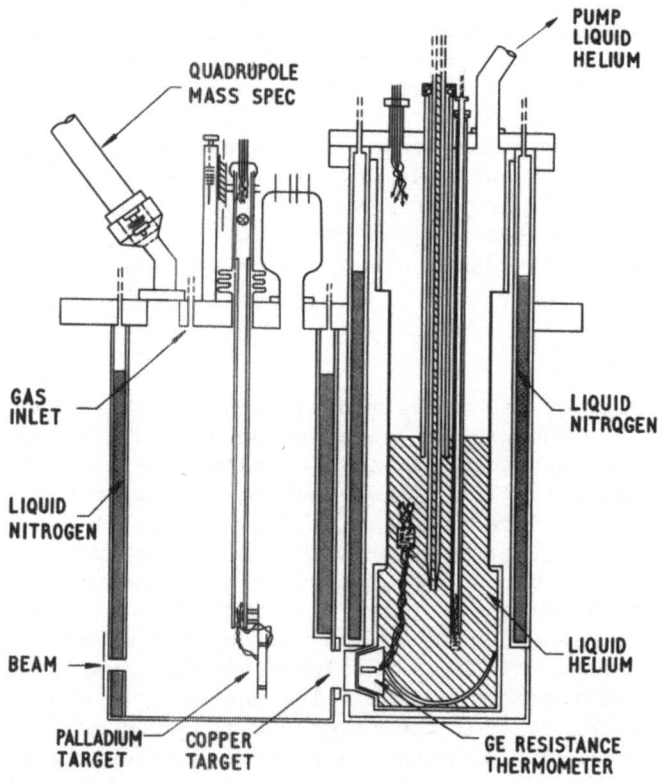

Fig. 1. Schematic diagram of target chamber showing He cryostat.

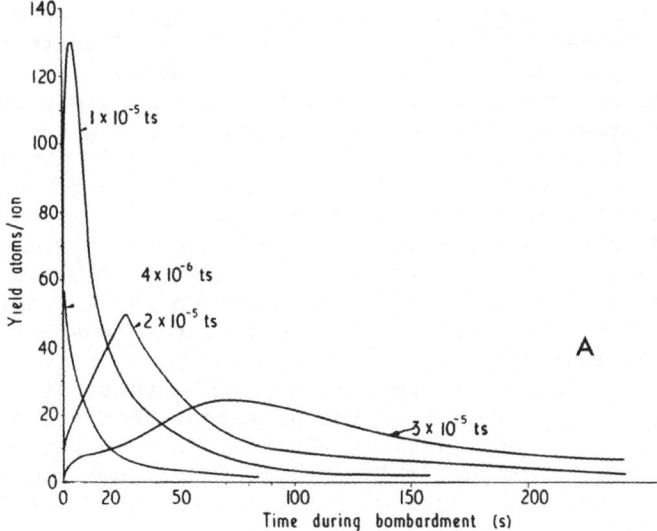

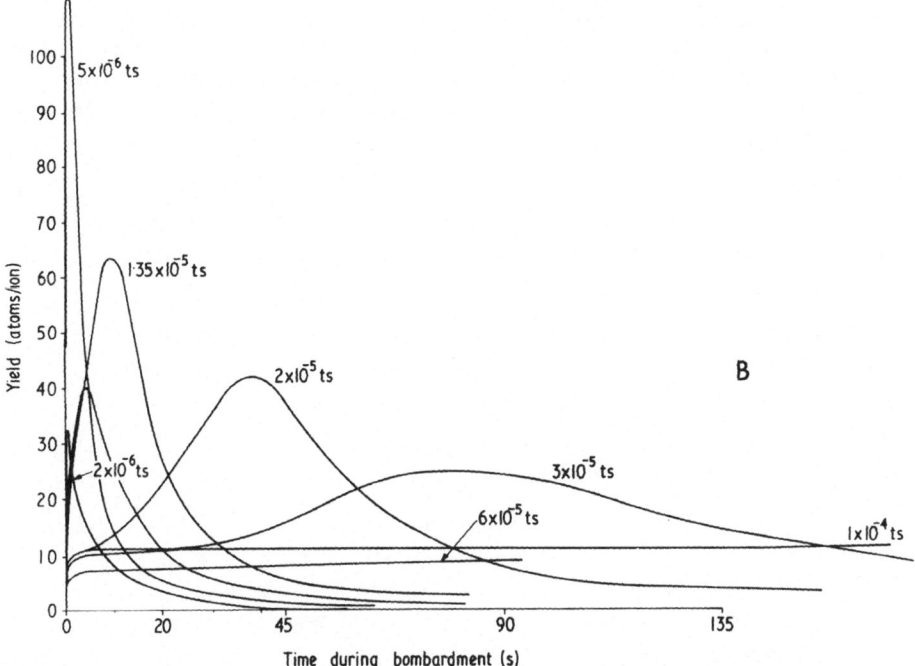

Fig. 2. Yield of gas desorbed from the target by 5 keV protons as a function of time for different initial thickness of condensed gas layer at 4.2 K. (a) Argon – 0.1 μA, 5 keV, $H^+ \to A$, 4.2° K. (b) Nitrogen – 0.1 μA, 5 keV, $H^+ \to N_2$, 4.2° K.

thereafter the yield decreases as the layer is eroded further.
This is illustrated more clearly in Figure 3 where the initial
yield is plotted as a function of the thickness of the layer from
a series of runs as shown in Figure 2. The yield increases approximately linearly with layer thickness reaching quite a sharp maximum and then decreasing. The maximum yield is 100 ± 20% atoms/ion
for nitrogen and argon and 50 ± 20% atoms/ion for carbon monoxide.
The absolute yields are measured relative to the hydrogen partial
pressure rise when the incident ion beam is allowed to impinge on
a saturated palladium target which under suitable conditions has
been found to have a yield of 1 atom/ion [4]. A correction is
applied for the variation in sensitivity of the mass spectrometer
to the various gases.

It is clear that the maximum yield occurs at a coverage of a
few monolayers. This is quite different from the case of solid
hydrogen. The peak yield is thus a thin film phenomenon, and at
greater thickness the yield rapidly approaches the yield of a
bulk material. The range of 5 keV protons in N_2 or CO is 1050 Å as
calculated from the data of Schiøtt [5]. This range corresponds
to ∿4 x 10^{17} molecules cm^{-2} or a gas exposure of 8 x 10^{-4} torr sec.
Judging from the nitrogen case, Figure 3b, which is the only one
with sufficiently thick layers, there is no marked change in yield
when the range of the incident ions is less than the thickness of
the condensed gas layer. This again is in contrast to the results
for hydrogen.

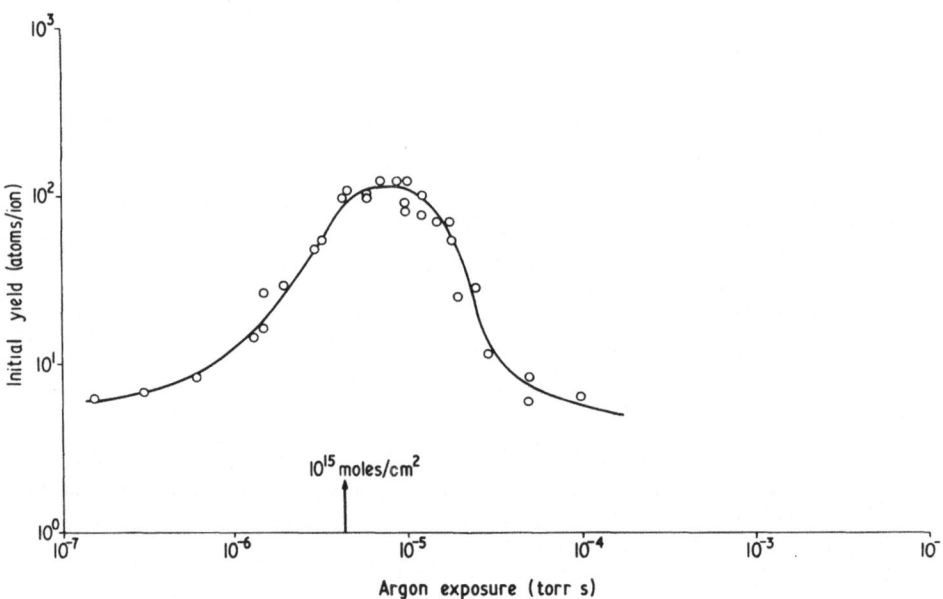

Fig. 3. Initial yield of gas as a function of condensed layer
thickness. 5 keV protons on condensed gas at 4.2K. (a) Argon –
0.1 μA, 5 keV, $H^+{\rightarrow}A$, 4.2° K.

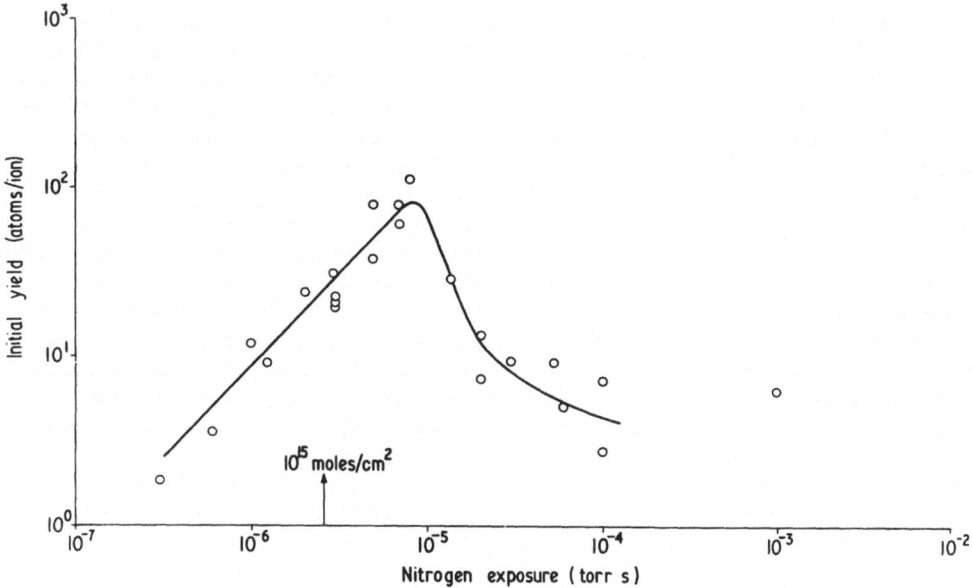

Fig. 3(b).   Nitrogen – 0.1 µA, 5 keV, $H^+ \rightarrow N_2$, 4.2°K

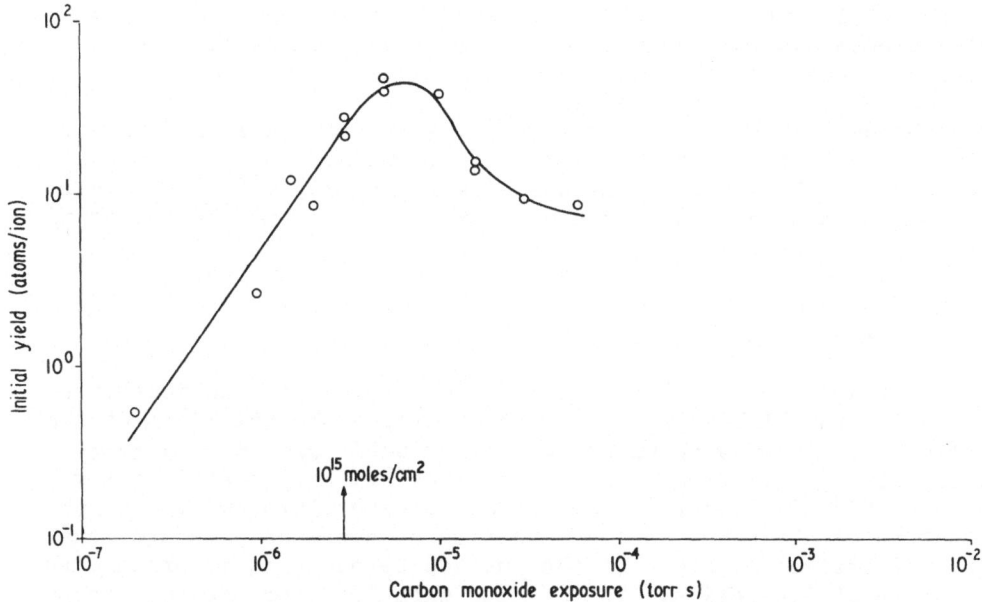

Fig. 3(c).   Carbon monoxide – 0.25 µA, 5 keV, $H^+ \rightarrow CO$, 4.2° K.

There are thus two main results to explain; the reason for the maximum yield of ∿100 atoms/ion and the yield from the thick layers, ∿5 atoms/ion. The first obvious check is to see whether the thermal spike mechanism, which was successful in explaining the high hydrogen yields, would apply in this case. However it is soon clear that there is no possibility of this mechanism explaining the yields for the heavier gases. For thin condensed gas layers the incident protons will slow down and give up their energy in the copper substrate, thus the calculation is identical to that described previously for hydrogen [2]. The spike radius is ≈ 250 Å, the temperature 30 K and the lifetime ∿10^{-10} sec. From vapour pressure data this leads to yields of ∿10^{-4} atoms/ion for nitrogen and less for Ar and CO. The yield from a conventional sputtering mechanism assuming a thick layer of condensed gas has also been calculated. Estimates from both the theory of Sigmund [6] (which is recognized not to apply too well for light ions and tends to give an overestimate) and also from the simpler theory of Pease [7] leads to yields from nitrogen in the range 0.1 to 1.0 atoms/ion. It is thus difficult to explain the observed yields by this process.

A further possibility is release of gas by dissociation or ionization in a way similar to the process of electron desorption of adsorbed gas [8]. There is then the possibility of cluster formation around ions as has been observed in the electron bombardment of solid hydrogen [9] and in the ion bombardment of water [10]. However, there is little information available to confirm whether such a mechanism can explain the observed yield. It is rather unlikely that such a mechanism can explain the sharp maximum in the yield curve. As discussed earlier this appears to be a thin film phenomenon and one possibility is that phonons produced in the copper by the incident ion cannot propagate across the boundary from the copper to the condensed gas when the gas layer is thick because the phonon spectra in the two materials are radically different. However, when the condensed gas film is thinner than the phonon mean free path then the condensed gas layer will not have a well defined phonon spectrum and thus energy transfer across the boundary may be allowed.

Helium

Because of the relatively high temperature of the substrate and hence the high vapour pressure of helium the thickness of the helium with increasing exposure will probably grow by physisorption only to a few monolayers before the sticking coefficient will become effectively zero. The surface concentration will then be constant with further increase in time at constant pressure, the rate of arrival of atoms at the surface being in equilibrium with the rate of desorption. The results with 10^{-8} torr helium present are shown in Figure 4a. It is apparent that this curve is much

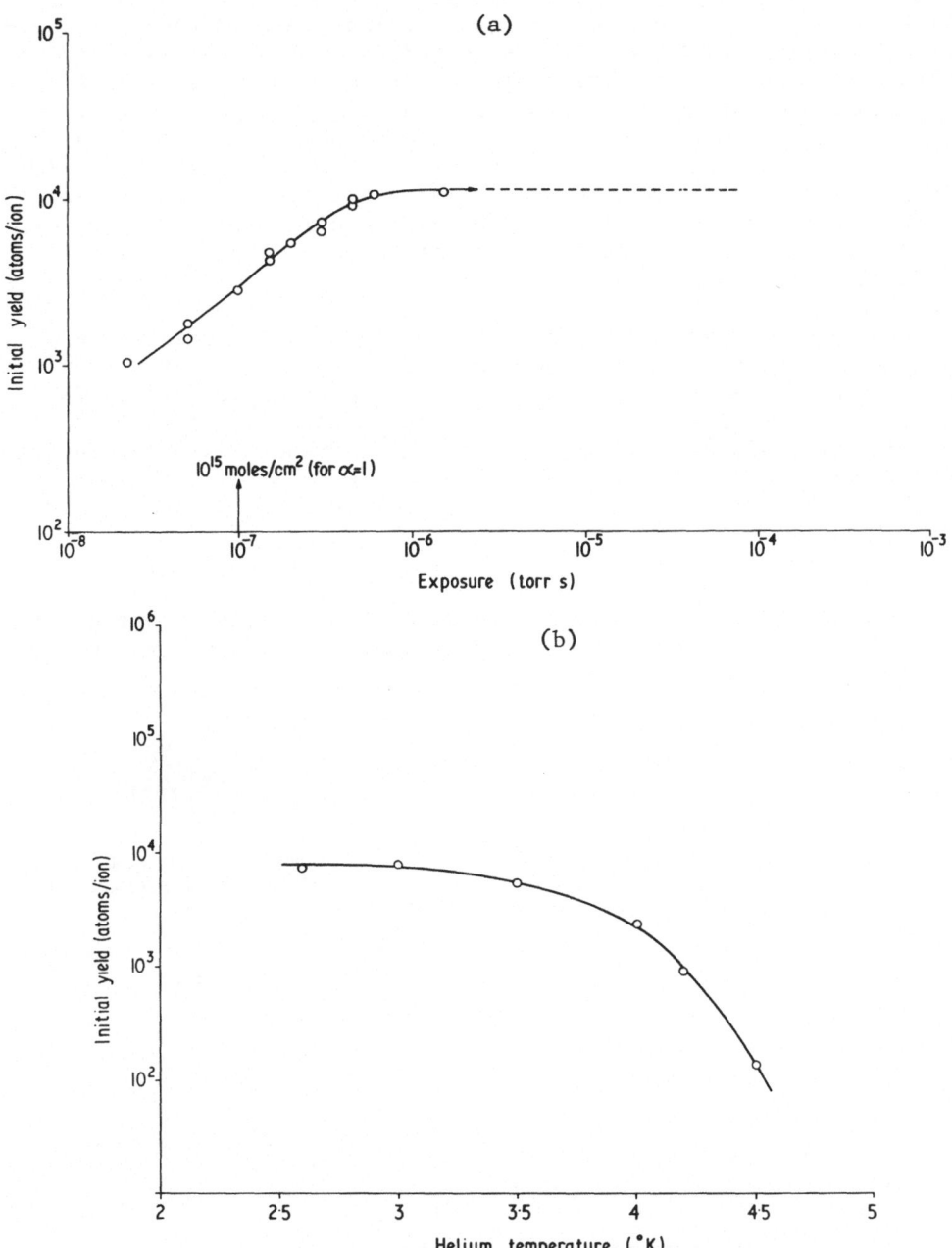

Fig. 4. Desorption of adsorbed helium by 5 keV protons. a) Yield
vs exposure time. Temperature = 3.0 K. Pressure during adsorption
5×10^{-9} torr. b) Yield vs temperature for constant exposure of 10^{-6}
torr sec. 100 secs of 1.5×10^{-3} μA $H^+ \rightarrow$ He·Cu. Pressure during
adsorption 10^{-8} torr He.

closer to the behaviour of the hydrogen results described earlier
[2] than to those of the heavier gases. The yield increases linearly to a maximum which is of the same order of magnitude as that for
hydrogen. Since the substrate temperature must again be ∿30° K
the vapour pressure of the helium liquid will be high'(since it is
atmospheric pressure at 4.2 K), and so the yield can be readily
explained by the thermal spike mechanism. The fact that the helium
yield is lower than hydrogen (which has a heat of sublimation about
10 times greater) may be due to the lower equilibrium coverage, or
to the physisorption of He on copper at these coverages having a
higher heat of adsorption than the heat of evaporation of bulk
hydrogen.

A further experiment with helium is shown in Figure 4b where
the yield is measured as a function of substrate temperature for a
constant exposure of 10^{-6} torr secs. No quantitative measure is
available of the layer thickness but the results illustrate the
decrease in yield as the surface coverage decreases with increasing
temperature.

Gas mixtures

Finally some measurements were made of the yield of one gas as
the thickness of another gas on top or underneath it was varied.

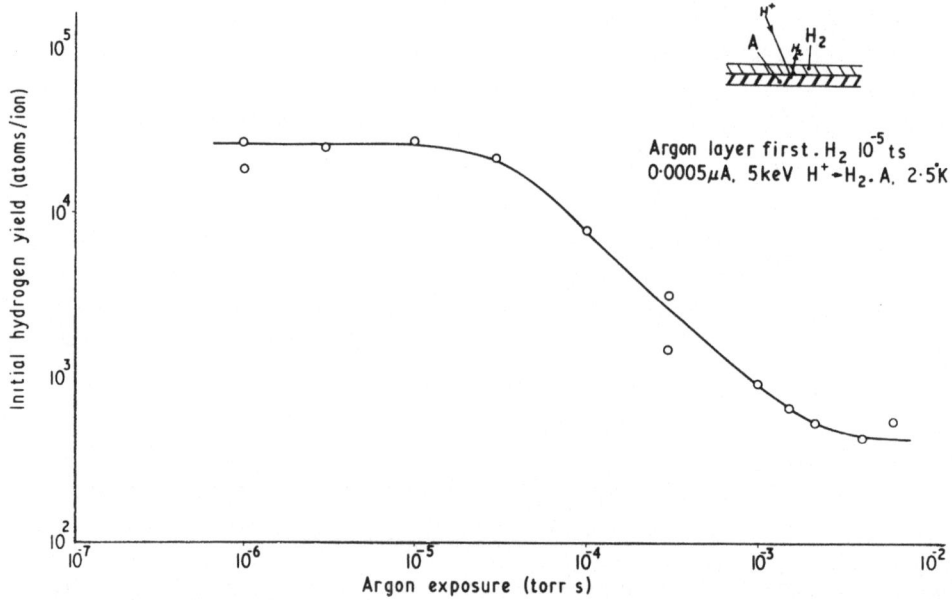

Fig. 5. Desorption of argon from a layer of condensed argon covered
by a layer of hydrogen of varying thickness by 5 keV protons.
Temperature 2.5 K.

In Figure 5 the results obtained for the yield from a fixed thickness of argon is shown as the thickness of a hydrogen layer on top of the argon is increased. The argon yield remained constant until the hydrogen layer on top of it increased to the point where the hydrogen layer thickness exceeded the range of the incident proton ($\sim 2 \times 10^{-4}$ t-s). It is interesting to note that the yield did not decrease significantly until this point, indicating that the argon desorption could take place through the hydrogen layer. The corresponding yield of hydrogen in a similar sandwich is shown in Figure 6. In this case in successive runs a constant thickness of hydrogen is condensed on top of various thicknesses of argon. It is found that the hydrogen yield is initially at the same level as obtained when it is condensed on a simple copper substrate. As the argon thickness is increased above 10^{16} molecules/cm^2 the hydrogen yield decreases. At this layer thickness some of the incident proton energy is being lost in the argon. However, no calculations have yet been made of thermal spikes in argon, so it is not yet clear whether the temperature spike is simply lower in argon than copper or whether the argon layer is acting as a thermal barrier to the energy which is deposited in the copper substrate. A further possible explanation is that the hydrogen actually diffuses into the argon rather than staying on top after condensation. Some evidence for this has been obtained when the condensed gas sandwich

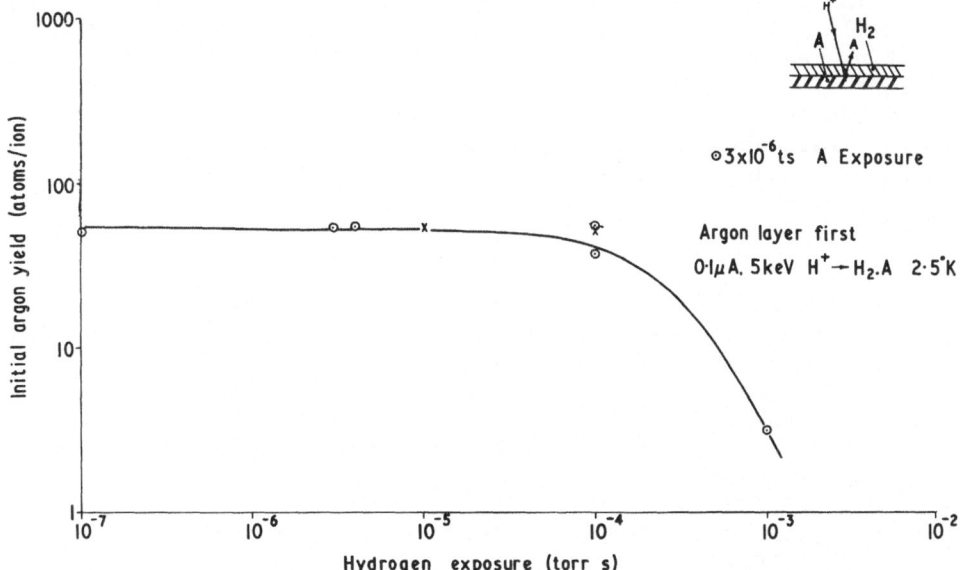

Fig. 6. Desorption of hydrogen from a layer of hydrogen condensed on top of a layer of argon of varying thickness by 5 keV protons. Temperature 2.5 K. Hydrogen layer thickness $\sim 2 \times 10^{16}$ molecules/cm^2.

is produced with nitrogen on top and hydrogen underneath. It is
found that hydrogen will diffuse through the nitrogen layer at a
detectable level in our apparatus at 2.8 K. The rate of diffusion
increases with temperature and a plot of the flow rate vs 1/T (for
constant nitrogen thickness of 10^{17} molecules/cm^2) indicated an
activation energy for diffusion of 170 cals/mole.

Conclusions

The desorption of heavy gases such as Ar, N_2 and CO by 5 keV
protons is a much less probable process than the desorption of the
light gases such as hydrogen and helium. This is in general agreement with the very large difference in their heats of sublimation.
From the present results helium appears to behave in a similar way
to hydrogen with yields of $\sim 10^4$ atoms/ion and this can be explained
qualitatively on the basis of the thermal spike mechanism previously
developed for hydrogen. However, a detailed explanation is not yet
possible because in the present experiments the helium is only
adsorbed rather than condensed, and its thickness has not been well
defined.

It appears that the yield of the heavier gases cannot be explained either by the thermal spike mechanism or by a conventional
momentum transfer sputtering theory. It has been put forward as
a hypothesis that an ionization or atomic excitation mechanism is
the explanation, but further experimental work, particularly on
the mass distribution of the desorbed species, would be necessary
to establish this hypothesis.

Finally, the measurements on gas mixtures have produced some
interesting results which may be useful in assessing practical applications of cryosurfaces in the presence of ions. Although no
clear explanation has yet been found for these results it is possible that this approach could be a useful method of further investigation of mechanisms since it is a simple way of looking at a
variety of substrates.

References

[1] S. K. Erents and G. M. McCracken, Vacuum 21, 257-260 (1971).
[2] S. K. Erents and G. M. McCracken, J. Appl. Phys. 44, 3139 (1973).
[3] J. N. Chubb, Private Communication, 1968.
[4] G. M. McCracken and J. H. C. Maple, Brit. J. Appl. Phys. 18,
    919-30 (1967).
[5] H. E. Schiøtt, Danske, Vidensk Selby verdi Math - Fys. Meddr.
    35, No. 9 (1966).
[6] P. Sigmund. Phys. Rev. 184, 383-416 (1969).
[7] R. S. Pease, Rendiconti Sci. Int. Fis. "Enrico Fermi" 13,
    158-165 (1960)
[8] D. Menzel and R. Gomer, J. Chem. Phys. 41, 3329 (1964).
[9] R. Clampitt and L. Gowland, Nature 223, 815-817 (1969).
[10] G. D. Tantsyrev and E. N. Nikolaev, Doklady Akad Nauk S.S.R.
    206, 151-154 (1972).

THE ANGULAR DISTRIBUTION OF FAST CHARGED PARTICLES REFLECTED BY THE SURFACE OF A SINGLE CRYSTAL

N. P. KALASHNIKOV

The Moscow Physical-Engineering Institute
Moscow, U.S.S.R.

ABSTRACT

The back scattered particles with the energy E_f close to the incident energy E_i escape the mono-crystal target from the thickness which is smaller or comparable with the inelastic scattering length. In the case of small energy losses $\frac{E_i-E_f}{E_i} \ll 1$, the angular distribution of the back-scattered particles corresponds to the usual Rutherford scattering if the incident and final directions of the particle momenta do not coincide with crystallographic axes. When the incident momentum direction and/or the back-scattered momentum direction coincides with any crystallographic axis, the blocking effect essentially reduces the scattering of the fast charged particles. When the energy losses are not small enough, the differential back-scattering coefficients of fast charged particles for a single crystal and for an amorphous medium coincide, provided that the incident and final directions are not parallel to crystallographic axes. In the opposite case the channeling effect for positive charged particles leads to the essential decrease of the differential back-scattering coefficient.

When a beam of fast charged particles interacts with a surface of monocrystal some of the particles are backward scattered. The energy loss (E_i-E_f) where E_i and E_f are the incident and final particle energies determines the target thickness from which the back-scattered particles with the energy E_f escape the target [1,2].

The effective length of the back scattering is characterized by the expression [2]

$$l_{eff} = (E_i - E_f)/\bar{\varepsilon} \tag{1}$$

where $\bar{\varepsilon}$ is the energy loss per the length unit. When the mean sqaure of the multiple-scattering angle accumlated over the effective length of the back scattering is smaller with unity, then the distribution function of the back-scattered particles is determined by the probability of the elastic scattering through a large angle, with the particle experiencing multiple small-angle elastic and inelastic scattering before and after the collision that leads to the large-angle scattering. First let us consider the case when the effective length of the back scattering is smaller or comparable with the effective channeling length L_o [3]

$$l_{eff} < L_o \lesssim p\kappa^{-2} \tag{2}$$

where p is the incident particle momentum and $\kappa = me^2\sqrt{z_1^{2/3}+z_2^{2/3}}$ (Z_1e and Z_2e are the atomic charges of the target and incident particle $\hbar = c = 1$).

When the fast positive charge particle passes through the monocrystal near the close-packing direction, the interaction with the atomic lattice leads to the well-known channeling effect [3]. The capture of the particles in the channeling regime takes place when the entrance angle, θ_o, i.e., the angle between the momentum of the incident particle and crystallographic plane or axis, is smaller than the critical Lindhard channeling angle, θ_c

$$\theta_o < \theta_c = \sqrt{\frac{2\pi Z_1 Z_2 e^2}{\kappa a^2 E_{kin}}} \tag{3}$$

where E_{kin} is the kinetic energy of the incident positive charged particle, "a" is a lattice constant. However, the establishment of the channeling regime occurs on the crystal thickness that is larger than L_o

$$L_o \simeq 2p[\kappa^2 + p^2\theta_o^2]^{-1} . \tag{4}$$

The value L_o characterizes the crystal thickness where the essential reconstruction of the particle wave function takes place from the incident plane wave into the channeling wave function [4].

Therefore, in the case $l_{eff} < L_o$, the angular distribution of the back-scattered particles corresponds to the usual Rutherford scattering (if the incident and final directions of the particle momenta do not coincide with crystallographic axes). When the incident momentum direction or the back-scattered momentum direction coincides with any crystallographic axis, the blocking effect essentially reduces the scattering of fast charged particles.

The motion of a particle in a monocrystal can be represented as motion in the total potential

$$U(\vec{r}) = \sum_a U_o(\vec{r} - \vec{R}_a) \tag{5}$$

where R_a is the position vector of the "a-th" atom and the sum is over all the atoms of the lattice. The scattering amplitude by the potential (5) can be written as

$$f(\vec{P}_f, \vec{P}_o) = -\frac{E}{2\pi} \int d^3r \; U(\vec{r}) \; \exp\{-i(\vec{P}_i \vec{P}_f)r$$

$$-\frac{i}{v}[\int_0^\infty ds \; U(\vec{r}-\vec{n}_i \cdot S) + \int^\infty ds \; U(\vec{r}+\vec{n}_f s)]\}, \quad (6)$$

where $\vec{n}_i = \dfrac{\vec{P}_i}{P_i}$, $\vec{n}_f = \dfrac{\vec{P}_f}{P_f}$.

Let us consider the case when the incident momentum direction $\vec{P}_i$ does not coincide with any crystallographic axis, but $\vec{P}_f$ is nearly parallel to the crystallographic axis OX, i.e., the angle between P_f and OX is small enough. In this case the scattering amplitude takes the form

$$f(\vec{q}) = -\frac{\alpha P}{\pi} \int d^2\rho \; K_o(\rho\sqrt{q_x^2 + 2}) \sum_{n=0}^{N-1} \exp\{-i \; q_x N_x - i \; \vec{q}_\perp(\rho+\xi_n)\}$$

$$\cdot \exp\{-2i\alpha \sum_{k=1}^{n} K_o(\kappa|\vec{\rho}+\vec{\xi}_k+\vec{\theta}_o X_k|)\}, \quad (7)$$

where X_k is the coordinate of the "K-th" atom along the crystallographic axis OX, $\vec{\xi}_k$ is the transverse thermal displacement of the "k-th" atom from the equilibrium position, and $K_o(Z)$ is a modified Bessel function of the second kind. By using this expression for the scattering amplitude (7), one can obtain the minimum yield along the crystallographic axis. The scattering amplitude takes the form

$$f(\vec{q})\Big|_{\theta_o =} = f_o(\vec{q}) \sum_{n=o}^{N-1} \exp\{-iq_x na - i2\alpha n\ell n(\frac{q_\perp}{\kappa\gamma})\}$$

$$\cdot \frac{\Gamma(1+in\alpha)\Gamma(\frac{1}{2} + i\frac{n\alpha}{2})}{\Gamma(\frac{1}{2} - i\frac{n\alpha}{2})} \quad (8)$$

and noncoherent part of the cross section of scattering through a large angle is equal to

$$\frac{d\delta}{d\Omega} = \frac{d\delta_o}{d\Omega} \sum_{n=o}^{N-1} |\Gamma(1+i\alpha n)|^2 = \frac{d\delta_o}{d\Omega} \sum_{n=o}^{N-1} \frac{\pi\alpha n}{sh(\alpha\pi n)} = \frac{d\delta_o}{d\Omega} \frac{\pi}{2} \cdot \frac{1}{\alpha} \quad (9)$$

These results are valid provided that

$$\alpha \equiv z_1 z_2 e^2 < 1 \tag{10}$$

and the mean square of the thermal displacement, $\overline{u^2}$, is infinites-
imally small. By taking into account the thermal vibrations, the
scattering amplitude is approximately given

$$f(\vec{q}) = f_o(\vec{q}) \sum_n \exp\{-i\, q_x na - i\vec{q}_\perp \vec{\xi}_n - i\, 2\alpha n \, \ell n \frac{q}{\kappa\gamma}\}$$

$$\cdot\; \Gamma(1+i n\alpha,\; q_\perp^2 \overline{u^2}) e^{q^2 \overline{u^2}} . \tag{11}$$

It follows that in the high temperature limit the blocking effect
is essentially decreased and

$$\frac{d\delta}{d\Omega} \quad \frac{d\delta_o}{d\Omega} \, N \tag{12}$$

Let us investigate the angular dependence of the back-scattered
particles' yield in respect to the angle θ_o, between the final
momentum $\vec{P}_f$ and the crystallographic axis OX. The result is

$$\frac{d\delta}{d\Omega} = N \frac{d\delta_o}{d\Omega} \left\{ \frac{\pi}{2N\alpha} + \frac{q^4(1- \frac{1}{Nqa\theta_o})\Theta(\theta_o - \frac{1}{Naq})}{[q_x^2 + (\vec{q}_\perp - 2 \frac{\vec{\theta}_o}{a\theta_o^2} \ell n N q a \theta_o)^2]^2} \right. \tag{13}$$

where $\Theta(x) = \{ \begin{smallmatrix} 1, & x>o \\ o, & x<o \end{smallmatrix} $.

The relative yield of the back-scattered particles is

$$\chi_o(\theta_o) = \frac{\frac{d\delta}{d\Omega}(\theta_o)}{N \frac{d\delta_o}{d\Omega}} = \frac{\pi}{2N} + \frac{1+(\frac{\tilde{\theta}_c}{\theta_o})^2}{\{1+2(\frac{\tilde{\theta}_c}{\theta_o})^2(1- \frac{2q_\perp^2}{q^2})+(\frac{\tilde{\theta}_c}{\theta_o})^4\}^{3/2}}$$

$$\cdot\; (1 - \frac{1}{Naq\theta_o})\Theta(\theta_o - \frac{1}{Naq}) , \tag{14}$$

where $\tilde{\theta}_c = 2\alpha \frac{\ln(2\alpha N)}{qa}$ \tag{15}

is the characteristic angle of the blocking dip of the scattering
cross-section [5].

It follows that the relative yield goes to unity when $\theta_o > \tilde{\theta}_c$

$$\chi_o(\theta_o) \simeq 1 + (\frac{\tilde{\theta}_c}{\theta_o})^2 . \tag{16}$$

These results are consistent with the available experimental data.

Finally it will be noted that one can observe the effect of "double shadow," when both the incident momentum direction and the back-scattered momentum direction coincide with crystallographic axes. When $\vec{P}_i \uparrow\uparrow OY$ and $\vec{P}_f \uparrow\uparrow OX$ in accordance with (6) and (9), the "double shadow" results in the more essential reduction of the back-scattered particle intensity

$$\frac{d\delta}{d\Omega} = (\frac{d\delta_o}{d\Omega}) \cdot N_Z \cdot \frac{1}{\alpha^2} \tag{17}$$

The crystallographic structure substantially changes the angular distribution of the back-scattered particles when the energy losses are not small enough. In this case the back-scattered particles escape from the large target thickness. The approximation of one event of large-angle scattering [1,2] is valid while the effective back-scattering length remains smaller than the transport length. If the incident and final directions are not parallel to crystal-lographic axes, then in the approximation considered [1,2] the differential back-scattering coefficients of the fast charged particles from a single crystal and from an amorphous medium coincide. Therefore, in this case the differential back-scattering is determined by

$$R(P_f, E_f; P_o, E_o L) = F(L) \cdot \frac{Z(Z+1)}{8\pi E_i \ln \frac{2E_i}{I_z}} \frac{m_e}{M}$$

$$\cdot \frac{\cos(\vec{n}_o \hat{} \vec{p}_i) \cos(-\vec{n}_o \hat{} \vec{p}_f)}{\cos(\vec{n}_o \hat{} \vec{p}_i) + (\frac{E_f}{E_i})^2 \cos(-\vec{n}_o \hat{} \vec{p}_f)}$$

$$\cdot \{[1 - \cos(-\vec{p}_i \hat{} \vec{p}_f)] + \ln \frac{E_i}{E_f} \frac{Z\ln(183Z^{-1/3})}{4\pi\ln(2E_i/I_z)}\}^{-2} , \tag{18}$$

where $(\vec{n}_o \hat{} \vec{p})$ is the angle between $\vec{p}$ and the normal to the target n_o and $F(L)$ is a function of the target thickness

$$F(L) \simeq (1 + \frac{\overline{\varepsilon^2}}{(\overline{\varepsilon})^2})^{-1/2} \exp(- \frac{E_i - E_f}{\overline{\varepsilon}L} , \tag{19}$$

where $$\overline{\varepsilon} = n_o \frac{2Z_i e^2}{E_i} \ln \frac{2E_i}{I_z}$$

$$\overline{\varepsilon^2} = 2\pi n_o Z_1 e^2 \tag{20}$$

and I_z is the ionization potential of the target. The function $F(L)$ allows us to evaluate the effective length of the back scattering or the so-called saturation length (1).

When the incident momentum direction or the back-scattered momentum direction coincides with any crystallographic axis the channeling effect for the positive charged particle leads to the essential decrease of the differential back-scattering coefficient. In the case of small entrance angles (3) the positive charged particle can be captured in the channeling regime when the effective length of the back-scattering is larger than the length L_0 of the channeling regime establishment. Therefore, the wave function of the channeling particles [4] is exponentially small near the atomic planes or atomic rows. That fact leads to the essential reduction of the process probabilities which are characterized by small impact parameters.

Let us consider the differential cross section for the elastic scattering of fast positive charged particles through large angle in the case of small entrance angle respect with the crystallographic axis YOZ.

By using the wave function of the channeling particles from [4,6] one can obtain

$$\frac{d\delta}{d\Omega} = N_{tot} \frac{d\delta_o(\vec{P}_i - \vec{P}_f)}{d\Omega} \cdot 2(\frac{\theta_o}{\theta_c})^2 \exp\{- \frac{2p\sqrt{\theta_c^2 - \theta_o^2}}{\kappa} \} \tag{21}$$

where $\theta_o < \theta_c$.

It should be noted that, when the entrance angle θ_o becomes larger than the critical Lindhard channeling angle θ_c, $\theta_o \gtrsim \theta_c$, the scattering cross section turns into the usual Rutherford scattering cross section.

References

[1] N. P. Kalashnikov and V. A. Mashinin, Soviet Physics JETP <u>32</u> (6), 1098 (1971).
[2] N. P. Kalashnikov and V. A. Mashinin, Zh. Teh. Fiz. <u>43</u> (11), 2229 (1973).
[3] J. Lindhard, Mat.-Fys. Medd. Dan. Vid. Selsk. <u>34</u>, No. 14 (1965).
[4] N. P. Kalashnikov, "The Proceedings of the Fifth Soviet Conference on the Interaction Physics of the Charged Particles with Single Crystals" (Moscow) (May 28-30, 1973).
[5] N. P. Kalashnikov and E. A. Koptelov, Soviet Physics, Solid State Physics <u>15</u> (6), 1668 (1973).
[6] N. P. Kalashnikov, "The Proceedings of the Fifth International Conference on Atomic Collisions in Solids,"(Gatlinburg, Tenn., U.S.A.) (1973).

SECTION VIII
CHANNELING

HYPERCHANNELING*

J. H. BARRETT, B. R. APPLETON, T. S. NOGGLE
Solid State Division, Oak Ridge National Laboratory
Oak Ridge, Tennessee 37830, U.S.A.

C. D. MOAK, J. A. BIGGERSTAFF
Physics Division, Oak Ridge National Laboratory
Oak Ridge, Tennessee 37830, U.S.A.

S. DATZ
Chemistry Division, Oak Ridge National Laboratory
Oak Ridge, Tennessee 37830, U.S.A.

and

R. BEHRISCH
Max-Planck-Institut für Plasmaphysik
D-8046 Garching bei München, Germany

Introduction

The distinction of the phenomenon of hyperchanneling is well illustrated by some of the computer calculations of the slowing down of energetic ions in single crystals from the 1963 paper by Robinson and Oen [1] which initiated the present day studies of channeling. Figure 1, taken from their paper, shows some calculated trajectories projected onto the (001) face of a hypothetical bcc Cu lattice for 5 keV Cu ions incident parallel to the [001] axis at the points marked by the crosses. In the terminology of present day channeling, all the trajectories in Figure 1 are representative of axially channeled ions; Robinson and Oen noted, however, that a limited class of axially channeled ions (such as those represented by the trajectory in the upper left hand corner of Figure 1) "are constrained by large numbers of very glancing collisions to move in regions of low potential surrounded by relatively closely packed atomic rows." It is this particular type of axial channeling which we call <u>hyperchanneling</u>. This phenomenon was also treated as a special case in analytical calculations by Lehman and Leibfried [2]. Despite this early recognition, it was not until 1966 that experimental evidence of this effect was reported by Eisen [3]. He

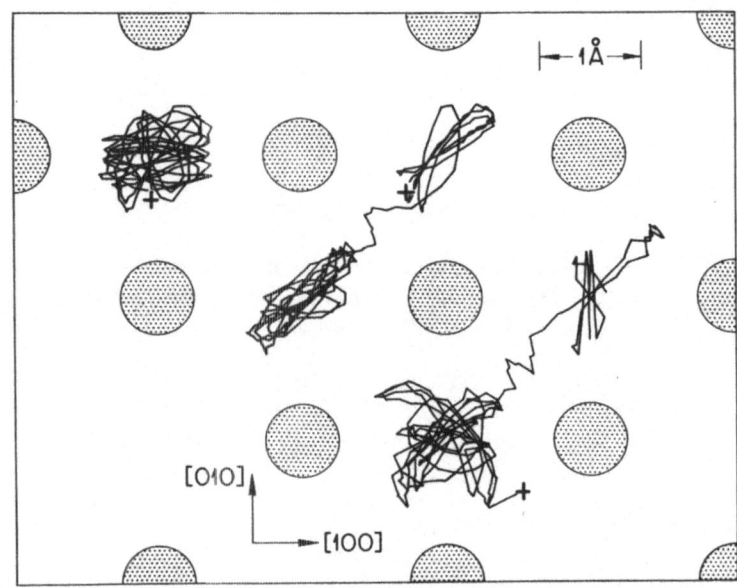

Fig. 1. Calculated trajectories projected onto the (001) face of
a bcc Cu crystal for 5 keV Cu ions incident parallel to [001].
Reference 1.

observed that a small fraction of 375 keV protons transmitted parallel to the [110] axis of a 2.2 µm thick Si single crystal emerged
with higher energies (lower energy loss rates) than those transmitted parallel to the (111) and (110) planes. He attributed this
high energy tail to protons which remained within a single axial
channel in transversing the crystal. No further investigations of
this phenomenon were reported until 1972 when Appleton, Moak,
Noggle, and Barrett [4,5] observed hyperchanneling for high energy
heavy ions in Ag. For 21.6 MeV I ions transmitted through thin Ag
single crystals in directions nearly parallel to the [011] axis,
they observed a distinct hyperchanneled group of ions with much
lower energy loss rates than regular axial channeling and with a
characteristic critical angle an order of magnitude less than the
axial critical angle. Because of the prominence of this effect for
high energy heavy ions, it is possible to study the details of the
hyperchanneling phenomenon. The results reported here are a continuation of this previous work. Although this effect has been
studied in Ag single crystals for 15-60 MeV I, 10-40 MeV O, 0.5-2.0
MeV He and 0.3-1.0 MeV H, only results for I ions, mostly at 21.6
MeV, will be discussed here.
      This paper discusses recent measurements of the hyperchannel-
ing phenomenon and presents the details of model calculations which
serve as a guide in interpreting the results. The measurements

reveal several new features which indicate that the hyperchanneling phenomenon is more complex than anticipated. Although it will be obvious from the results presented here that this effect can be utilized to investigate the phenomena of multiple scattering, radiation damage, and ion-atom interactions at large impact parameters in solids, the emphasis will be on presenting a perspective view of the present observations and their conciliation with the model calculations.

Experimental

A very schematic representation of the experimental apparatus used in these measurements is shown in Figure 2. A beam of heavy ions from a Tandem Van de Graaff Accelerator was collimated to a full angle of divergence of $\lesssim 0.01°$ onto a thin Ag single crystal held in a three axis goniometer. The energy distributions of the heavy ions transmitted through the oriented crystal were measured by a time of flight apparatus which gave a typical energy resolution of gaussian shape with a full width at half maximum of 180 keV for 21.6 MeV I ions. The energy detecting system had an acceptance angle of ± 0.012°. A beam monitor system was used to normalize various measurements with respect to one another.

Although the details of the experimental apparatus and specimen preparation are essentially the same as those reported earlier [4,5,6], the precision and versatility of the apparatus has been greatly improved, and procedures have been developed for characterizing the specimens at various stages in the experiments. A third degree of freedom was added to the goniometer to facilitate scans of the [001] axis of the [001] Ag single crystals. The entire beam line was made more rigid by supporting it from three independent concrete pillars. The ability to perform emergent scans was acquired by supporting the goniometer from one milling machine which

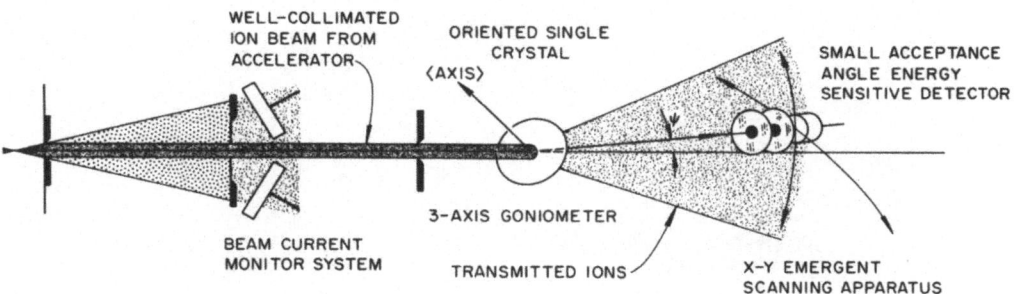

Fig. 2.   Schematic representation of experimental apparatus.

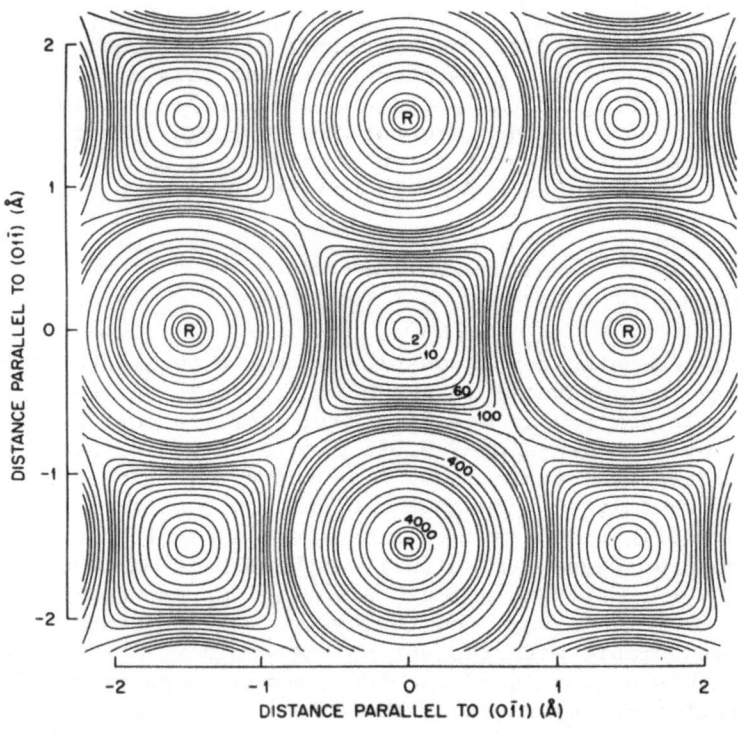

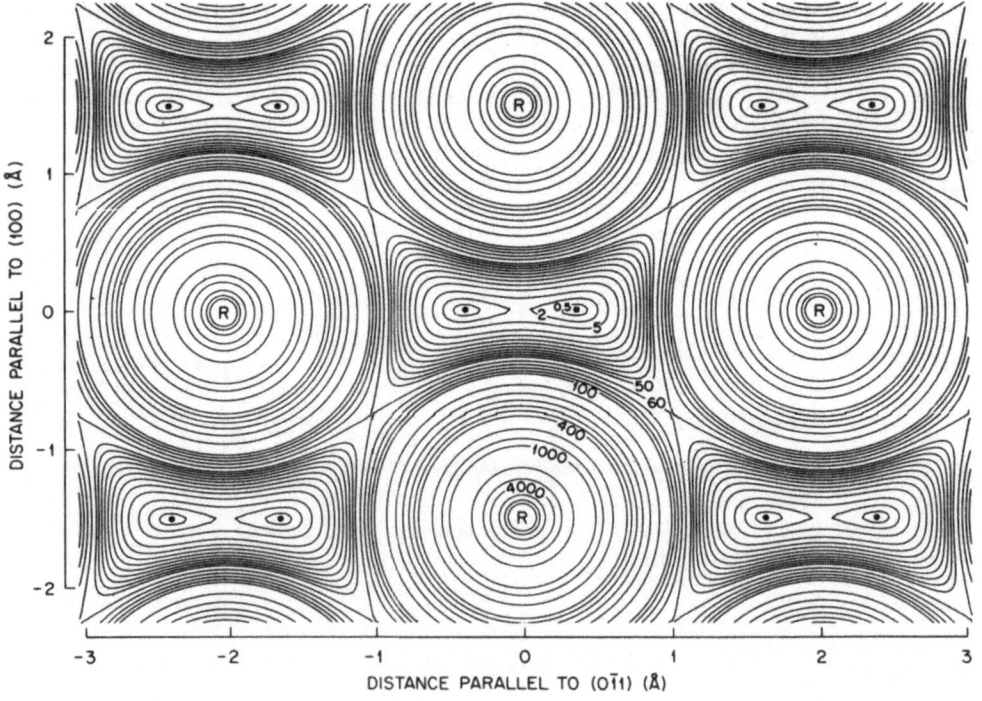

Fig. 3. Potential energy contours in [100] channels (above) and
[011] channels (below) for I in Ag.

◄───

also served as a pivot point for the time-of-flight apparatus which
was mounted on a second milling machine. This allows us to scan
the transmitted ion distribution in either the vertical or horizontal plane over a range of $\pm 10°$ with an accuracy of 5×10^{-3} degrees.
No emergent scan results will be reported in this paper because of
space limitations. The mosaic spread of the thin Ag single crystals
used were determined before and after irradiation by x-ray rocking
curve measurements. During the hyperchanneling experiments, the
mosaic spread at the position of the ion beam on the specimen could
be inferred from the x-ray measurements. It was found that the mosaic spread was not changed by the handling and irradiation involved
in a given measurement, and systematic mapping of the mosaic spread
of a particular specimen indicated that variations as much as 20%
occurred only in the region near the mounting edge of the thin self-
supporting films. The mosaic spreads appropriate to each measurement are given in the text. Typically, beam spot sizes of 0.5 mm
were employed while the crystal diameter was 3 mm. Low energy electron diffraction and Auger electron analysis of some Ag samples were
performed <u>after</u> a measurement and it was found that less than 1 - 5
monolayers of contaminant were present on the surfaces [7]; the contaminants were primarily C and O. Measurements of the mosaic spread
and surface contamination are important for the interpretation of
the hyperchanneling measurements since they can contribute significantly to the detailed spectra observed.

Statistical Equilibrium Theory

When an ion is moving nearly parallel to a principal axis in a
crystal, its motion can be separated into a component parallel to
the axis and components transverse to the axis. The parallel com- ·
ponent is influenced primarily by the stopping power of the crystal,
and the transverse components can be treated as if governed by the
continuum average potentials of the atomic rows parallel to the
axis. The treatment can be further simplified if it is assumed that
the transverse components have achieved statistical equilibrium.
For two dimensional motion, statistical equilibrium implies that an
ion has equal probability of being found at any point within the
area accessible to it under conservation of energy [8]. Although
some indications will be given below that statistical equilibrium
is not a valid assumption, it offers the simplest basis for constructing a theory and the resulting theory provides an explanation
of the main features observed in the experiments.
Sets of potential energy contours constructed from continuum
row potentials for iodine ions in silver are shown in Figure 3 for
two axes. All energies are in eV and are relative to the minimum

value, which is at the center for [001] and at the positions marked
by dots for [011]. The positions of the rows are denoted by the
leter R. The potential used was one determined by Robinson [9]
from planar channeling experiments for 21.6 – 60 MeV iodine ions in
Ag [10]. In constructing a theory, one needs to know the fraction of
area accessible to an ion of a transverse energy ε, which is the
fraction of area in a basic cell lying below the maximum potential
energy contour allowed, that is, $\phi = \varepsilon$; this fraction is given by

$$f\,(\phi) = A\,(\phi)\,/\,A_{cell} \quad .$$

One also wishes to know the fraction of area lying between the contour for ϕ and that for $\phi + d\,\phi$; this is

$$df = f'\,(\phi)\,d\,\phi \quad .$$

The values of $f(\phi)$ and $f'(\phi)$ determined for the [011] contours in
Figure 3 are shown in Figure 4. For the [011] contours, there is
a saddlepoint at the center of the four rows and another at the
point midway between any pair of rows. As first pointed out by Van
Hove [11], such critical points in the potential lead to logarithmic
singularities in f'; the positions of these singularities are indicated by the arrows in Figure 4. The calculations have not been
done in sufficient detail to show the structure of the lower singu-
larity. However, only the higher one (at ~ 60 eV) is likely to be
observable.

In addition to the notation already introduced of ε for transverse energy and ϕ for transverse potential energy, let τ be the
transverse kinetic energy. At times use will be made of the standard relationship between τ, the total energy E, and the angle ψ of
a trajectory to the axis: $\tau = E\psi^2$. Quantities associated with the
incoming and outgoing portions of the trajectory will be denoted by
subscripts i and o respectively.

It is convenient to specify the behavior of a trajectory by
specifying its incoming kinetic energy τ_i, the addition ε_a to its
total energy that is acquired from multiple scattering as it tra-
verses the crystal, its outgoing kinetic energy τ_o, and its outgoing
total energy ε_o. The fraction of all trajectories entering with
kinetic energy τ_i and acquiring additional energy ε_a that emerge
with energy between ε_o and $\varepsilon_o + d\varepsilon_o$ is

$$f'\,(\varepsilon_o - \tau_i - \varepsilon_a)\,d\varepsilon_o \quad .$$

The fraction of these trajectories that emerge with kinetic energy
between τ_o and $\tau_o + d\tau_o$ is

$$f'\,(\varepsilon_o - \tau_o)\,d\tau_o\,/\,f\,\,\,(\varepsilon_o) \quad .$$

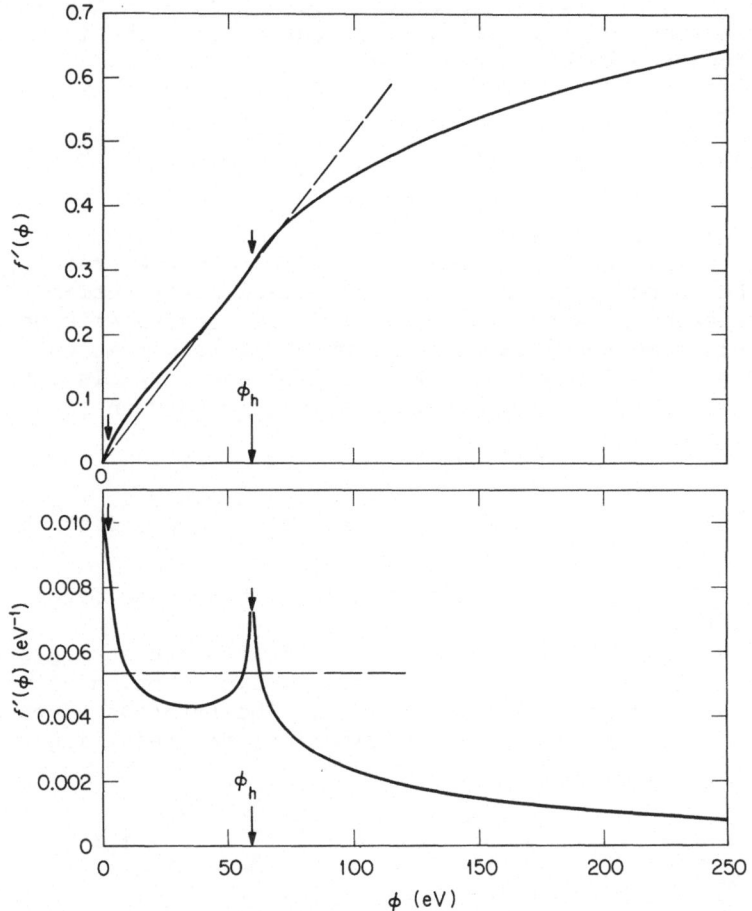

Fig. 4. Plots of $f(\phi)$ and $f'(\phi)$, where f is the fraction of area of the basic cell for [011] in Figure 3 that lies below ϕ.

Combining the two fractions just obtained gives the fraction of all trajectories entering with kinetic energy τ_i and acquiring additional energy ε_a that emerge with kinetic energy between τ_o and $\tau_o + d\tau_o$ and total energy between ε_o and $\varepsilon_o + d\varepsilon_o$; this fraction is

$$q_a (\varepsilon_o, \tau_o, \tau_i, \varepsilon_a) \, d\tau_o \, d\varepsilon_o$$

$$= \frac{f'(\varepsilon_o - \tau_o) \, d\tau_o}{f(\varepsilon_o)} \, f'(\varepsilon_o - \tau_i - \varepsilon_a) \, d\varepsilon_o . \tag{1}$$

For some purposes, one is interested not only in the differential distribution function of Eq. (1), but also in the integral distribution function which is given by:

$$g_a (\varepsilon_o, \tau_o, \tau_i, \varepsilon_o) = \int_{\tilde{\varepsilon}}^{\varepsilon_o} q_a(\varepsilon', \tau_o, \tau_i, \varepsilon_a) \, d\varepsilon' \qquad (2)$$

where $\tilde{\varepsilon}$ is the greater of τ_o and $(\tau_i + \varepsilon_a)$. For the present, attention will be concentrated on the differential distribution.

The detector in any given experiment is located at some angle ψ_o and covers some total solid angle which it is convenient to express as $\pi \psi_{det}^2$. The probability per incident particle that a particle with total outgoing transverse energy between ε_o and $\varepsilon_o + d\varepsilon_o$ is detected is

$$Q_a (\varepsilon_o, \tau_o, \tau_i, \varepsilon_a) = q_a (\varepsilon_a, \tau_o, \tau_i, \varepsilon_a) \, \tau_{det} \, d\varepsilon_o \qquad (3)$$

where $\tau_{det} = E\psi_{det}^2$; τ_{det} is not an actual energy but merely a convenient notation.

The additional transverse energy does not have a single value for all trajectories. Instead, there is a distribution of values given by a distribution function $p(\varepsilon_a)$. The fraction given by Eq. (1) must be averaged over this distribution; the resulting distribution function is

$$q_A(\varepsilon_o, \tau_o, \tau_i, \varepsilon_A) = \frac{f'(\varepsilon - \tau_o) \, f'_A (\varepsilon - \tau_i; \varepsilon_A)}{f(\varepsilon_o)} \qquad (4)$$

where ε_A is the average value of ε_a for the distribution $p(\varepsilon_a)$ and

$$f'_A(\varepsilon - \tau_i; \varepsilon_A) = \int_0^\infty p(\varepsilon_A) \, f' (\varepsilon - \tau_i - \varepsilon_o) \, d\varepsilon_a \quad .$$

The corresponding detection probability is

$$Q_A (\varepsilon_o, \tau_o, \tau_i, \varepsilon_A) = q_A(\varepsilon_o, \tau_o, \tau_i, \varepsilon_A) \, \tau_{det} \, d\varepsilon_o \qquad (5)$$

The use of subscript a on q or Q will denote a quantity specified for a particular value of ε_a, and use of subscript A will denote a quantity averaged over the distribution $p(\varepsilon_a)$. Using results

given by Lindhard [8,12], it may be estimated that the contribution
of nuclear multiple scattering will be much smaller than that of
electronic multiple scattering. It also appears that the contribution of defects and surface contaminants will be unimportant.
Therefore, ε_A will be taken to be that produced by electronic multiple scattering. Lindhard's results [12] allow an upper limit to be
estimated as

$$\varepsilon_A \leq \frac{1}{2} \ (m/M) \quad \Delta E \tag{6}$$

where m and M are the masses of an electron and an ion, respectively, and ΔE is the average energy loss of an ion as it passes through
the crystal.

Thin single specimens of the type used in this experiment contain dislocations which cause changes in the orientation on a microscopic scale. These misorientations along with deviations from
ideal planar geometry combine to give a range of orientations present in a given volume sampled by the ion beam. In order to measure
this "mosaic" spread, x-ray rocking curves have been obtained using
an x-ray beam of the same size as the ion beam. The distribution of
orientations as determined by the rocking curves is approximately
gaussian [6] and for the analysis reported here is taken to be
gaussian. For the experiments being reported here, ψ_o is always the
same as ψ_i. With such an arrangement, a gaussian mosaic spread
causes the detection probability to be

$$\overline{Q}_A \ (\varepsilon_o, \ \tau_i, \ \tau_i, \ \varepsilon_A) = \int^{\infty} \exp \left[\frac{-(\psi' - \psi_i)^2}{2\psi_M{}^2} \right]$$

$$\exp \left[\frac{\psi_i \psi'}{\psi_M{}^2} \right] I_o \left(\frac{\psi_i \psi'}{\psi_M{}^2} \right) Q_A(\varepsilon_o, \ \tau', \ \tau', \ \varepsilon_A) \frac{\psi' d\psi'}{\psi^2} \tag{7}$$

where $\tau_i = E\psi_i{}^2$, $\tau' = E\psi'^2$, ψ_M is the variance of the mosaic spread
(= 0.425 FWHM), and I_o is a modified Bessel function. Equation (7)
may be applied to integral as well as differential distributions.

A trajectory of given transverse energy samples the impact parameters with regard to the surrounding rows that lie within the
contour for that energy. This distribution of impact parameters
will produce a value of stopping power dependent on the transverse
energy of the trajectory. Since the energy varies during traversal
of the crystal because of multiple scattering, the most accurate
dependence would be on the total transverse energy averaged throughout the crystal. Formulation of the theory so as to implement such
a dependence would be very complicated. Consequently, to keep the
theory manageable, the dependence will be taken to be on ε_o only.

It is hoped that the range of ε_a is sufficiently small that the error in this assumption is small. Under this assumption, the stopping power is a function $S(\varepsilon_o)$ of the outgoing total transverse energy. Distributions in terms of transverse energy can be converted to distributions in terms of stopping power by considering ε_o as $\varepsilon_o(S)$ and replacing $d\varepsilon_o$ by $(dS/d\varepsilon_o)^{-1} dS$.

The final factor that must be incorporated into the theory is the energy resolution of the detector. As stated above, this is approximately gaussian in shape with a full width at half-maximum of 180 keV. This can readily be folded into the result of Eq. (7) with ε_o considered to be a function of stopping power or total emergent energy of an ion.

Two further approximations will be needed or useful. The first involves the form of $p(\varepsilon_a)$. Although the theory has been stated in terms of a general form, calculations can be greatly facilitated if the form is simple. The form that will be assumed is

$$p(\varepsilon_a) = \varepsilon_A^{-1} \exp(-\varepsilon_a/\varepsilon_A) \quad . \tag{8}$$

This form is superior to assumption of a constant value and appears to be the best approximation that will keep the mathematics tractable. Computer simulations of the type done by one of the authors [13] show such an exponential distribution to be a reasonably good approximation when the total transverse energy is sufficiently small. The second approximation is not so essential, but is convenient for making simple analyses of experiments using integral distributions. This involves approximating $f(\phi)$ by a straight line. This is the dashed line shown in the upper portion of Figure 4; the dashed line in the lower portion is the corresponding value of $f'(\phi)$. Specifically, $f(\phi)$ will be approximated as

$$f(\phi) = m\phi = (f_h/\phi_h)\phi \quad , \tag{9}$$

where ϕ_h is the energy of the contour passing through the upper saddle point and f_h is the fraction of the cell area enclosed by this contour. With this approximation the integral distribution corresponding to Eq. (5) with $\tau_i = \tau_o$ is

$$G_A(\varepsilon_o, \tau_o, \tau_o, \varepsilon_A) = f_h \left(\frac{\psi_{det}}{\psi_h}\right)^2 \left\{ \ln\left(\frac{\varepsilon_o}{\varepsilon_A}\right) \ln\left(\frac{\tau_o}{\varepsilon_A}\right) \right.$$
$$\left. - \exp\left(\frac{\tau_o}{\varepsilon_A}\right) \left[E_1\left(\frac{\tau_o}{\varepsilon_A}\right) - E_1\left(\frac{\varepsilon_o}{\varepsilon_A}\right) \right] \right\} \tag{10}$$

where E_1 is the exponential integral.

In the analysis [4,5] of previous experiments for 21.6 MeV iodine ions directed along [011] in silver, it was found that the best agreement of theory with experiment was obtained with $\phi_h = 120$ eV and $\varepsilon_A = 54$ eV rather than with the estimated values. The value of ψ_{det} was 0.012°, that of ψ_m was 0.022°, and f_h can be found from Figure 3 to be 0.31. Further analysis of the previous experiments suggested a linear relationship between S and ε_o. This provides all of the information needed to use the above theory to calculate distributions of stopping power as a function of tilt angle of the crystal. The results are shown in Figure 5. The four sections correspond to the various stages of the theory: Eq. (3), Eq. (5), Eq. (7), and Eq. (7) with the detector resolution function folded in. The final stage of the theory reproduces the main features observed [4,5] in the distribution of energy losses for orientations near [011] in silver except for the planar losses that occur at angles of a few times ψ_h when the tilting is done into a major planar channel. However, the planar losses are due to ions that have escaped from hyperchanneling to become planar channeled even though they are still within the axial critical angle. The above analysis does not apply to such particles. New results to be presented below show a relationship between S and ε_o that is non-linear. Although this will affect the detailed shape of curves such as in Figure 5, it will not affect the basic features. Other new results to be given below suggest that statistical equilibrium is not completely valid, particularly when the beam is directed near [001]. This takes the form of additional features in the energy loss distributions; however, the basic features of the statistical equilibrium theory are still present.

One point that develops out of the distributions of energy loss shown in Figure 5 is the clear identification of the secondary peak for the $\psi_i = \psi_o = 0$ distribution with the energy ϕ_h in Figure 4. This allows the limiting transverse energy for hyperchanneling to be associated with a definite stopping power.

The model that has been developed should be equally applicable to all axes and all crystal structures as long as statistical equilibrium and the other assumptions on which it is based are satisfied. The parameters of the theory may vary, of course. The two sets of contours shown in Figure 3 indicate that ϕ_h will be about twice as large for [001] as for [011] in silver; the same should be true in general for fcc metals. Other calculations suggest that ϕ_h will be large in the diamond structure, particularly for [011], and that it will be small in the bcc structure.

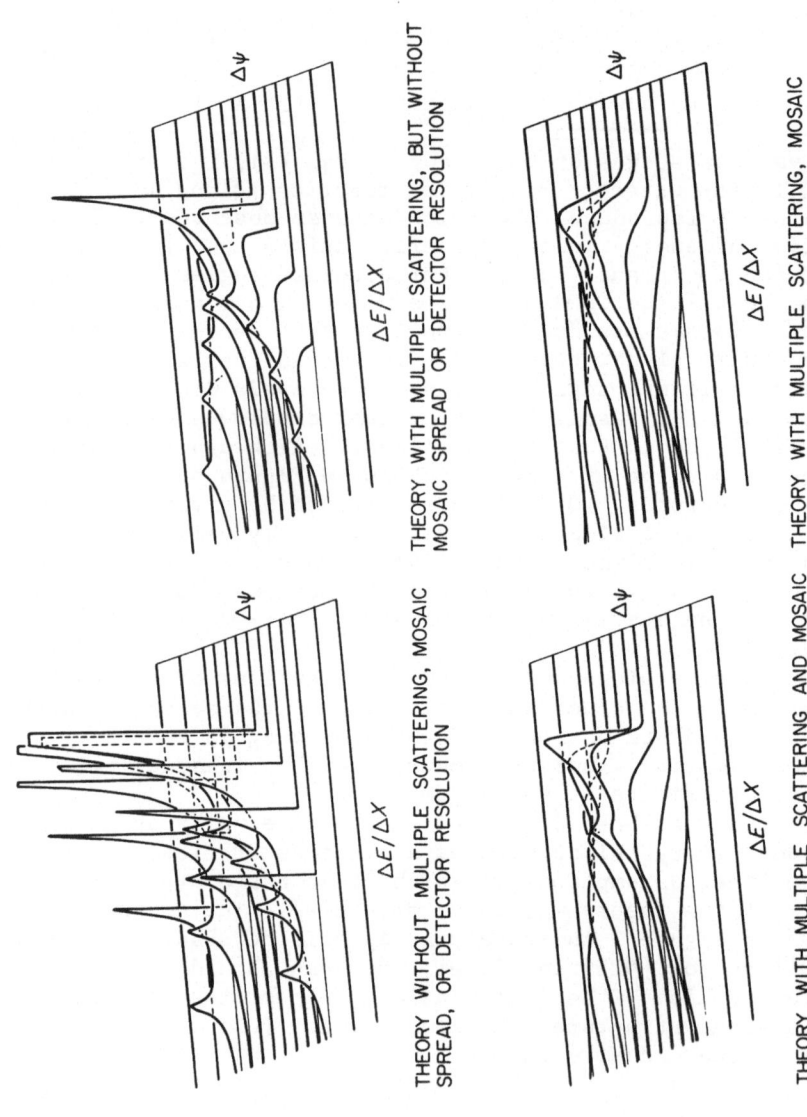

Fig. 5. Plots of the distribution of stopping power for different tilt angles for the stages of the statistical equilibrium model discussed in the text.

Results and Interpretation

   Measurements for [011]. A series of normalized transmitted
energy distributions are shown in Figure 6 for 21.6 MeV I ions incident at various angles to the [011] axis in the (100) plane. For
these measurements the energy detector was kept in line with the incident beam and the angle of incidence was varied by tilting the
crystal. In the remainder of this paper we will refer to such

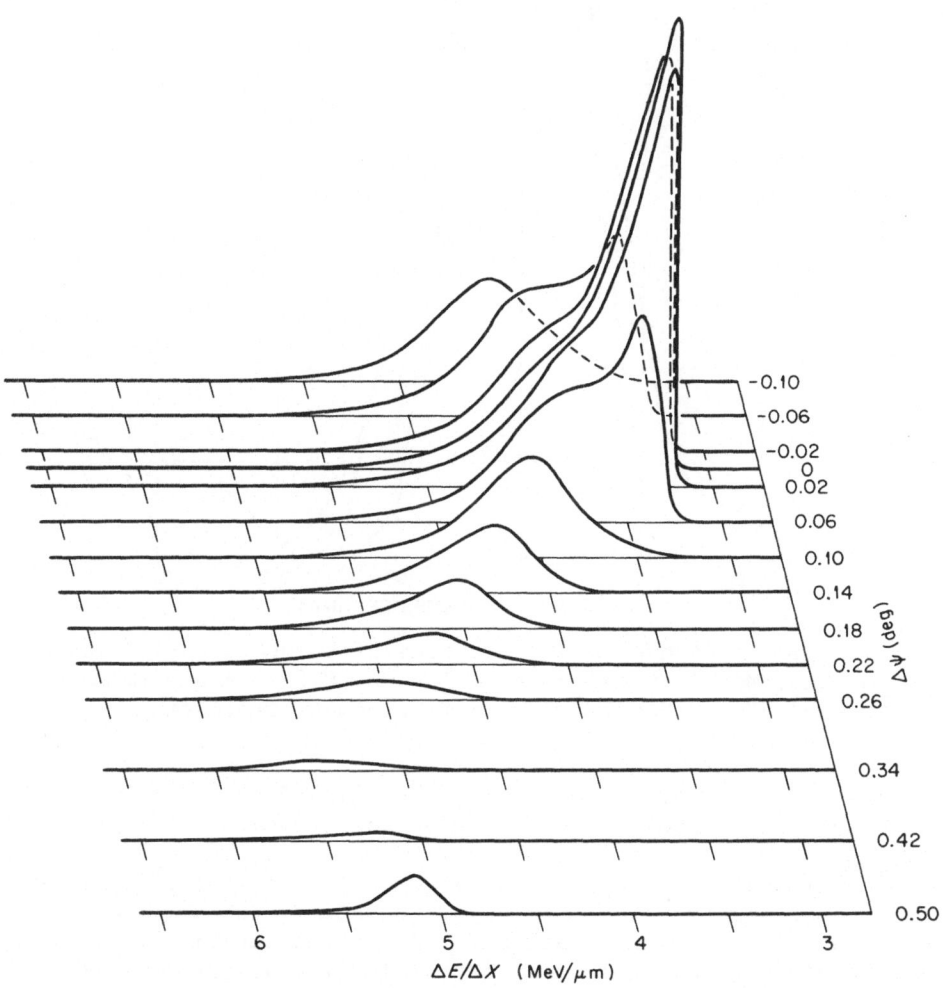

Fig. 6. Normalized transmitted energy-loss distributions for 21.6
Mev I in Ag incident at various angles to [011] in the (100) plane.
The path length was 1.18 μm and mosaic spread was 0.055° FWHM.

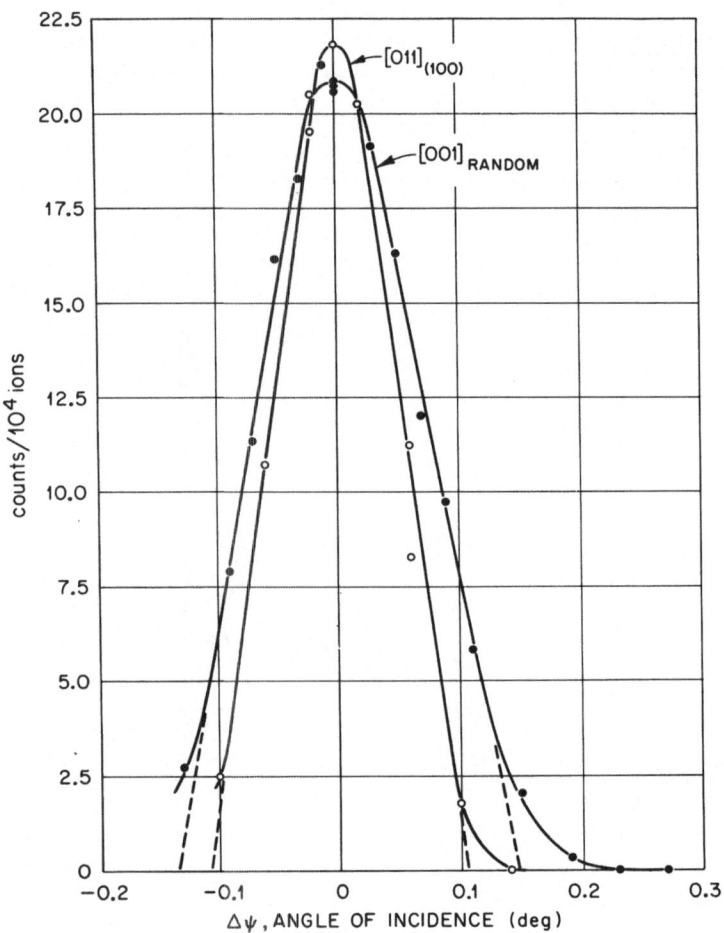

Fig. 7. Fraction of hyperchanneled ions as a function of incidence angle for 21.6 MeV I in Ag; path length was 0.835 μm and mosaic spread 0.055° FWHM. See text for definition of fraction.

measurements as crystal or goniometer scans, and the axis being scanned and tilt plane will be given by the notation scheme $[011]_{(100)}$.

A comparison of Figure 6 with Figure 5 shows that all the general features observed for [011] hyperchanneling are adequately reproduced by the model calculations. Identification of the secondary peak in the energy spectra as a distinctive feature associated with those ions which are making the transition from hyperchanneling to regular axial channeling allows us to quantify the hyperchanneled fraction. As an example we have plotted the fraction of ions hyperchanneled during the $[011]_{(100)}$ scan versus the angle of incidence in Figure 7. The companion plot for $[001]_{Random}$ will be discussed presently. For [011] this figure shows the total fraction of detected ions with energy losses $\leqq 3.88$ MeV/μm, which is the energy loss corresponding to the secondary peak for $\Delta\psi = 0$. The angular width taken from the data is a measure of the critical angle for hyperchanneling or equivalently the critical transverse energy. The magnitude of the distribution represents the surviving hyperchanneled fraction and contains valuable information about the effects of multiple scattering. As was the case in our earlier work on hyperchanneling [4,5], curves calculated from Eqs. (7) and (10) with the estimated values of the parameters reproduce the shapes of the distributions like these in Figure 7 but are narrower and have a higher peak than the experimental ones. It was found that good agreement with experiment could be found with the values listed in Table 1. The base angle of the [011] curve in Figure 7 has been used in conjunction with calculation utilizing Eqs. (7) and (10) to obtain the critical angle for hyperchanneling ψ_h, which is $(\phi_h/E)^{1/2}$. The peak height of the curve is used with Eqs. (7) and (10) to obtain the value of ε_A. Different regions of the crystal give slightly different peak heights and hence slightly different values of ε_A. These differences may be due to variations in mosaic spread and to varying amounts of radiation damage produced in assuring that the best alignment with [011] has been obtained. As in the earlier work, the values of ϕ_h and ε_A that give best agreement with the measurements are larger than the estimated values. However, the values of ϕ_h and ε_A inferred from the present experiments do not differ as much from the estimated values as did the values inferred from the earlier experiments. We believe the greater rigidity of the present beam line to be responsible for this improvement.

Table 1

Crystal	Axis and Tilt Plane	ψ base (degrees)	ψ_h (degrees)	ϕ_h (eV) Meas.	Th.	$\varepsilon_A/1$ (eV/μm) Meas.	Th.
Ag 9-2 A	$[011]_{(100)}$	0.10_7	0.12_3	92	60	16	8
$t_o = 0.835$ μm							
$\psi_M = 0.025°$	$[001]_{Random}$	0.14_2	0.17_0	176	120	14	10
	$[001]_{(100)}$	0.13_6	0.16_1	158	120	23	10
Ag 7-4	$[011]_{(100)}$	0.15_4	0.17_4	184	120	32	10
$t_o = 0.775$ μm							
$\psi_M = 0.038°$							

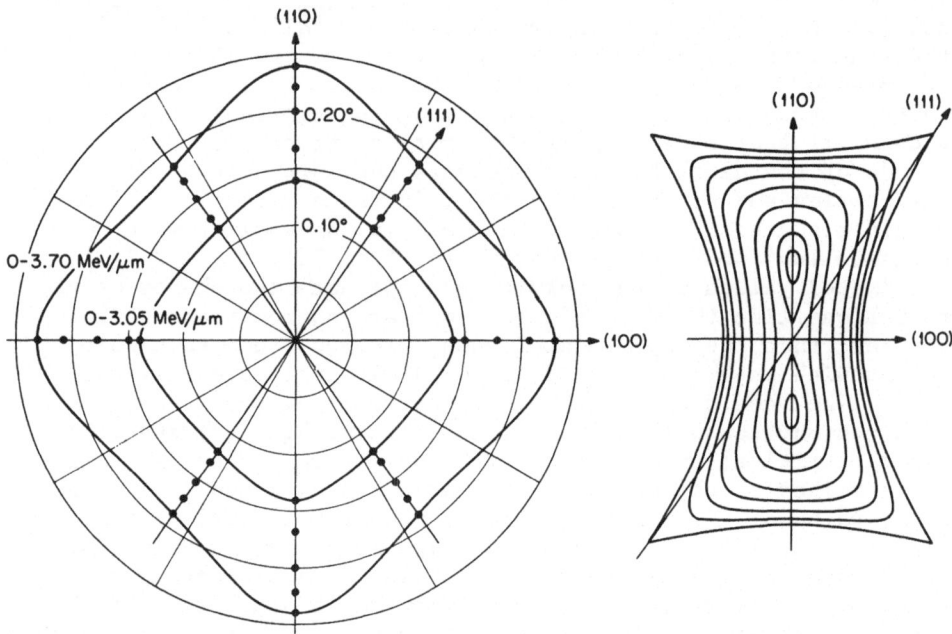

Fig. 8. At the left is shown the base angles of hyperchanneled fraction curves as in Figure 7 for 21.6 MeV I in Ag with different major crystallographic planes used as tilt planes through [011]. At the right are shown some of the energy contours for this axis.

One of the central assumptions made in the model calculations is that statistical equilibrium is achieved by the channeled ions. A test of the validity of this assumption was made for the [011] axis by performing crystal scans in several different planes passing through the axis. If statistical equilibrium is achieved then the measured widths for the hyperchanneled ions should be independent of the direction in which the crystal is tilted. The results of these measurements are presented in the left-hand portion of Figure 8, which shows a polar plot of some hyperchanneling angles extracted from $[011]_{(110)}$, $[011]_{(100)}$ and $[011]_{(111)}$ crystal scans. These angles are the base widths obtained from plots like Figure 7. Note that all the data are for hyperchanneled ions having energy losses smaller than that of the transition energy loss, 3.88 MeV/μm. The marked asymmetry for even the lowest energy losses indicates that statistical equilibrium is probably not fully realized by the hyperchanneled ions. Some of the observed differences at larger energy losses may result from the fact that initially hyperchanneled ions can "leak out" more easily when tilts occur toward the more open planes, (111), than the less open ones, (100) and (110).

    <u>Measurements for [001]</u>. We observed even stronger evidence
that statistical equilibrium was not achieved for our particular
conditions when we investigated the behavior of ions hyperchanneled
along [001]. A [001]_{Random} crystal scan is shown in Figure 9. The
crystal was tilted so that the beam was incident at various angles
to [001] in a plane 30° from (100) in order to approximate a random
tilt direction and avoid the appearance of any planar-related effects at larger tilt angles. Two characteristic features of this
scan are similar to ones observed in the [011] scans just discussed.
The prominent peak at $\approx$ 3.75 MeV/μm for $\Delta\psi \cong 0$ is due to hyperchanneled ions, and the secondary peak at $\approx$ 5.25 MeV/μm results from
ions escaping from the [001] hyperchannel. The surprising new feature is the appearance of group structure between these two peaks.

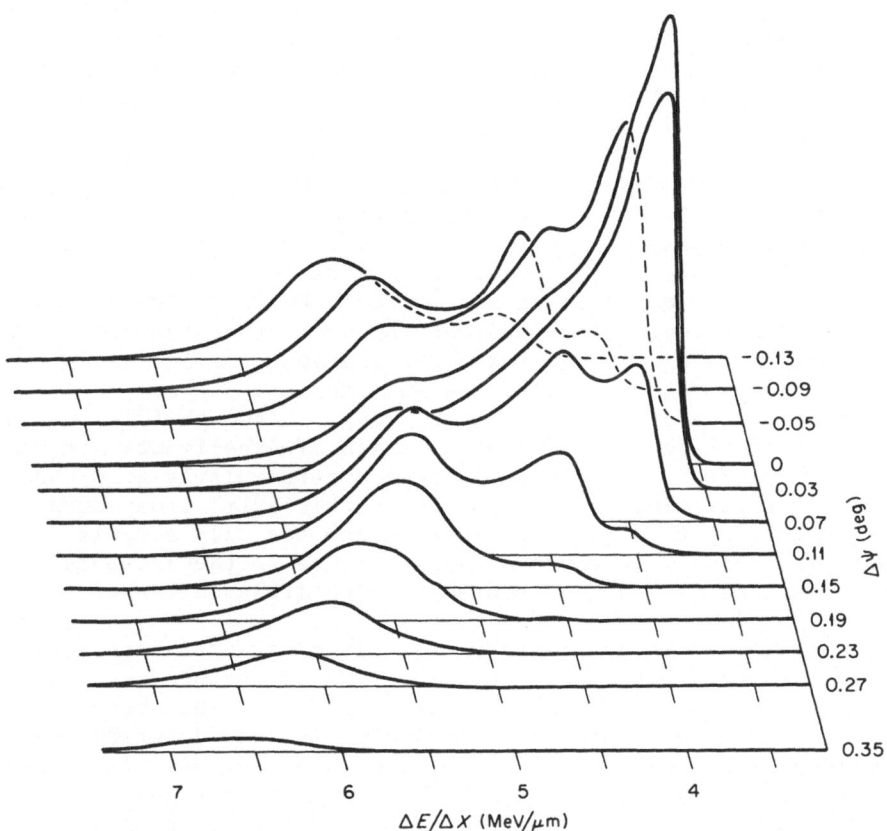

Fig. 9. Normalized transmitted energy-loss distributions for 21.6
MeV I incident on an Ag crystal at various angles to [001] in a
random tilt plane. The path length through the crystal was 0.835
μm and the mosaic spread was 0.055° FWHM.

Since this intermediate structure in the energy loss spectra always occurs at losses less than that for the transition peak it is clearly a hyperchanneling phenomenon. This structure <u>cannot</u> be explained on the basis of model calculations like those presented above. As discussed there, such calculations for [001] will have the same qualitative features as those shown for [011] in Figure 5.

The origin of this intermediate group structure is not well understood at the present time, although a number of possibilities exist. The most likely appears to be that one or more discrete energy loss groups are supported within the [001] hyperchannel in a manner analogous to the discrete wavelength groups which exist within planar channels for these heavy ions [14,15]. A set of measurements was formulated to determine if these groups have such a wavelength behavior and the results are shown in Figure 10. The experimental set-up was identical to that used in the crystal scans. The crystal was tilted so that the ion beam was incident at 0.06° to the [001] axis. At 21.6 MeV this tilt angle maximized the population of the intermediate group structure. Then, keeping this crystal orientation, the incident energy was varied from 21.6 MeV to 31.25 MeV and the series of transmitted energy distributions shown in Figure 10 were recorded. In the event that the intermediate structure observed in the 21.6 MeV spectrum results from a group of ion trajectories within the hyperchannel having a definite wavelength one should be able by increasing the velocity of the ions to re-tune this same group (same energy loss rate) when the ions have made one fewer oscillation [15]. The various spectra shown are not properly normalized to each other so no conclusions should be drawn about relative intensities. It is evident, however, from the complexity of the changes in the energy loss distributions for increasing incident energies in Figure 10 that the intermediate structure cannot be attributed to a single wavelength group. Although (011) was not used as a tilt plane in the present series of experiments, it would appear from Figure 3 that this might be a particularly good choice for future explorations of the nature of this extra group structure. Without a more systematic investigation one can only speculate whether the features in these spectra are due to the presence of several discrete wavelength groups or to entirely different mechanisms.

Although the extra group structure is of great interest, it appears as a perturbation on the basic form of the statistical equilibrium model. Consequently, one should still be able to apply the model to the basic features of the energy loss distributions to test values of ϕ_h and ε_A or to infer the values by matching calculated curves to the measurements. The perturbations may be expected to introduce some small uncertainty into such comparisons, but the comparisons still retain most of their validity. Accepting this argument, we can characterize the hyperchanneled fraction for [001] as we did for [011]. A representative result is given in Figure 7 which shows a plot of the fraction of detected ions with

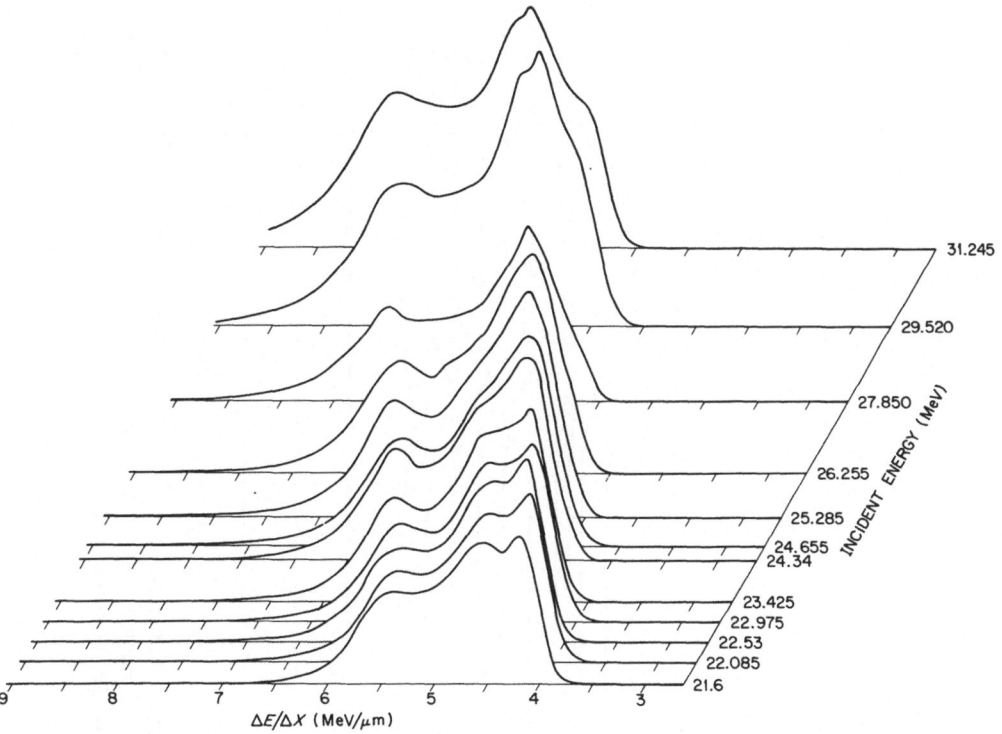

Fig. 10. Transmitted energy loss distribution for I ions with energies from 21.6 MeV to 31.25 MeV incident in the (100) plane at 0.06°
to the [001] direction. Path length = 0.835 µm, mosaic spread =
.055° FWHM.

energy losses less than 5.25 MeV/µm versus the angle of incidence
for a [001]$_{Random}$ crystal scan. Several measurements for [001] are
listed in Table 1 along with the results discussed earlier for [011].
The parameters that yield the best agreement for [001] are shown for
two tilt planes in one crystal and for one of these tilt planes in
a second crystal of larger mosaic spread. The inferred values of
ϕ_h and ε_A are larger than the estimated ones just as for [011].
They do, however, have just about the relation to the [011] values
that one expects from the theory.

    Comparisons Between [001] and [011]. The results discussed so
far show some interesting differences between hyperchanneling phenomena for [001] and [011]. These two cases are contrasted again in
Figure 11 which shows [001]$_{(100)}$ and [011]$_{(100)}$ crystal scans for
21.6 MeV I in Ag. The hyperchanneling model indicates that ions
with transverse energies slightly greater than the critical value
will tend to leak out of the hyperchannel most readily in directions

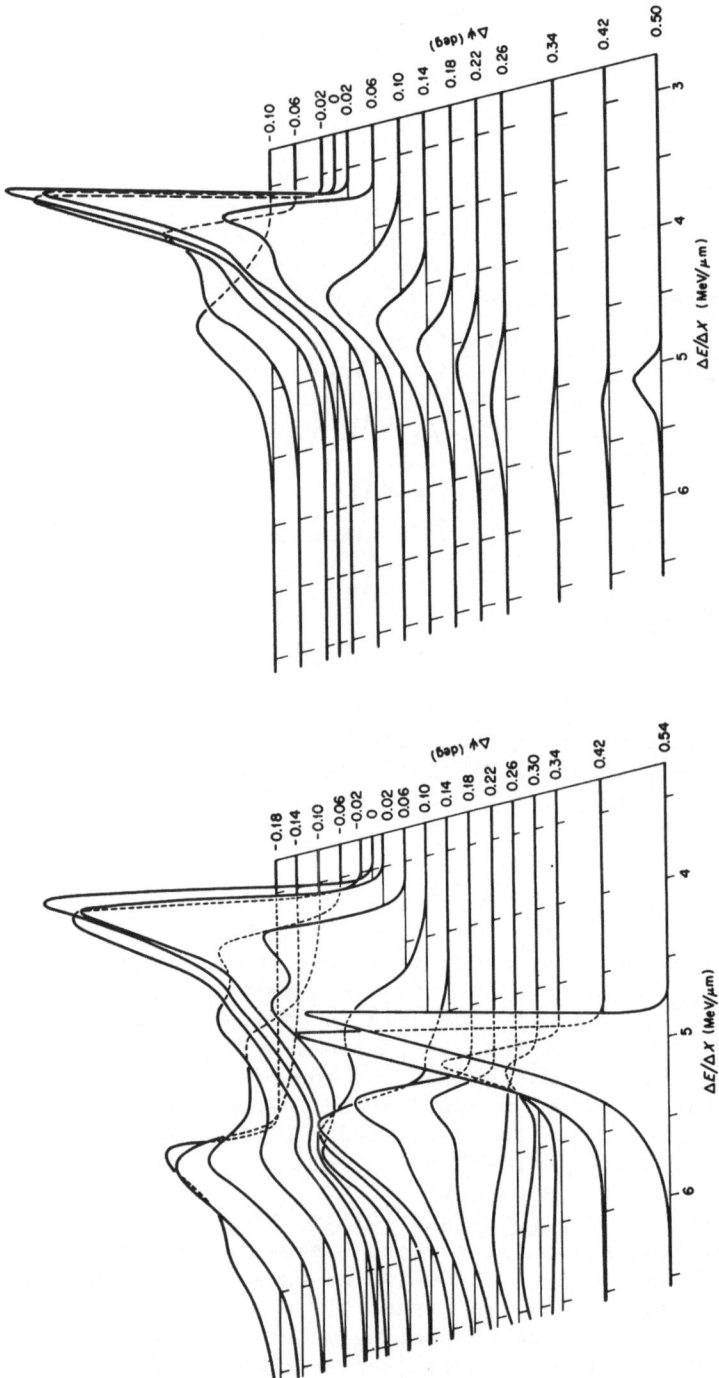

Fig. 11. Comparison of the normalized transmitted energy-loss distributions for 21.6 MeV I ions incident at various angles to [001] in (100) (pathlength = 0.835 μm mosaic spread = 0.055° FWHM) and to [011] in (100) (pathlength = 1.18 μm, mosaic spread = 0.055°).

which present the easiest path across the potential barrier. These
are the (111) type directions for the [011] hyperchannel and the
(100) for [001]. It can be argued that these ions will have trajectories and energy losses characteristic of (111) and (100) planar
channeling respectively. This criterion was in fact the first argument made as a means of identifying hyperchanneled ions before the
significance of the transition peak was known [3,4,5]. The energy
loss information compiled in Table 2 can be used to judge the validity of this assumption for the cases being considered by comparing
the energy losses corresponding to the transition peak for each

axis, $\frac{dE}{dx}\big)_t$ to that of the most open plane intersecting this axis.

These results would indicate that for [011], $\frac{dE}{dx}\big)_t$ is very nearly

the same as the leading edge value of the energy loss for (111),

$\frac{dE}{dx}\big)_{L.E.}$. For [001], $\frac{dE}{dx}\big)_t$ is more nearly equal to the peak of the

(100) planar energy loss distribution, $\frac{dE}{dx}\big)_{peak}$. Actually one can

understand the order of these differences by considering in detail
the nature of the trajectories taken by ions making the transition
from [011] into ($1\bar{1}1$) and from [001] into (100). The most obvious
difference between [001] and [011] is the intermediate group structure in the [001] case. If this structure is due to discrete oscillations supported within a hyperchannel as we speculated earlier
there are two possible reasons why it is observed for [001] and not
for [011]. The first is associated with the nature of the energy
loss that characterizes the ions which make up the transition peak
and can be understood with the aid of the information in Table 2.
The larger the difference in energy loss for a particular axial

direction between the best hyperchanneled ions, $\frac{dE}{dx}\big)_{L.E.}$, and the

transition ions, $\frac{dE}{dx}\big)_t$, the better are the chances of observing any

intermediate group structure with a finite energy resolution, which
in our case is 180 KeV. One can see from the table that the difference between these two bounds for [001] is more than 2.5 times that
for the [011].

Table 2

Axis	$\frac{dE}{dx}\big)_{L.E.}$ (MeV/μm)	$\frac{dE}{dx}\big)_t$ (MeV/μm)	Plane	$\frac{dE}{dx}\big)_{L.E.}$ (MeV/μm)	$\frac{dE}{dx}\big)_{peak}$ (MeV/μm)
[001]	3.5_6	5.2_5	(100)	4.9_0	5.1_7
[011]	3.2_1	3.8_8	(111)	3.8_8	4.0_7

The second reason why groups of ions with discrete oscillatory trajectories are more likely for [001] can be deduced from the shapes of the potential energy contours for the two axes shown in Figure 3. The flat sides of the hyperchanneling contours parallel to (01$\bar{1}$) and (0$\bar{1}$1) in the [100] projection are more conducive to supporting recurring trajectories than the more intricate [011] contours. The shape of the [011] contours would tend to mix trajectories thus causing a more rapid approach to statistical equilibrium.

Hyperchanneled ions with a particular transverse energy will have trajectories which occupy the space within the potential contour corresponding to that transverse energy. Thus ions with the smallest transverse energies will have the lowest energy losses. By taking differential slices of energy loss distributions like those shown in Figure 11, one can obtain intensity distributions as a function of the angle of incidence for ions emerging with a particular energy loss [5]. Using the angular extent of these plots to determine the corresponding transverse energy, it is possible to extract the functional dependence of $\frac{dE}{dx}$ versus transverse energy. Two representative results are shown in Figure 12 for [001]$_{Random}$ and [011]$_{(100)}$. A reasonable approximation for the cases investigated is that $\frac{dE}{dx}$ varies linearly with $(\varepsilon_\perp)^{1/2}$. Although the transverse energy associated with a particular $\frac{dE}{dx}$ changed slightly depending on the scan direction for a particular axis, the functional dependence remained unchanged. You will recall that a linear dependence of $\frac{dE}{dx}$ on $\varepsilon_\perp$ was assumed in the model calculations. Incorporating the square root dependence observed here will make some changes in the calculated distributions but will not affect the basic features; for example, it will not account for the intermediate structure observed for [001]. These results nevertheless show how such measurements can indicate the directions future changes in the model should take.

Conclusion

The experiments reported above give improved measurements of [011] hyperchanneling in Ag. They also give the first results for hyperchanneling along a second direction, [001], in Ag. The theory outlined above is based on the assumption of statistical equilibrium and has been shown to provide an adequate basis for understanding the general features of hyperchanneling. These general features include the main peak, the transition peak associated with ϕ_h, the variation of these peaks as the crystal is tilted, and the relative values of ϕ_h for the [001] and [011] axes. For [001] hyperchanneling, additional group structure appears in the energy loss spectra in addition to the basic structure described by the statistical

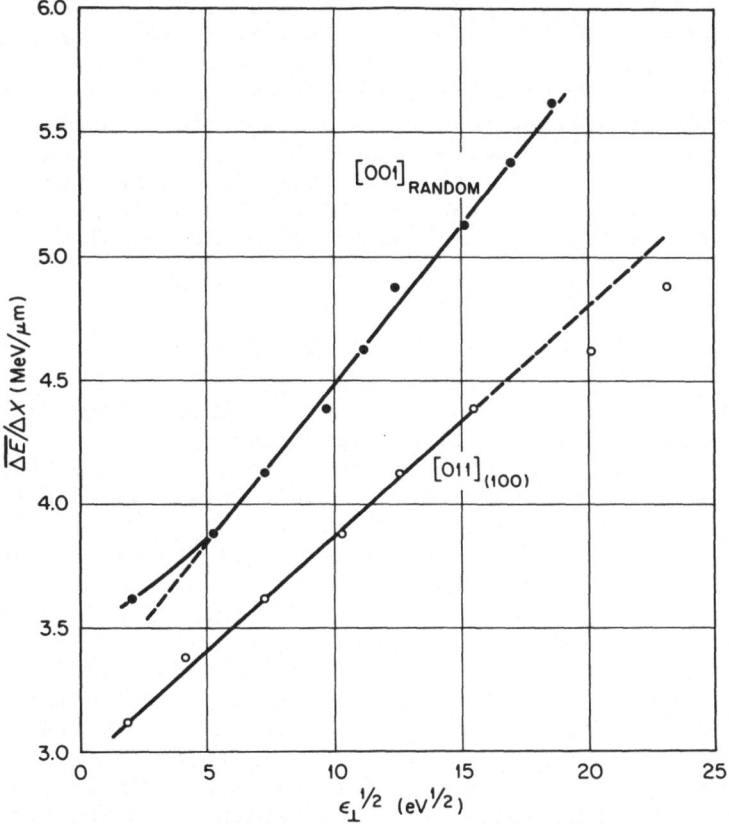

Fig. 12. Average rate of energy loss versus transverse energy as extracted from $[001]_{Random}$ and $[011]_{(100)}$ goniometer scans in a manner described in the text.

equilibrium theory. The most likely explanation of the extra groups appears to be the wave effects of the kind observed for planar channeling. Further experiments using tilts in (011) and variations of E in an attempt to understand the nature of the extra group structure are highly desirable.

The values of the two parameters, ϕ_h and ε_A, that have been adjusted to give best agreement of theory with experiment are higher than the estimated values. However they are closer to the estimated values than those obtained by a similar adjustment for best fit to earlier experiments [4,5]. We attribute this to improvements in the experimental apparatus, measurement procedures, and specimen evaluation. Efforts should continue to improve both the estimated values and the values obtained by fits to experimental results. Also, the effect of radiation damage should be studied to determine its

influence on the measured peak heights as well as to evaluate the
great sensitivity of hyperchanneling to this factor. Preliminary
measurements made by scanning the emergent angle with the incident
angle held fixed show considerable promise of improved results.
Such measurements with $\psi_i = 0$ are as susceptible of analysis by the
present model as the goniometer scans; the accuracy of such scans
is better than 5×10^{-3} degrees; the radiation damage problem associated with the analyzing beam is considerably diminished and easier
to interpret; and finally the beam normalization problem is simplified.

References

*Research sponsored by the U. S. Atomic Energy Commission under
   contract with Union Carbide Corporation.
[1] M. T. Robinson and O. S. Oen, Phys. Rev. 132, 2385 (1963).
[2] C. Lehmann and G. Leibfried, J. Appl. Phys. 34, 2821 (1963).
[3] F. H. Eisen, Phys. Letters 23, 401 (1966).
[4] B. R. Appleton, C. D. Moak, T. S. Noggle, and J. H. Barrett,
   Phys. Rev. Letters 28, 1307 (1972).
[5] B. R. Appleton, J. H. Barrett, T. S. Noggle and C. D. Moak,
   Radiation Effects 13, 171 (1972).
[6] T. S. Noggle, Nucl. Instr. and Meth. 102, 539 (1972).
[7] L. H. Jenkins and D. M. Zehner, private communication.
[8] J. Lindhard, Kgl. Dan. Vidensk. Selsk, Mat. - Fys. Medd. 34,
   No. 14 (1965).
[9] M. T. Robinson, "Interatomic Potentials and Simulation of
   Lattice Defects," edited by P. C. Geblen, J. R. Beeler, Jr.,
   and R. I. Jaffee (Plenum, N.Y. 1972), p. 281.
[10] T. S. Noggle, C. D. Moak, S. Datz, and B. R. Appleton, unpublished data.
[11] L. Van Hove, Phys. Rev. 89, 1189 (1953).
[12] J. Lindhard, Kgl. Dan. Vidensk. Selsk. Mat. - Fys. Medd. 28,
   No. 8 (1954).
[13] J. H. Barrett, Phys. Rev. B 3, 1527 (1971).
[14] B. R. Appleton, S. Datz, C. D. Moak, and M. T. Robinson, Phys.
   Rev. B 4, 1452 (1971, and references quoted therein.
[15] S. Datz, C. D. Moak, B. R. Appleton, M. T. Robinson, and O. S.
   Oen, "Atomic Collision Phenomena in Solids," ed. D. W. Palmer,
   et al. (North-Holland Publ. Co. 1970), p. 374.

CHANNELING STUDIES OF ALKALI HALIDES

P. B. PRICE* and J. C. KELLY
Materials Irradiation Group
School of Physics, University of N.S.W.
Australia

Introduction

The susceptibility of the alkali halides to radiation damage by light charged particles of MeV energies appears first in the high minimum yield observations of aligned crystals by Matzke [1]. Measurements of the effect of doses up to a few times 10^{15} ions cm^{-2} of MeV protons in various alkali halides were made by Morita et al. [2] and Ozawa et al. [3]. A saturation in the normalized initial slope of the aligned backscattered spectrum $\dfrac{d\chi_{min}}{dz}$, was observed by Morita et al. [2] for 1.5 MeV protons in KBr. In a study of a number of alkali halides with 1.5 MeV protons, Ozawa et al. [3] failed to observe a saturation in $d\chi_{min}/dz$. Using a 1.0 MeV He$^+$ beam, Hollis [4] has found a recovery in χ_{min} for NaCl at a dose of about $5 \cdot 10^{15}$ ions cm^{-2} at both 300°K and 77°K.

Measurements by Matzke [1] of widths of angular scans about the <100> axis for KCl and NaCl showed reasonable agreement with estimates of $\psi_{1/2} = \alpha\psi_1$. However, no estimate of the dose used in the measurements was given and the amount and form of the filling in of the dips is not known.

Angular scans of a number of alkali halides have been carried out using low (<10^{13} ions cm^{-2}) doses for each point to give approximately 'zero dose' $\psi_{1/2}$ values. The effects of increasing dose on χ_{min} are reported for beams of H$^+$, D$^+$ and He$^+$ on a number of alkali halides up to maximum doses of about 10^{17} ions cm^{-2}.

Experimental Procedure

The experimental arrangement, Fig. 1, consisted of the usual collimating system followed by a chamber containing a goniometer. Beam collimation was effected by using one (in the case of He$^+$ ions) and two (in the case of H$^+$ and D$^+$) apertures giving angular divergences of less than 0.1° and 0.03° respectively. For angular

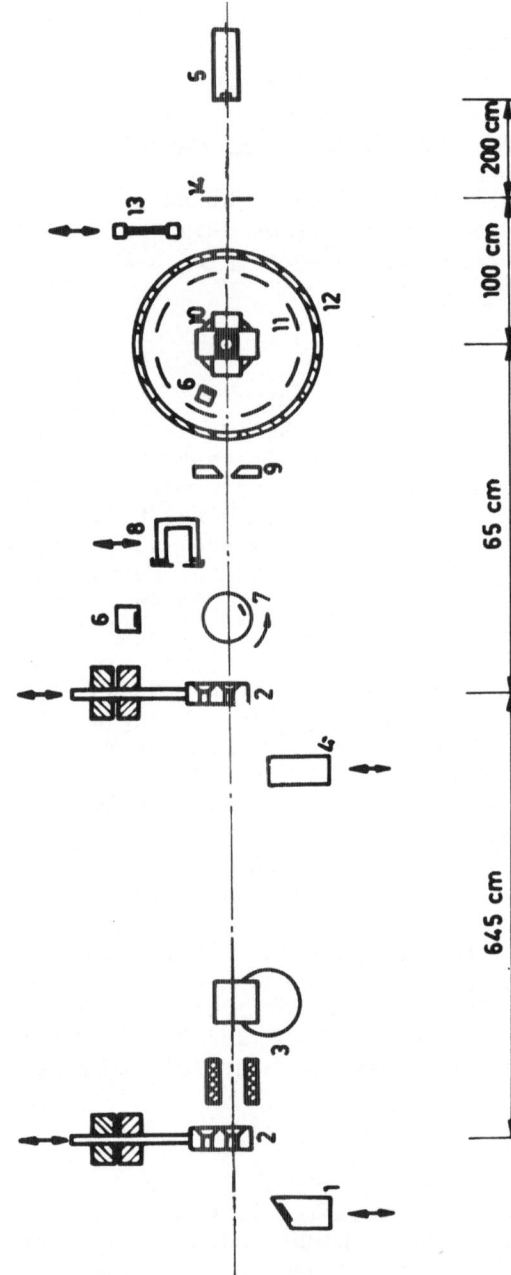

Fig. 1. Schematic diagram of channeling rig. 1 & 4, beam viewers; 2, adjustable apertures; 3, steering magnet; 5, laser; 6, detectors; 7, beam monitor; 8, moveable Faraday cup; 9, anti-scattering aperture; 10, crystal on goniometer; 11, liquid nitrogen cooled shield; 12, chamber wall; 13, moveable phosphor screen; 14, auto-collimation screen for laser alignment of crystal.

scans using proton beams, the horizontal tilt increment was 1 minute of arc.

A beam chopper was used to monitor the beam current after the final aperture. Calibration was by means of a removable Faraday Cup. The beam uniformity was visually checked before each aperture and after the goniometer. The goniometer was surrounded by a liquid nitrogen cooled shield and the targets could be heated to 450°C.

Harshaw alkali halide crystals[†] were cleaved normal to the <100> and chosen to be visually free of cleavage steps. Alignment of the crystals was effected by using a laser beam, previously made colinear with the ion beam [5], reflected from the (100) face of the crystal.

Results

In order to introduce as little damage as possible during scans the dose per scan point was kept to about $5 \cdot 10^{12}$ ions cm^{-2}, and as few points as possible were used. This meant that statistical errors varied from 5% at the shoulder to about 15% in the bottom of the dip. Values of $\psi_{1/2}$ were measured with proton beams of varying energies for LiF, NaF and NaCl with the results shown in Table 1.

Table 1

Comparison of predicted and experimental values of α, β and $\psi_{1/2}$ where $\psi_1(\text{axial}) = (2Z_1Z_2e^2/Ed)^{1/2}$, $\psi_1(\text{planar}) = (Z_1Z_1e^2d_p\text{Na}_{TF}/E)^{1/2}$ and α_{calc}. is obtained from Fig. 5 of Andersen (1967). The calculated $\psi_{1/2}(\text{planar})$ values are obtained from Lindhard (1965) $\psi_1^*/2\bar{Z}_2^{1/6}$.

Crystal	Direction	Energy MeV	Calculated ψ_1	α	$\psi_{1/2}$	Experimental $\psi_{1/2}$[(C)]	α
NaCl	<100>	0.7	0.82	0.90	0.74	0.73	0.89
LiF	<100>	1.3	0.47	1.15	0.54	0.57[(B)]	1.22
NaF	<100>	0.8	0.67	1.08	0.72	0.67[(A)(B)]	1.00
NaCl	(100)	0.7	0.14		0.26	0.20	β=1.4
NaF	(100)	0.8	0.14		0.26	0.23[(B)]	β=1.6

(A) Measured after a dose of 2×10^{15} ions cm^{-2}.
(B) Values obtained by extrapolation of $\psi_{1/2}(z)$.
(C) Maximum error about 10%.

Agreement with values predicted from the theories of Lindhard [6] and Anderson [7] for the axial angular half widths is quite good and is within the maximum error estimates for the results.

Both the (100) planar scans (0.7 MeV H^+ on NaCl and 0.8 MeV H^+ on NaF) gave results lower than the predictions of Lindhard using $\psi_{1/2} = \psi_1{}^*/2\bar{Z}_2{}^{1/6}$. Picraux et al. [8] have given for the planar case

$$\psi_{1/2} = \beta(Z_1 Z_2 e^2 N d_p a_{TF}/E)^{1/2}$$

with the value of β expected to be about unity. In the case of 0.7 MeV protons on NaCl and 0.8 MeV protons on NaF we have found the value of β to be 1.4 and 1.6 respectively. Alexander [9] has suggested the value of β should be between 1.5 and 2.0. In all of the above cases the yield from the heavier element was measured, though the dechanneling would have been due to both the target elements.

Surface values of $\chi_{min}(z)$ have been obtained by extrapolation of χ_{min} versus depth curves. Calculated values for axial directions have been obtained using Lindhard's formula

$$\chi_{min} = \pi N d(u_\perp{}^2 + a^2{}_{TF})$$

and Barrett's formula

$$\chi_{min} = \pi N d(Cu_\perp{}^2 + C^1 a^2{}_{TF})$$

with $C = 3.0$ and $C^1 = 2.0$ and 0.0, and where $\psi_{1/2}$ has been measured from Picraux et al. [8]

$$\chi_{min} = \pi N d(1/4a_{TF})^2 \left(\frac{\psi_1}{\psi_{1/2}}\right)^4 .$$

Planar values are compared with the expression $\psi_{min}(\text{planar}) \simeq 2a_{TF}/d_p$ for a static lattice from Davies et al. [10] where d_p is the spacing between the planes. Table 2 compares the calculated values with the extrapolated experimental values. The agreement between calculated and experimental values is quite good in view of the small sample size.

For small values of $\chi_{min}(\simeq 0.01)$ the absolute error in the extrapolation could give rise to errors of 25% - 50% for the proton case. This may account for the unusual observation of $(\chi_{min})_{expt} < (\chi_{min})_{theor}$ found for NaCl and NaF when the results are compared with Barrett's formula. Except in the case of LiF, $a^2{}_{TF} \sim u^2$ and the sum of these two terms used in Lindhard's expression agrees fairly well with the results for NaCl and NaF while the observed result for KCl is anomalously high. For LiF, $a_{TF}{}^2 > \bar{u}^2$ and the Lindhard expression with only a_{TF} is probably more correct, though here again the observed proton result is somewhat greater than predicted. In two cases (KCl and LiF) the results for the helium observations are

Table 2

Comparison of observed values of χ_{min} with various theoretical estimates

Crystal	Direction	Projectile	χ_{min} (Lindhard)			χ_{min} (Barrett)		χ_{min} (A)	χ_{min} (observ.)
			$\pi N d \bar{u}_\perp^2$	$\pi N d a_{TF}^2$	$\pi N d (u_\perp^2 + a_{TF}^2)$	$C=3, C'=.2$	$C=3, C'=0$		
KCl	<100>	H^+	0.014	0.008	0.022	0.044	0.042	–	0.04
		He^+	0.014	0.008	0.022	0.044	0.042		0.02
NaCl	<100>	H^+	0.015	0.012	0.027	0.048	0.046	0.012	0.015
		He^+	0.015	0.012	0.027	0.048	0.046		0.01
NaF	<100>	H^+	0.012	0.021	0.033	0.040	0.036	0.006	0.01
LiF	<100>	H^+	0.013	0.039	0.052	0.047	0.039	0.006	0.04
		$He^\pm$	0.013	0.034	0.047	0.045	0.039		0.01
KCl	<110>	H^+	0.020	0.012	0.032	0.062	0.060		0.09
								$2a_{TF}/d_p$	
KCl	(100)	H^+						0.11	0.22
NaCl	(100)	H^+						0.13	0.11

(A) Estimated from $\chi_{min} = \pi N d (1/4\pi a_{TF})^2 (\psi_1/\psi_{1/2})^4$ (see Picraux et al., 1969).

lower than the corresponding proton results though they also agree
fairly well with the predicted values.

Only static lattice estimates have been made for (100) proton
minimum yields from KCl and NaCl. Agreement with the results is
good for NaCl but not at all good for KCl. Similarly there is a
quite large discrepancy between the predicted and observed results
for the <110> axis results for KCl. The <100> results should only
differ from the <110> result by the ratio $d_{<110>}/d_{<100>} = 1.41$
whereas the observed ratio is 2.25. It is possible, though unlikely, that the crystal was considerably misaligned to this extent in
each of the three KCl observations but not in any of the others.

The effect of increasing the ion beam dose on the crystal was
monitored by observing the change in the normalized back-scattered
yield. In earlier runs the irradiation was interrupted and a sample
spectrum taken, whereas for later runs sample spectra were taken
while the crystal was continuously irradiated.

Initial observations were made by irradiating NaCl and KCl
single crystals in the <100> direction with a 1.0 MeV deuteron
beam, Fig. 2. Shown for comparison are the results of Hollis [4]
for 1.0 MeV He$^+$ on NaCl. No recovery similar to that observed by
Hollis was found though a plateau is observed for KCl.

Proton irradiations were carried out in four alkali halides
at various energies viz. LiF (1.3 MeV), NaF (0.8 MeV, NaCl (0.7 MeV)
and KCl (1.0 MeV) up to a total dose of about 10^{16}H$^+$cm^{-2}, Fig. 3.
In each of these cases χ_{min} was measured at a depth of 0.64 μm.
All of the curves show a point of inflection between a dose of 10^{14}
and 10^{15}H$^+$cm^{-2}, though the effect is quite small close to the surface.

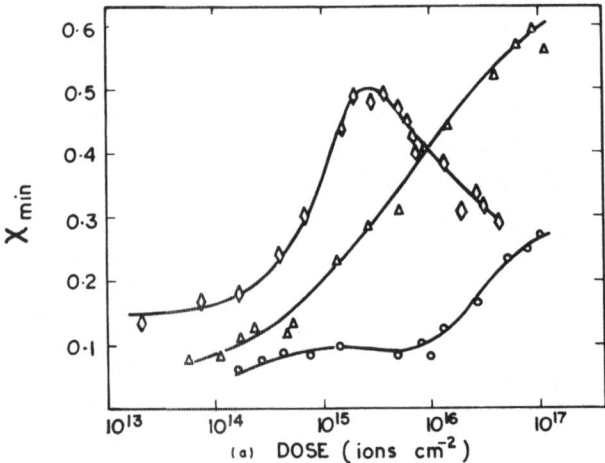

Fig. 2. χ_{min} as a function of dose. Diamonds, 1 MeV He$^+$ on NaCl
(Hollis). Triangles, 1 MeV D$^+$ on NaCl. Circles, 1 MeV D$^+$ on KCl.

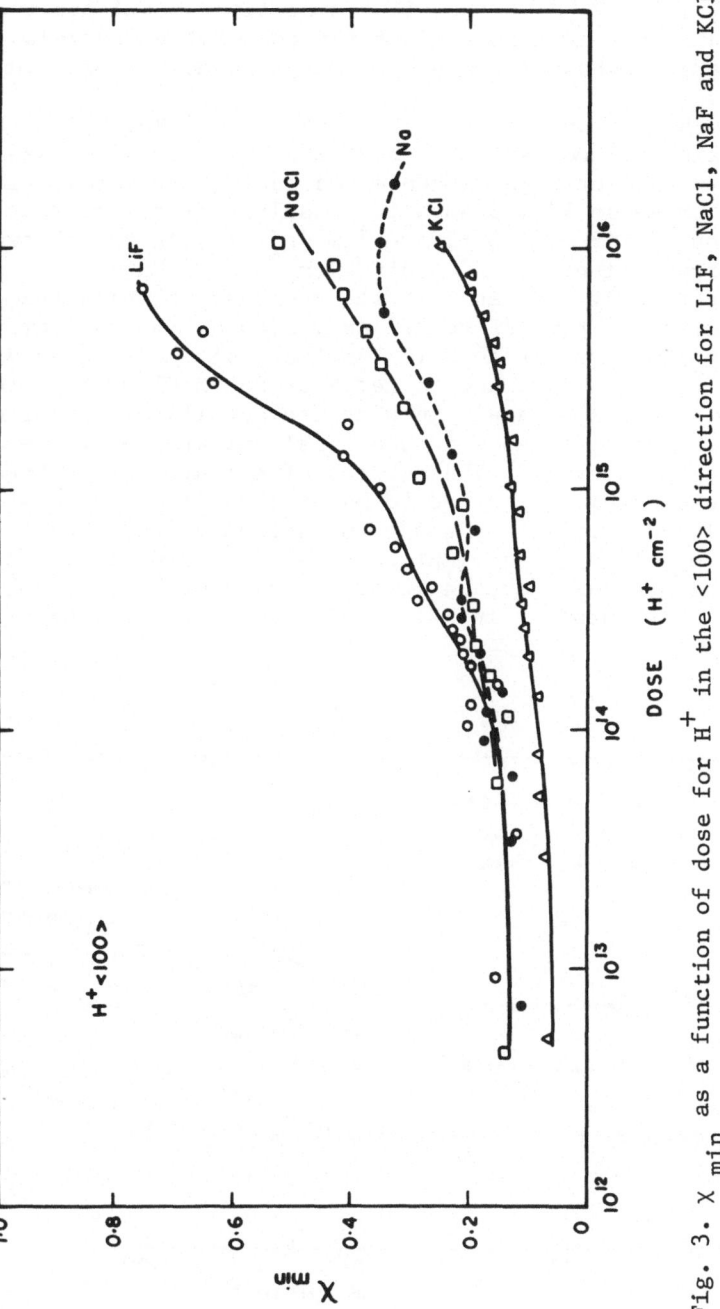

Fig. 3. χ_{min} as a function of dose for H^+ in the <100> direction for LiF, NaCl, NaF and KCl.

With increasing depth the effect is more pronounced. Fig. 4 shows the curves corresponding to depths of 0.64 µm, 1.6 µm and 2.5 µm for 1.0 MeV H^+ in the <110> direction in KCl for two crystals. A distinct hump is evident for one of the crystals. Both crystals are Harshaw crystals but are believed to come from different blocks.

No such plateau exists for any of the 1.0 MeV He^+ irradiations. The crystals studied were LiF, NaCl and KCl. In all cases a recovery in the value of χ_{min} was observed during the irradiations.

The curves in Fig. 5 show the results for an interrupted irradiation of LiF at $4.6 \cdot 10^{12} He^+ cm^{-2} s^{-1}$ together with two continuous irradiations at $4.0 \cdot 10^{12} He^+ cm^{-2} s^{-1}$ of NaCl and KCl. Two trends are readily apparent; first, the dose at which recovery occurs increases from LiF to KCl and, second, the maximum value of χ_{min} decreases from LiF to KCl. Further, at a dose greater than $10^{17} He^+ cm^{-2}$, χ_{min} continued to decrease for NaCl whereas for KCl it suddenly increases again in what are actually a series of steps. A corresponding difference in the final appearance of the irradiated areas is also observed. The surface of KCl appears to break into flakes whose edges correspond to major crystal directions.

He^+ irradiations in a random direction do not produce an earlier recovery in χ_{min} as might be expected and which would correspond to the greater increase in $d\chi_{min}/dz$ observed by Ozawa et al. [3] for 1.5 MeV protons in KCl. This together with the effect of

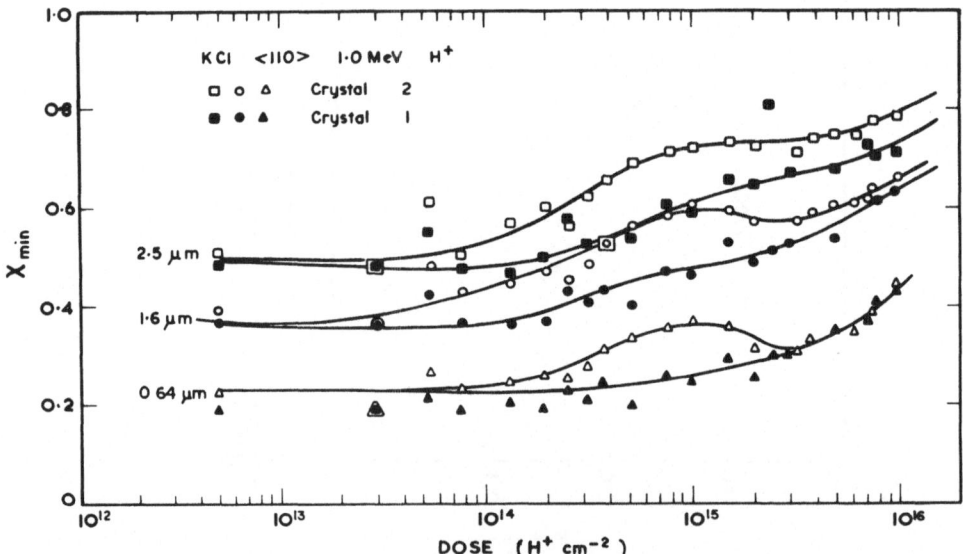

Fig. 4. χ_{min} as a function of dose at various depths for 1 MeV H^+ in the <110> direction for two KCl crystals.

increasing dose rate can be seen for the case of LiF in Fig. 6.
The effect of increasing current is to raise the saturation value
of χ_{min} rather than to decrease the dose required. Between each of
the irradiations shown the crystal was annealed for 35 minutes at
400°C with an aligned spectrum taken midway through the anneal.
The displacement of the curve for a beam current of 30 nA appears
to be due to an incomplete annealing of the damage or to a slight
misalignment as a similar irradiation gave a recovery at the same
dose as that found with a lower current beam.

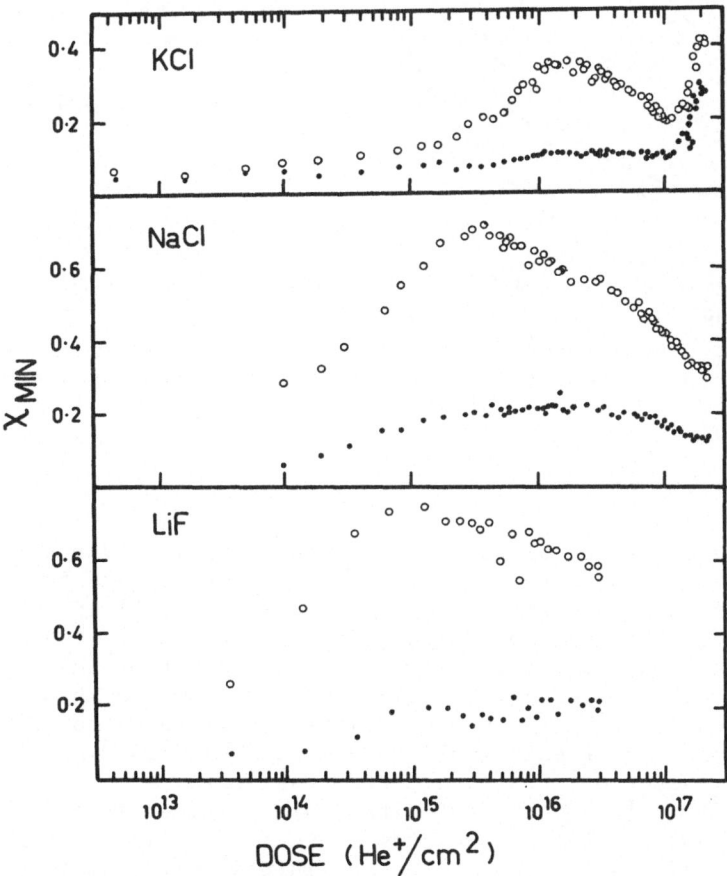

Fig. 5. χ_{min} as a function of dose for 1 MeV He[+] in the <100>
direction for KCl, NaCl and LiF.

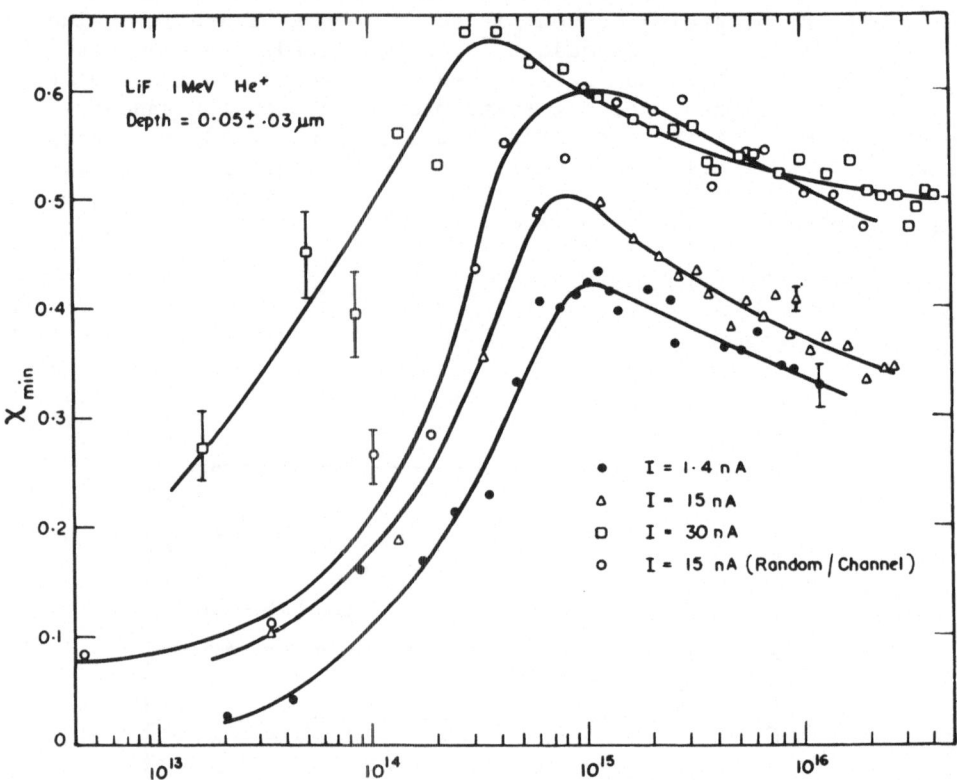

Fig. 6. χ_{min} as a function of dose for <100> 1 MeV He$^+$ irradiation
at several dose rates, together with an irradiation in a random
direction.

Discussion

        Estimates of the displaced atom concentration over the depth
range 1750 Å – 3500 Å for He$^+$ irradiations can be made from both
theories of Bøgh [11] and of Feldman and Rogers [12] and also from
the theory of Khinchin and Pease [13]. After a dose of $5 \cdot 10^{13}$He$^+$cm^{-2}
in NaCl the number of displaced atoms estimated from the χ_{min} values
is $1.8 \cdot 10^{21}$cm^{-3} compared with $3.22 \cdot 10^{18}$cm^{-3} from the Khinchin-Pease
theory which is probably an overestimate, Billington and Crawford
[14]. Channeling effects would reduce the Khinchin-Pease estimate
by up to a factor of 10 at the depth considered.
        Thus it would appear that direct displacement effects alone
cannot account for the damage curves observed. Indirect displacements are believed to account for the initial high damage rate
observed with the alkali halides.

Pooley [15] has observed a saturation in the F-center concentration at a dose of about $10^{16}H^+cm^{-3}$ in KCl at 400 KeV. A similar saturation in the indirect displacement yield occurring at a slightly lower dose would account for the plateau observed in the proton irradiations. The higher energy density deposited in the He^+ irradiations would imply an earlier saturation in the F-center concentration. We have observed just such a saturation at a dose of $10^{14}He^+cm^{-2}$ in NaCl.

Although directional effects do not seem to be important in the case of the 1.0 MeV He^+ irradiations of Fig. 6 we have observed a shift of the plateau in the proton irradiations to a lower dose as the incident direction changes from the <100> and <110> axial direction to the (100) planar direction suggesting that H^+ irradiations may be more sensitive to directional effects.

Acknowledgements

This work was supported by the Australian Institute for Nuclear Science and Engineering and the Australian Research Grants Committee.

References

*Present address: Department of Physics, Queen's University, Kingston, Ontario, Canada.

†The NaF was from Optran through the courtesy of Dr. S. Dryden of the National Standards Laboratory, Sydney.

[1] H. J. Matzke, Phys. Stat. Sol. (a) 8, 99 (1971).

[2] K. Morita, K. Tachibana, and N. Itoh, Phys. Lett. 33A, 257 (1970).

[3] K. Ozawa, F. Fujimoto, K. Komaki, M. Mannami, and T. Sakunai, Phys. Stat. Sol. (a) 9, 323 (1972).

[4] M. J. Hollis, Phys. Rev. B 8, 931 (1973).

[5] P. B. Price, M. J. Hollis, and C. S. Newton, Nuc. Instr. and Method. 108, 605 (1973).

[6] J. Lindhard, Mat. Fys. Medd. Dan. Vid. Selsk. 34, No. 14 (1965).

[7] J. U. Andersen, Mat. Fys. Medd. Dan. Vid. Selsk. 36, No. 7 (1967).

[8] S. T. Picraux, J. A. Davies, L. Eriksson, N. G. E. Johansson, and J. W. Mayer, Phys. Rev. 180, 873 (1969).

[9] R. B. Alexander, A.E.R.E. - R 6849 (1971).

[10] J. A. Davies, J. Denhartog, and J. L. Whitton, Phys. Rev. 165, 345 (1968).

[11] E. Bøgh, Can. J. Phys. 46, 653 (1968).

[12] L. C. Feldman, and J. W. Rogers, J. Appl. Phys. 41, 3776 (1970).

[13] G. H. Khinchin and R. S. Pease, Repts. on Progress in Physics 18, 1 (1955).

[14] D. S. Billington and J. H. Crawford, "Radiation Damage in Solids," Princeton University Press.

[15] D. Pooley, Brit. J. Appl. Physics 17, 855 (1966).

MOLECULAR ION TRANSMISSION THROUGH A MONOCRYSTALLINE THIN FILM

M. J. GAILLARD, J. -C. POIZAT and J. REMILLIEUX
Institut de Physique Nucléaire, Université Claude Bernard Lyon-I
(Institut National de Physique Nucléaire
et de Physique des Particules)
43, bd du 11 novembre 1918, 69621 Villeurbanne, France

ABSTRACT

We have observed the channeling effects on
the transmission of H_2^+ molecular ions through
thin gold crystals. Energy and thickness dependences have been tested in planar channeling conditions. Similarly, the production of H_2^+ ions
from an incident H_2^+ beam has been studied. All
these observations are consistent with a molecular
recombination taking place at the back surface of
the crystal.

Introduction

Previous experiments using H_2^+ and H_3^+ molecular ion beams
have exhibited the consequences of the break-up process on the
interactions of the dissociation products with a solid target.
Bacher et al. [1] have shown that the Coulomb repulsion of protons
issued from a molecular beam is partly responsible for a broadening
of a narrow nuclear reaction resonance. Caywood et al. [2] and
Eisen and Uggerhøj [3] have shown the repulsion effects on the
transverse energy distribution and the channeling properties of an
incident molecular beam. However, we have first observed [4] that
H_2^+ ions can be transmitted through thin carbon foils. More recently
we have shown [5] that the transmission probability of H_2^+ ions
through a thin gold single crystal is strongly enhanced in channeling
conditions. In order to reach some understanding of this phenomenon,
we have performed more extensive channeling experiments in which the
influence of various parameters, such as incident energy, crystal
thickness, channeling direction, has been tested. We shall describe
these experiments and discuss these results which have been found to
be in agreement with a molecular recombination process at the back
surface of the target.

Experimental Set-Up

The experimental set-up, shown in Fig. 1, is a more sophisticated version of the arrangement used in previous experiments [5].
H^+, H_2^+ and H_3^+ beams provided by our 2 MeV Van de Graaff are collimated to a half-width angle $\leq 0.03°$. The beam spot size on the
crystal is 0.5 mm. The self-supporting [111] oriented gold single
crystals, obtained by the technique described in ref. 6-7 are carefully selected to be free of holes. Their thickness is measured
by comparing the widths of energy spectra of α-particles backscattered respectively from these crystals (in random incidence) and
from amorphous gold foils of known thicknesses.

The incident beam is measured by means of a beam-chopper. The
various components transmitted through the crystal (molecular ions,
neutral and charge dissociation products) are separated by a magnetic analysis. Here two difficulties arise: first, the transmitted H_2^+ fraction is very small ($10^{-4} - 10^{-7}$), second there is a D^+
(or DH^+) component in the incident H_2^+ (or H_3^+) beam which cannot
be separated from the transmitted H_2^+ component. Therefore the
following precautions are taken: a collimator limits the angular
dispersion of particles to be analyzed to $\psi_a = 1.2°$ and an electrostatic quadrupole lens is used, before the magnet, to focus the H_2^+
and D^+ transmitted beam on an annular detector and a gold foil.
Backscattered protons and deuterons are easily identified by their
energy. The detection efficiency is typically $\sim 10^{-5}$ and remains
constant (within $\sim 2\%$) whatever the particle position is when it
goes through the annular detector. It is important to note that
any H_2^+ ion transmitted through the collimator will impinge on the
gold foil. All along the experiments, the quadrupole can be visually
adjusted by replacing the detection system by a ZnS screen.

Experimental Results and Discussion

We are interested in measuring the probability P for an incident H_2^+ ion to be transmitted through the crystal in a molecular
state. Unfortunately, we are prevented from measuring P by the
collimation downstream from the crystal; we can only measure
$T_{H_2^+/H_2^+}$ called the transmission yield, which is the probability
for an H_2^+ ion to be transmitted through the crystal and the collimator. T is obviously smaller than or equal to P. In fact we have
observed in some particular cases that T is very weakly dependent
on the size of the limiting aperture in the range of size used in
all the experiments ($\psi_a = 1.2°$). So we can assume that T is very
close to P.

On the other hand, each determination of T was associated with
another measurement in which the probability T_{H^+/H^+} for an incident
proton with the same velocity to be transmitted through the crystal
was measured in the same detection set-up. Let f be the ratio of
$T_{H_2^+/H_2^+}$ and T_{H^+/H^+}

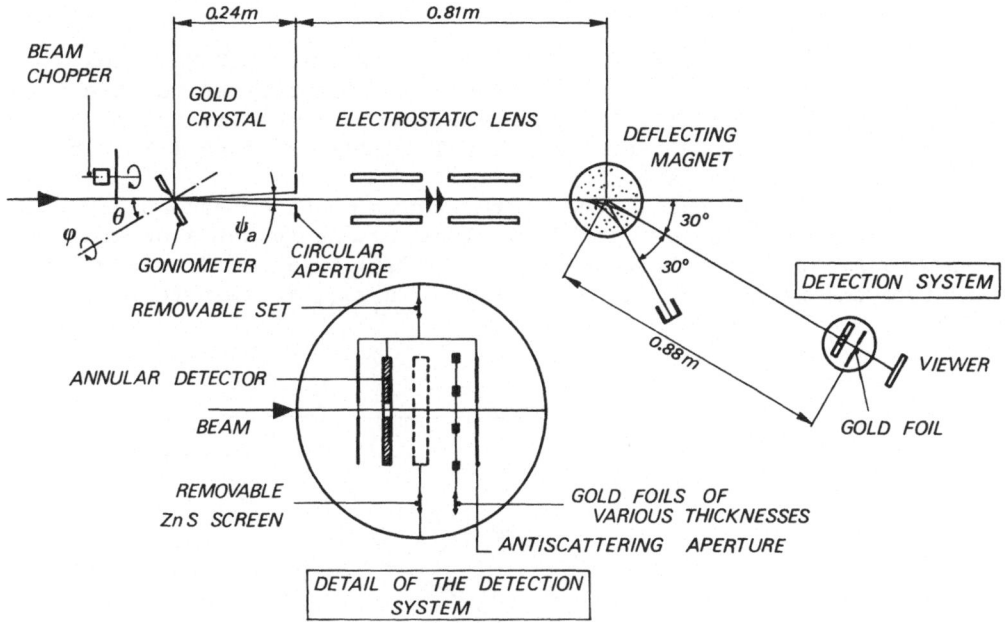

Fig. 1.   Experimental set up.

$$f_{H_2^+/H_2^+} = \frac{T_{H_2^+/H_2^+}}{T_{H^+/H^+}}$$

f would be equal to P if the angular dispersion due to multiple
scattering (correlated or not) were the same for the transmitted
H_2^+ fraction and for the transmitted proton beam. In fact it is
reasonable to assume that the angular dispersion of the transmitted
H_2^+ ions is much smaller. So f is larger than P. (It must be
noted that the determination of the proton transmission with an
incident H_2^+ beam would be less significant since repulsion effects
downstream the crystal would broaden the angular dispersion).
Finally P is not measured, but closely surrounded by two measured
values T and f.

We shall also describe experiments in which the H_2^+ production
from a H_3^+ incident beam is studied. In this case the same technique is used to obtain $T_{H_2^+/H_3^+}$, T_{H^+/H^+} and $f_{H_2^+/H_3^+}$.

The ion dose required for a transmission measurement along a
given direction is a few $\mu C/mm^2$. If many measurements are performed
with the same crystal, a decay of the transmission yields can be
observed, due to carbon build-up and/or radiation damage. In such
cases, a small correction is needed for the experimental transmission yields.

H_2^+ transmission along planar directions

Transmission yields, T and f, for 1.0 and 1.3 MeV H_2^+ ions
incident in various planar directions have been plotted vs. the
planar spacing d_p on Fig. 2. The pathlength in the crystal is
540 Å for all planar directions. The T random value at 1 MeV is
also shown for comparison.

One can notice that T and f are pretty close to each other and
their variations are very similar. That is due to the slow and
monotonous variation of T_{H^+/H^+} with d_p (T_{H^+/H^+}, as said before, is
equal to the ratio of $T_{H_2^+/H_2^+}$ over f). T_{H^+/H^+} is seen to vary

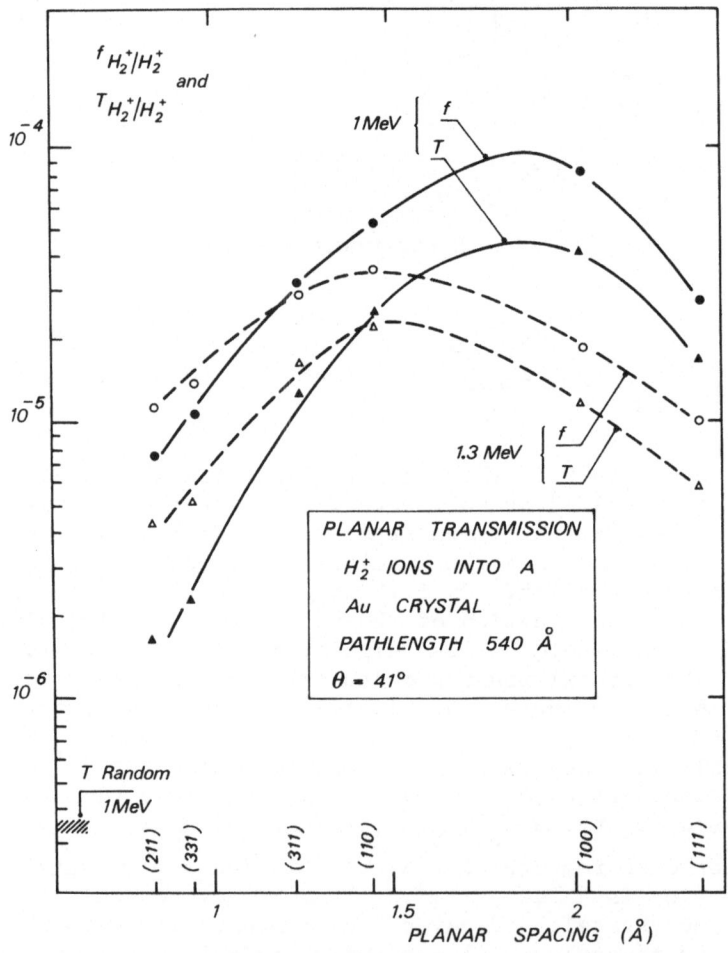

Fig. 2. Planar spacing dependence of the H_2^+ transmission yields
T and f, for 1 and 1.3 MeV H_2^+ ions. Pathlength 540 Å.

from $\sim$65% in the (111) direction to $\sim$25% as $(1 - \chi_m)$ does (χ_m being
the normalized scattering yield in planar alignment [6]. That is
not surprising since the half-acceptance of the limiting aperture
(0.6°) is at least equal to the $\psi_{1/2}$ angle (half-width of scattering
dip at midway between aligned and random values) for the (111) direction. So it is reasonable to assume that the channeled particles,
and the transmitted H_2^+ ions, are able to go through the aperture,
and then that P, surrounded by T and f is closer to T than to f.
In any case the most important points are that the d_p-dependence of
P is very similar to those of T and f, and that these observed dependences are free from any geometrical effects due to the tight
collimation. The striking feature of these results is the maximum
observed and the shift of this maximum towards lower d_p when incident energy increases from 1 to 1.3 MeV.

Fig. 3 shows the variation of T vs. the planar spacing d_p for
1 MeV H_2^+ ions incident on two crystals, with respective pathlengths
of 340 and 540 Å. It was impossible to associate the (100) direction
with a pathlength as short as 340 Å which could be obtained only with
a small tilt angle (23°), the (100) direction alignment with a
<111> orientated crystal requiring a larger tilt angle ($\sim$41°). So
a maximum is observed but its shift towards lower d_p for shorter
pathlengths is only suggested.

Fig. 4 shows the variations of the transmission yield T along
the (110) direction; (a) with incident energy for a given pathlength (320 Å), (b) with the pathlength for a given energy (1 MeV).
Experimental points shown with solid circles were obtained by
varying the tilt angle of the same crystal.

Fig. 5 shows the pathlength dependence of the transmission
rate, normalized to unity for the pathlength 340 Å, of 1 MeV H_2^+
ions incident along four planar directions (experimental data for
pathlength 940 Å have been obtained and published previously [5]).

We have now to examine which physical processes can be found
compatible with these experimental results. We can first imagine
that the transmitted H_2^+ ions are survivors which remained in a
molecular state all the way through the foil. However, this possibility must be ruled out because, as pointed out by Brandt and
Sizmann [8] for hydrogen atomic states, electronic screening is too
strong for a bound molecular H_2^+ state to exist inside a solid. In
such a case indeed, it would be reasonable to expect an increase
of the transmission rate with the planar spacing d_p, which is in
opposition with our results. On the other hand, the electron loss
cross-section for an atomic ion decreases with increasing energy;
so, expecting a similar behaviour from a molecular ion, we should
observe an increase of the transmission rate with increasing energy
for any planar direction, in opposition with the results of Fig. 4a.

The only possibility left lies in a recombination process at
the exit surface of the target. Then every incident molecular ion
breaks up at the entrance surface, and the two protons follow more
or less independent trajectories. If certain kinematic conditions
are satisfied, recombination of the two protons into a molecular

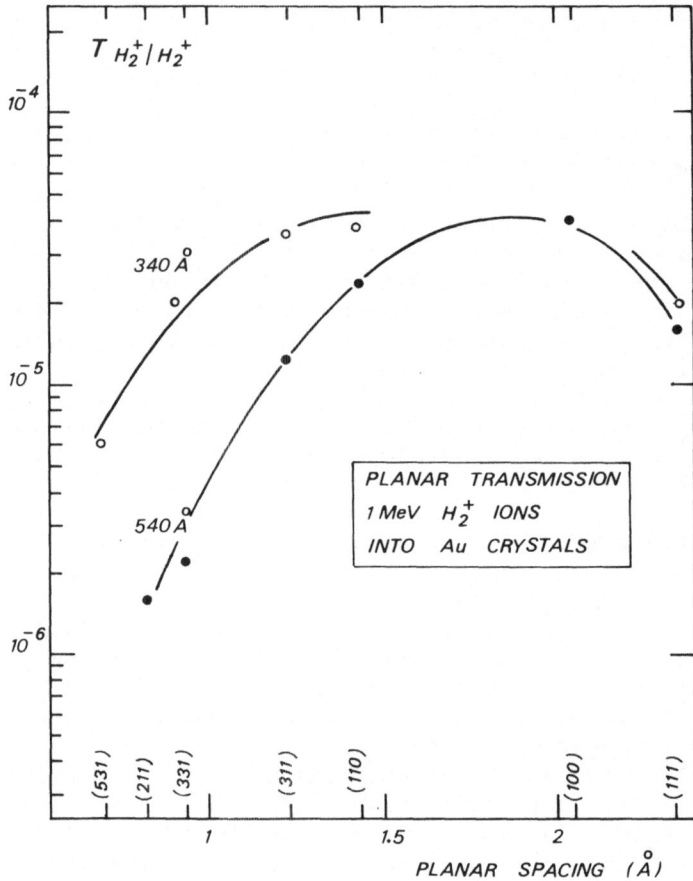

Fig. 3. Planar spacing dependence of the H_2^+ transmission yield T
for 1 MeV H_2 ions. Pathlengths 340 and 540 Å.

ion can take place at the exit surface, in conjunction with an electronic capture, either by one of the protons, or by both of them
directly into a molecular state. The parameters to be considered
are those which have an influence on the distance between the two
protons and their impulses in the center of mass when they come out
of the foil, and also the electronic capture cross-section.

One can first suppose that two incident protons to be transmitted through the foil in a H_2^+ molecular state must be channeled
(whatever the transmission process is). Assuming the incident
molecular orientation to be isotropic, and initial proton separations given by the distribution of vibrational levels, one can
easily show that the probability for the two protons to be channeled
i) in the same planar channel increases with the interplanar spacing d_p and ii) in two adjacent planar channels reaches a maximum

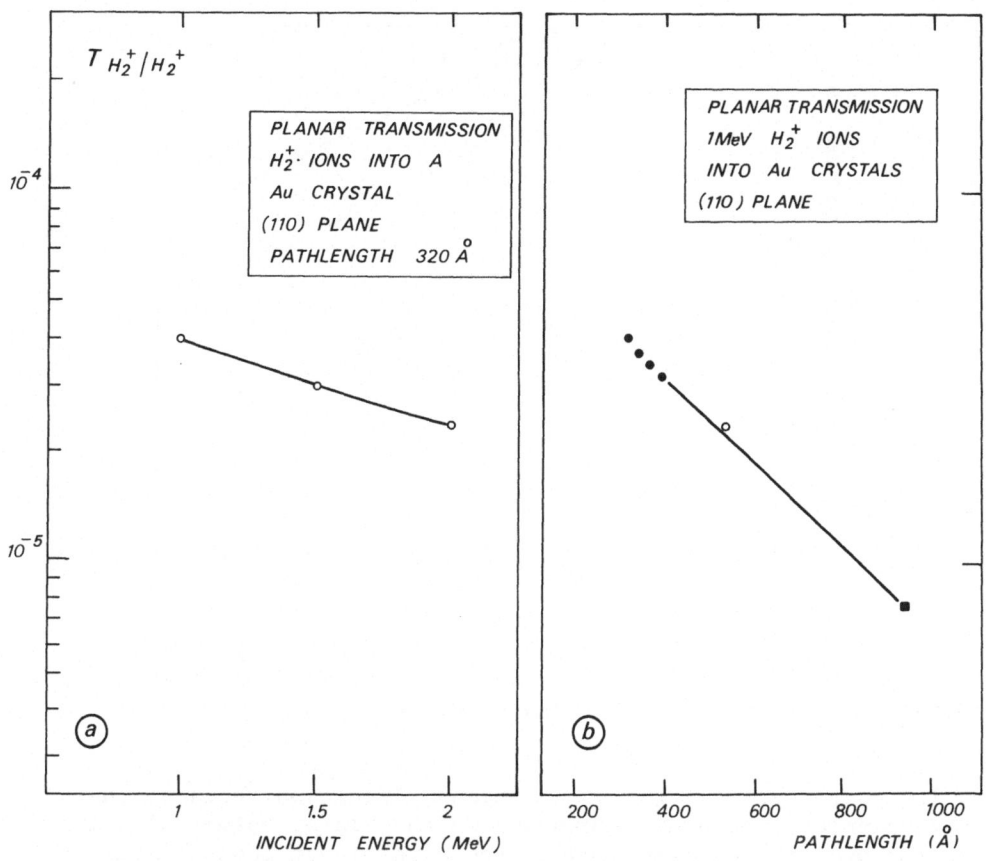

Fig. 4. H_2^+ transmission yields T for H_2^+ ions incident along the (110) planar directions: (a) energy dependence for a pathlength of 320 Å; (b) pathlength dependence for 1 MeV H_2^+ ions.

for $d_p \sim 1.1$ Å. Drawing any definite conclusions from this remark would be hazardous. However one must keep in mind these two possibilities when discussing the second stage of the transmission.

This second stage deals with the interactions suffered by protons in the bulk of the crystal. They suffer electronic and nuclear multiple scattering, as usual, plus a mutual Coulomb repulsion. Although Coulomb repulsion is strongly screened in solids, its effect probably cannot be neglected and increases the distance between the two protons. So foil thickness and incident energy, in addition to their influence on multiple scattering effects, determine the transit time during which repulsion applies in the foil (in the energy and thickness ranges used, unscreened repulsion could increase the initial distance by a factor ~ 2). In planar channeling conditions we can expect the repulsion to be effective mainly

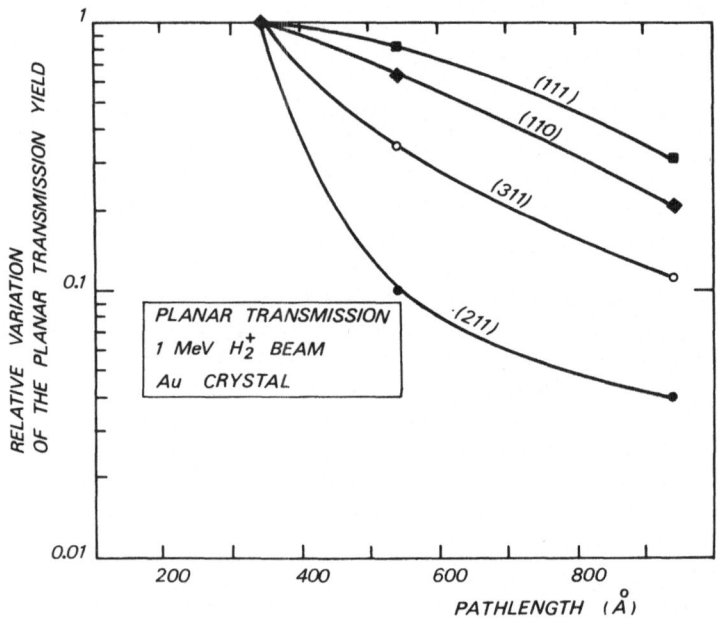

Fig. 5. Pathlength dependence of H_2^+ transmission yields T for
1 MeV H_2^+ ions incident along some planar directions (T values are
normalized to their values for pathlength 340 Å).

by its component parallel to the planar direction, because the
perpendicular component is opposed by the planar potential. If
the two protons are channeled in the same planar channel this last
component, however, induces an increase of transverse energies and
then an increase of oscillation amplitudes (and extra dechanneling
[2]). If the two protons are channeled in two adjacent planar
channels, Coulomb repulsion is much more screened, and this geometry
could be favorable for two protons to keep at a short distance from
each other in minor planar directions.

The last state is evidently the recombination itself at the
back surface of the foil. The recombination probability depends
first on the kinematic conditions (relative distance of protons and
impulses in the center of mass) resulting from the previous stages.
For given kinematic conditions, it depends also on the electronic
capture cross-section. Though being not known in the case of capture into a molecular state, this cross-section should decrease
with increasing energy at least as rapidly as for capture in an
atomic state ($\sim E^{-3}$).

We can now make a comparison of our experimental data with
these predictions. The maxima observed for the d_p dependence of
transmission yields (Fig. 2 and 3) could be explained as follows:
when d_p is large, in the (111) direction particularly ($d_p = 2.35$ Å),
molecular transmission associated with protons channeled in two

adjacent channels becomes unlikely, because their mean separation is at least equal to d_p. If they are channeled in the same planar channel, the volume available is such that the two protons can keep at mean distances larger than those required for recombination. So the transmission rate cannot increase when d_p becomes much larger than the mean H_2^+ interatomic distance, say 1 Å. Fig. 3 also shows that repulsion does not play a significant role in the (111) direction since the transmission yield is very little affected by increasing the pathlength from 340 to 540 Å. On the other hand, transmission along minor planes is clearly seen to be very sensitive to repulsion and/or dechanneling.

Energy dependence is shown on Fig. 2 and 4b. Fig. 2 shows that transmission rate variations are in opposition in minor and major planar directions (and are small along (110), as shown also in Fig. 4b for a larger energy range). In the (111) direction, experimental data obtained at 1 and 1.3 MeV are compatible with a E^{-4} dependence, not too far from the E^{-3} dependence which can be expected if repulsion and dechanneling effects are small. On the contrary these effects are determinative along minor planes and, for the same change of energy, produce a slight increase of the transmission rate.

Then we can conclude by saying that energy and pathlength dependences are very sensitive to the planar spacing d_p, but are explainable, at least qualitatively, by molecular recombination.

H_3^+ beam experiments along planar directions

Studying a possible production of H_2^+ ions downstream a thin crystal from an incident H_3^+ beam was thought to be of great interest as a complement to the H_2^+ beam experiments. Naturally the first move was to make sure that no H_3^+ ion is transmitted for any crystal orientation, and this was verified.

But the production of H_2^+ ions was found to be nearly as large for an H_2^+ beam. The same experimental procedure was used, and measurements of T and f led us to the same conclusion about geometrical effects. Fig. 6 shows the d_p dependence of H_2^+ transmission yields T from a H_3^+ beam, along with previously presented H_2^+ transmission yields T (Fig. 3) measured for the same crystal and the same proton velocities. It can be seen that T values for equal proton velocities respectively with H_2^+ an H_3^+ incident ions are very close to each other and follows the same d_p dependence. This similarity appears to be one more experimental argument against a H_2^+ transmission without dissociation: this assumption would imply that H_3^+ dissociation is always associated with the creation of a H_2^+ ion, which is very unlikely.

Then the physical processes leading to production of H_2^+ ion seems to be very similar in both cases, except for the following differences due to the presence of three protons instead of two: i) the proton separation is slightly different, as well as the initial probability for two protons to be channeled; ii) the

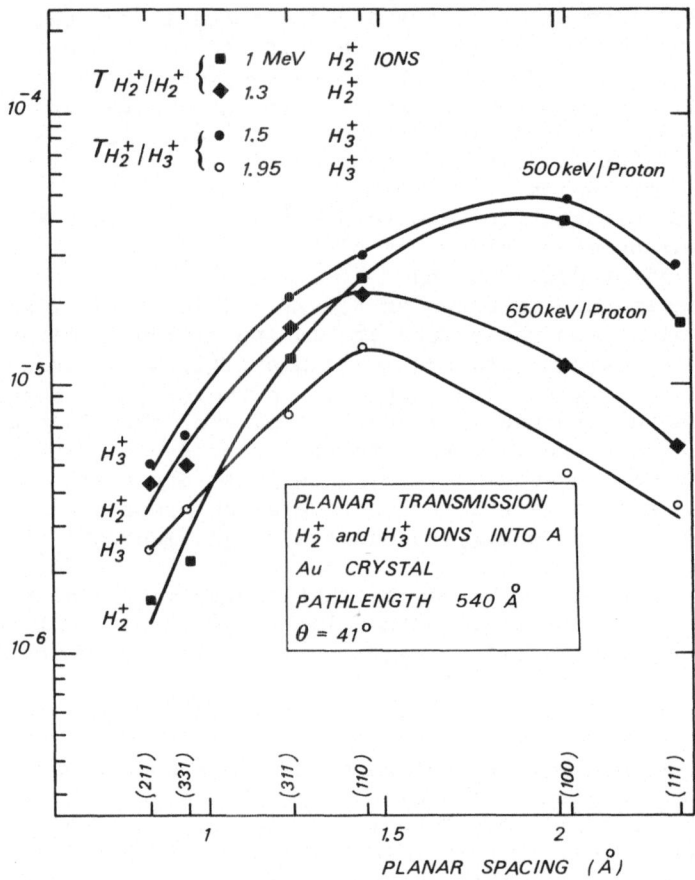

Fig. 6. Comparison of planar spacing dependences of H_2^+ production probabilities for same velocity H_2^+ and H_3^+ incident ions (500 and 650 keV/proton). The pathlength in the crystal is 540 Å.

repulsion effects are also different, due to i) and to the third proton (dechanneling effects are known to be larger with H_3^+ ions than with H_2^+ ions [2]); iii) then the proton separation distribution at the back surface is also different. On the other hand three protons instead of two are candidates to the creation of a H_2^+ ion. However, the effect of the third proton on electronic capture by the two others is not known.

These differences, acting in opposite ways, are seen on Fig. 6 to have only a small influence on the H_2^+ production probability. However, the energy dependence of the H_2^+ production is much stronger in the case of H_3^+ incident ions, which is yet unexplained.

Angular acceptance for H_2^+ transmission along planar directions.

The angular width of the peak of H_2^+ transmission probability has been measured, with H_2^+ and H_3^+ incident ions and for various planar directions, by varying the incident direction with respect to the plane orientation. These angular widths have been found, in both cases, to be 20 - 30% smaller than the angular dips of Rutherford scattering dips. This indicates that the H_2^+ production is strongly enhanced even with not very well channeled particles. Again H_2^+ and H_3^+ beams give similar results.

H_2^+ transmission along axial directions

The H_2^+ transmission probability has been measured also along various axial directions. Table I gives experimental T values measured along five directions (all of them are contained in a (110) plane) with a 315 Å thick gold crystal and 1 MeV H_2^+ incident ions. In spite of small differences in pathlengths (due to tilt angle changes), it appears that T does not exhibit any clear dependence on d, the interatomic distance in the row. The maximum value, observed in the [211] direction, is close to the value found for the (110) planar direction (similar results have been obtained with H_3^+ incident ions, again with indications of a stronger energy dependence). These results, which are in opposition with all other channeling effects, must be explained by saying that, compared to planar channeling conditions, the volume accessible to channeled particles is larger and repulsion effects can apply more easily. Then, although the channeling probability is higher, the mean distance between the two protons can be larger. We have been prevented from measuring accurately angular widths by geometrical difficulties related to the collimation of the detection system. However, the angular acceptance seems to be relatively smaller than in planar conditions. This indication, along with the lack of definite d-dependence of T, could suggest some

Table I

H_2^+ transmission yield along axial directions

Axis	[110]	[211]	[411]	[332]	[311]
d(Å)	2.88	4.99	8.65	9.57	13.52
Pathlength (Å)	385	330	385	320	360
$10^5 \times T_{H_2^+/H_2^+}$	2	3.8	3.6	2.1	0.6

connection with hyperchanneling: potential barriers midway between
the rows of atoms which form the channel are dependent on the
geometrical arrangement of rows in each axial direction and then
oppose more or less efficiently to the transverse component of repulsion.

Conclusion

The experimental results presented in this paper have shown
that the production of H_2^+ molecular ions downstream a thin crystal
from H_2^+ and H_3^+ incident ions can be explained in terms of a
molecular recombination. However, we have just reached a qualitative agreement, and much more experimental and theoretical work is
still needed. Measurements of the polarization of the internuclear
axis of transmitted ions would be particularly useful and will be
attempted in the near future.

Acknowledgments

We acknowledge enlightening discussions with Prof. W. Brandt.

References

[1] A. D. Bacher, E. A. McClatchie, M. S. Zisman, T. A. Weaver and
    T. A. Tombrello, Nucl. Phys. 181A, 453 (1972).
[2] J. M. Caywood, T. A. Tombrello and T. A. Weaver, Phys. Lett.
    37A, 350 (1971).
[3] F. H. Eisen and E. Uggerhøj, in Atomic Collisions in Solids IV
    (Gordon and Breach, London, 1972) p. 181.
[4] J. -C. Poizat and J. Remillieux, Phys. Lett. 34A, 53 (1971).
[5] J. -C. Poizat and J. Remillieux, J. Phys. B5, L94 (1972).
[6] J. -C. Poizat and J. Remillieux, Journal de Phys. 33, 1013 (1972).
[7] R. Kirsch, D.E.S. Univ. of Lyon (1972).
[8] W. Brandt and R. Sizmann, Phys. Lett. 37A, 115 (1971)(and in
    references therein).

A COMBINATION OF DECHANNELING AND ENERGY MEASUREMENTS OF PROTONS IN THIN SILICON SINGLE CRYSTALS

G. GÖTZ, K. D. KLINGE and U. FINGER
Bereich Angewandte Physik II
der Friedrich-Schiller-Universität
69 Jena, DDR

ABSTRACT

A special backscattering technique has been used to measure dechanneling and energy loss of protons in thin monocrystals of silicon (1-4µm) simultaneously.

At the back of the crystalline targets a very thin layer of carbon or LiF was deposited. Backscattering of the protons on these layers was used for the measurement of the energy-loss ratio $\alpha = (dE/dt)_{hkl}/(dE/dt)_{random}$ in the energy range $E_p = 0,7$ to 1.6 MeV. We found $\alpha = 0.69 \pm 0.03$ in <111> and $\alpha = 0.52 \pm 0.03$ in <110> direction.

In connection with this, the energy loss distribution and the dechanneling at incidence angles of 0° - 1° in relation to the <111> and <110> axis were measured.

Also investigated was the variation of de-channeling and energy loss distribution with the amount of radiation damage, introduced by implantation of argon ions.

The experimental results for dechanneling along the <110> and <111> axis are in relatively good agreement with calculations in which we took into consideration nuclear contributions of the second order and used a more realistic potential.

Introduction

The backscattering technique has been widely used in connection with the channeling effect to investigate the lattice location of dopant and the radiation damage in ionimplanted semiconductors.

To determine the depth profile of dopants and lattice disorder it
is necessary to convert the energy scale of the backscattering
spectra into an equivalent depth scale. For this, the stopping
power S(E) for the ingoing (channeled) and outgoing (non-channeled)
ion trajectories must be known.

In thin perfect crystals the channeling phenomena of positive
particles can be described on the assumption of conservation of
transverse energy [1]. In this case the impinging beam can be
divided into two parts, one aligned and one random, and the channeled stopping power can be obtained from the random stopping power
by multiplying the latter with an energy dependent factor:

$$S(E)_{hkl} = \alpha S(E)_r \tag{1}$$

α is in the order of 0.3...1.0 [2,3,4,5,6]. For thicker crystals,
especially for crystals with lattice defects, the distribution in
transverse energy of an aligned beam will change appreciably with
depth because of scattering processes by electrons, vibrational and
disordered lattice atoms. To calculate the depth it is necessary
to know the fractions of the aligned and random beam and consequently
to determine a mean stopping power.

In this paper we have measured simultaneously the dechanneling
and the stopping power of protons in thin silicon crystals for
<111>, <110> and random direction. We also studied the dependence
of the dechanneling and stopping power from the incidence angle
$(0 \leq \psi_i \leq 1°)$ and from the amount of radiation damage, generated
by implanatation of argon ions.

Experimental

Experimental technique

The meausrements have been performed by the backscattering
technique using a proton beam from the 2 MV-Van de Graaff accelerator of JENA Friedrich-Schiller-University. The beam was energy
analyzed with a magnetic analyzer which was stabilized by a N.M.R.
set to $\Delta B/B \leq 10^{-4}$. The feedback stabilization maintained the
energy of the beam at ± 1.2 keV. The proton beam, collimated to
an angle ≤ 0.03° full width, was kept at a low enough intensity
($\leq$ 5 nA on a 1 mm^2 spot) to reduce radiation damage, heating and
pile up to a negligible value. The beam current at the time of
the experiment was measured on the target by a current integrator.
An annular electron suppressor (bias - 200 V) was placed in front
of the target in order to return secondary electrons emitted from
the target by proton impact. The dose during orientation of the
crystal and during the measurement was of the order of 1 μC. At
this dose no change in the aligned spectra could be seen.

The crystals could be oriented relative to the incident beam
by a two-axis goniometer which had an accuracy and reproducibility

better than ±0.03°. The scattered particles were measured by a
conventional surface-barrier detector and their energies analyzed
with a 256-channel pulse height analyzer. The energy resolution
of the whole detector system was 8 keV for 1.4 MeV protons.

The implantations were done on a 400 kV Cockroft-Walton accelerator. The samples were mounted on a variable temperature target
holder which could be heated up to 400°C. The temperature was
measured by a thermocouple with an accuracy of ± 10°C. In order to
ensure a uniform implantation, the beam was scanned over the area
of the sample. To avoid channeling effects, the crystals were
oriented 8° off the mean axis. The current density was $0.25\mu A \cdot cm^{-2}$,
and the ion energy was 200 keV. The dose could be determined within
an accuracy of ± 20%.

The annealing of the crystals was carried out in vacuum
($\leq 10^{-5}$ torrs) for 15 minutes.

Target preparation

The preparation of the silicon crystals has been described
elsewhere in detail [7,8]. The fact of selective electrochemical
etching is used for removing the low resistivity n^+-substrate
($10^{-2}\Omega \cdot cm$) from an n-type ($1\Omega \cdot cm$) epitaxial film. This way, a
large area (3 mm diameter), homogeneous (deviation in thickness
less than 5%), thin (1.0 - 6.0 μm) single crystal silicon film
supported by a frame of thicker material was obtained. Backscattering measurements produced the same minimum yield as for bulk silicon
(3%).

On the backside of the crystal a very thin layer of carbon or
LiF($\leq 700\text{Å}$) was evaporated. The thickness was chosen so as to produce an energy loss of the protons in the layer much lower than
the energy resolution of the detection system.

Experimental method

Usually the stopping power of energetic ions has been measured
in transmission experiments [5,6]. In the present work we have
used a backscattering technique in which the scattering from the
crystal and the scattering from the backlayer have been utilized.
The advantages of this method are:
   - dechanneling and energy loss can be measured simultaneously;
   - it is not necessary to determine the thickness of the crystal.
The principle of the method is shown in Figure 1. A monoenergetic beam (particle mass M_1, atomic number Z_1, energy E_0) enters
the crystal (atomic mass M_2, atomic number Z_2) at an angle ψ_1
relative to the surface normal. Particles scattered through the
angle θ (laboratory system) and leaving the surface at an angle
ψ_2 to the surface normal are energy-analyzed. Particles which penetrate the crystal can be scattered by the atoms of the backlayer
(atomic mass M_3, atomic number Z_3). Because there is a large difference in the collision energy loss between particles scattered

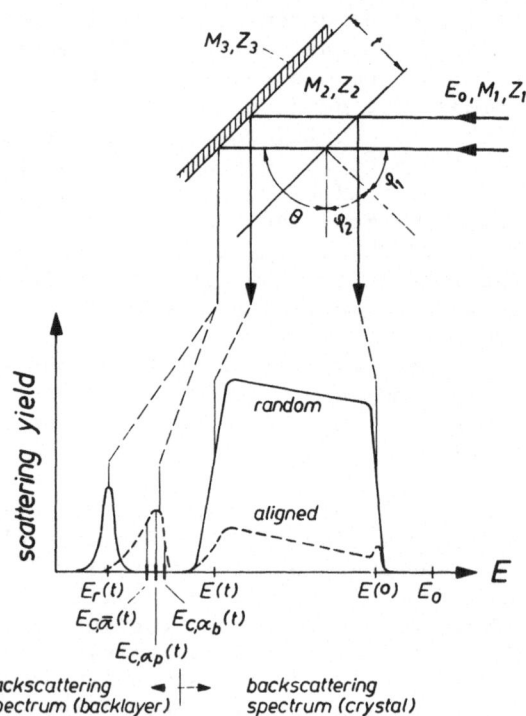

Fig. 1. Schematic diagram of the experimental arrangement for
measuring dechanneling and energy loss in thin single crystals.
Scattering process and energy spectra at aligned and random incidence.

from M_2 and those scattered from M_3 it is possible for not too
thick crystals to separate the corresponding spectra. In Figure 1
the spectrum from M_2 is denoted by "backscattering spectrum from
the crystal" and from M_3 "backscattering spectrum from the layer."
The random spectrum from the backlayer has a Gaussian shape with
the mean energy $E_r(t)$. The shape of the aligned spectrum from
the backlayer is caused by the different energy loss of the channeled particles penetrating the depth t. It must be noted that
there is a broadening of the spectra due to the multiple scattering
of the particles because of the penetration of the crystal after
the scattering processes. We chose such conditions that the energy
width due to multiple scattering is in the order of the energy
resolution of the detecting system. In the aligned spectrum from
the backlayer three characteristic energy values can be observed:

$E_{c,\bar{\alpha}}(t)$ the mean energy of ions incident in a channeling direction,

$E_{c,\alpha_p}(t)$ = the most probable energy of the channeled particles,

$E_{c,\alpha_b}(t)$ = the energy of the best channeled particles.

$\bar{\alpha}$, α_p and α_b are the corresponding ratios between the channeling and random stopping powers.

Particles scattered at the depth t have the following energies:

random incidence

$$E_r(t) = k\left[E_o - \int_o^{t/\cos\phi_1} NS(E)dt\right] - \int_{t/\cos\phi_2}^o NS(E')dt \qquad (2)$$

aligned incidence

$$E_c(t) = k\left[E_o - \int_o^{t/\cos\phi_1} N\alpha(E)S(E)dt\right] - \int_{t/\cos\phi_2}^o NS(E'')dt \qquad (3)$$

k is the well known collision factor [9] and N the atomic density.

If the energy loss in the crystal is very low, then the mean random values of the energy can be used in equations (2) and (3):

$$\bar{E} = E_o - 1/2 \int_o^{t/\cos\phi_1} NS(E)dt \qquad (4)$$

$$\bar{E}' = k\left[E_o - \int_o^{t/\cos\phi_1} NS(E)dt\right] - 1/2 \int_{t/\cos\phi_2}^o NS(E')dt \qquad (5)$$

$$\overline{E''} \simeq \overline{E'} + (E_c - E_r) \qquad (6)$$

The stopping power for this energy values can be obtained from experimental or theoretical results [10,5,11,12,13]. Under these approximations from Eqs. (2) and (3), $\alpha(\bar{E})$ is determined.

$$\alpha(\bar{E}) = \frac{k \cdot E_o - E_c}{k \cdot E_o - E_r} \left(1 + \frac{S(\bar{E}')}{k \cdot S(\bar{E})} \frac{\cos\phi_1}{\cos\phi_2}\right) - \frac{S(E'')}{k \cdot S(\bar{E})} \frac{\cos\phi_1}{\cos\phi_2} \qquad (7)$$

This formula does not contain the thickness t of the crystal and α can only be calculated from the measured values E_r and E_c.

Typical spectra for the backscattering of 1.4 MeV protons on a 4.0 μm thick silicon crystal with a 700 Å carbon backlayer are shown in Figure 2.

The yield in the channel region 170-250 corresponds to the backscattering spectrum from the silicon crystal for random, <111> and (110) incidence. The backscattering spectra from the carbon backlayer (channel 120-150) are clearly separated. For the <111> axis the energy values E_r, $E_{c,\bar{\alpha}}$, E_{c,α_p} and E_{c,α_b} are plotted.

The random spectra are measured rotating the crystal about the beam direction 5° off the <111> axis.

It is clearly seen that the strong dechanneling in the (110) plane leads to a big random portion in the backscattering spectrum from the carbon layer. By contrast, the low dechanneling along the <111> axis corresponds to a big aligned portion in the backscattering spectrum.

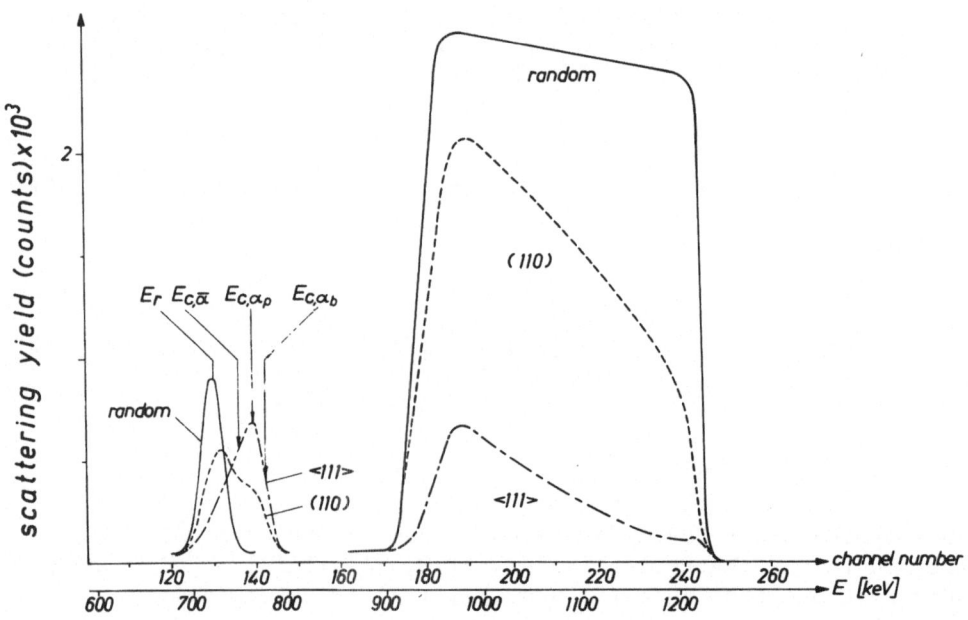

Fig. 2. Typical energy spectra of backscattered protons (E_0=1.4 MeV) incident at random, <111> and (110) direction on a silicon crystal (t=4.0 μm).

Results

Dependence on the incidence angle

An angle of incidence ψ_i (relative to a crystal axis or plane) different from zero caused a change of the initial distribution in transverse energy $E_\perp$. The number of particles with a higher value $E_\perp$ was increased. That corresponds to an increased dechanneling with increasing ψ_i. The increased dechanneling is related to an increased mean stopping power characterized by α.

Dechanneling and energy loss of 1.4 MeV protons were studied in the <111>, <110> and (110) channels of silicon in the range of $0° \leq \psi_i \leq 1°$. Typical energy spectra of protons backscattered from a 4 μm thick crystal near the <111> axis are presented in Figure 3. The figure also presents the depth scale t_r for random incidence. For the backscattering spectra from the backlayer a scale for α, calculated from Eq. (7) is represented.

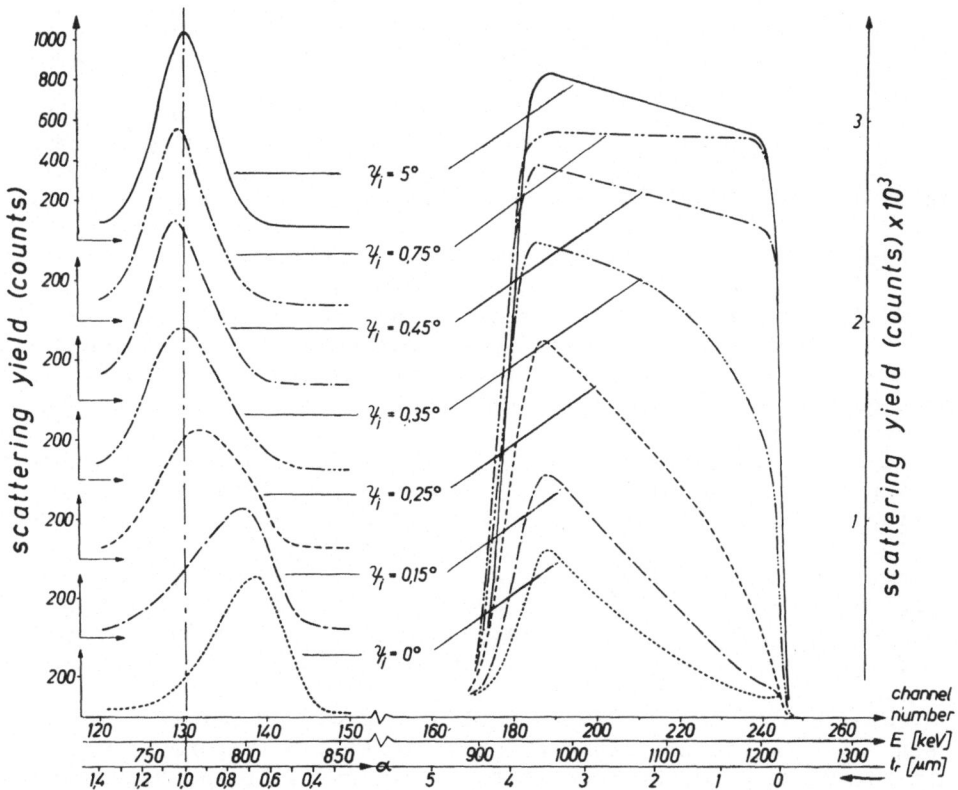

Fig. 3. Energy spectra of backscattered protons (E_0=1.4 MeV) for different angles of incidence ψ_i to the <111> axis.

In the energy spectra from the crystal the increased dechannel-ing with increasing ψ_i is shown. There is also a change in the shape of the spectra. The energy loss of the protons scattered on the backside of the crystal (left edges of the spectra) for inci-dence angles in the region $0° \leq \psi_i \leq 0.35° = \psi_{1/2}$ is lower than in the random case. $\psi_{1/2}$ is the half-width of the channeling dip. For incidence angles larger than $\psi_{1/2}$ the energy loss is somewhat larger than in the random case. For angles larger than 2° the random value is reached.

Interesting are the energy spectra from the carbon backlayer. As was to be expected the channeled fraction of the particles for $\psi_i = 0°$ is very large, but an inclination of 0.15° caused already a broadening of the spectrum due to an increase in the random fraction. A maximum of the energy width is reached at $\psi_i = 0.25°$. At $\psi_i = 0.45°$ the energy width corresponds to the random value. For $\psi_i > \psi_{1/2}$ the mean stopping power is greater than for random incidence.

In accordance with the results from the energy spectra from the crystal the stopping power for the random case is obtained at $\psi_i \geq 2°$.

The results for the <110> channel are exactly the same. Also Gibson et al. [14] described a similar effect for channeling exper-iments using thin single gold crystals.

As can be seen from Fig. 4 for the (110) channel there is no increased stopping power. But the yield due to scattering processes near the surface of the crystal is increased for incidence angles $0.18° \leq \psi_i \leq 0.38°$ as shown in the energy spectra from the crystal.

Dependence from radiation damage

If defects are present in the crystal, due to radiation damage or lattice strain, an increased dechanneling by the defect centers is caused. Connected with this is an increased energy loss for channeled particles penetrating the damaged crystal in comparison to a nonimplanted crystal.

At different temperatures (T_I = 20° and 200°C) silicon crystals were implanted by argon ions. The implantations were made with different doses in order to get a wide range of damage up to amorph-isation. For this, at T_I – 20°C doses from $5 \cdot 10^{13}$ cm^{-2} to $1.5 \cdot 10^{15}$ cm^{-2} were used. The energy of 200 keV corresponds to a pro-jected range of 2350 Å [15]. To avoid dimpling effects the thick-nesses of the crystals must be large enough ($t \geq 3$ μm for a dose of $1.5 \cdot 10^{15}$ cm^{-2} at T_I – 20°C). The annealing of the crystals has carried out at T_A = 400, 600, 800, 1030°C.

Some results for the implantation of a 4.0 μm thick silicon crystal at room temperature are shown in Fig. 5 Figure 5a repre-sents the backscattering spectra from the crystal (also represented are the depth scales t_r and $t_{<111>}$ for random and <111> incidence), and Fig. 5b the backscattering spectra from the carbon layer (also plotted one scale for α).

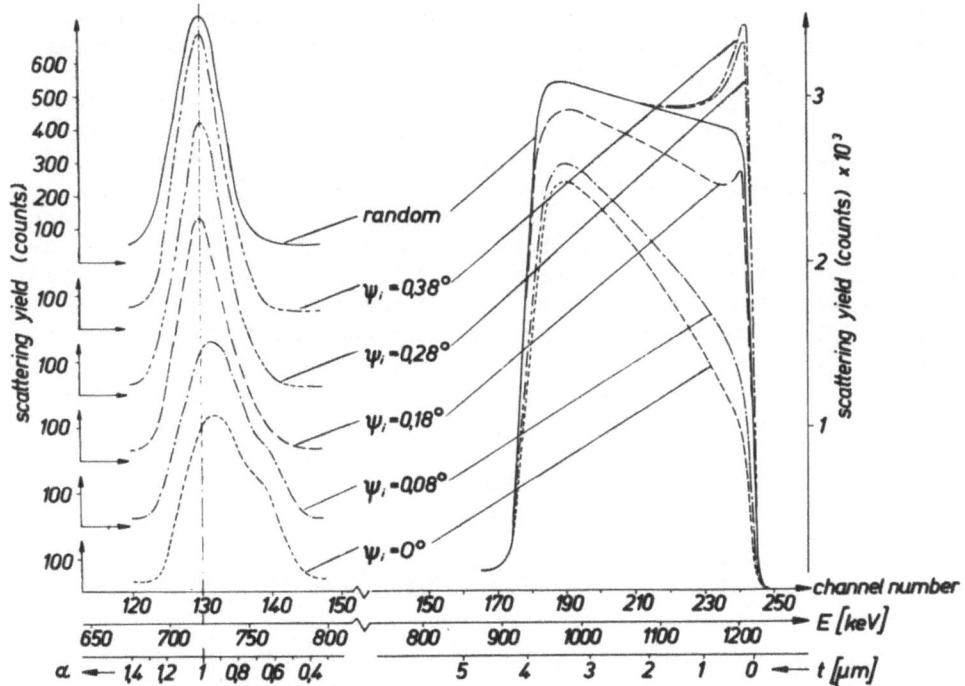

Fig. 4. Energy spectra of backscattered protons (E_o = 1.4 MeV) for different angles of incidence ψ_i to the (110) plane.

In Fig. 5a the peak close to the spectrum edge (channel 230-240, <111>) corresponds to the damaged layer of the crystal. Behind this layer the random fraction of the beam is increased, owing to dechanneling on defect centers. At a dose of $5 \cdot 10^{13}$ cm^{-2} no dechanneling is measured in spite of a low radiation damage. Amorphisation is obtained between $5 \cdot 10^{14}$ and $1 \cdot 10^{15}$ cm^{-2} (not shown in Fig. 5a). At a dose of $2 \cdot 10^{14}$ cm^{-2} annealing at T_A = 400°C considerably reduces damage, but the shape of the aligned spectrum is preserved generally. As we are interested in the influence of damage centers on the dechanneling and stopping power we will principally not concern ourselves here with annealing effects.

In agreement with the results from the backscattering spectra of the crystal the spectra from the backlayer show for a dose of $5 \cdot 10^{13}$ cm^{-2} no measurable broadening (dechanneling) but there is a change at increasing doses in the shape of the energy spectra. For the dose range between $1 \cdot 10^{14}$ and $1.5 \cdot 10^{15}$ cm^{-2} the spectra were unfolded into a random and an "aligned" fraction. The term "aligned" fraction is not quite correct, because here the random fraction due to the minimum yield is contained. It is remarkable that for all spectra the energies E_{c,α_p} for the "aligned" fraction respectively E_r for the random fraction, have the same value irrespective of the amount of radiation damage. Only for the highest damage ($1.5 \cdot 10^{15}$ cm^{-2}) there is a minor change in the shape of the "aligned" fraction of the spectrum. This points to the dechanneling of the "aligned" beam, penetrating the crystal behind the damaged layer. This is caused by the multiple scattering in the amorphized silicon layer leading to an increased beam divergence.

The random fractions in the spectra from the carbon layer due to the dechanneling of the protons by the damage centers (displaced lattice atoms) are proportional to the difference in the scattering yields (yield from the damaged minus yield from the undamaged crystal) direct behind the damage peak in the spectra from the crystal.

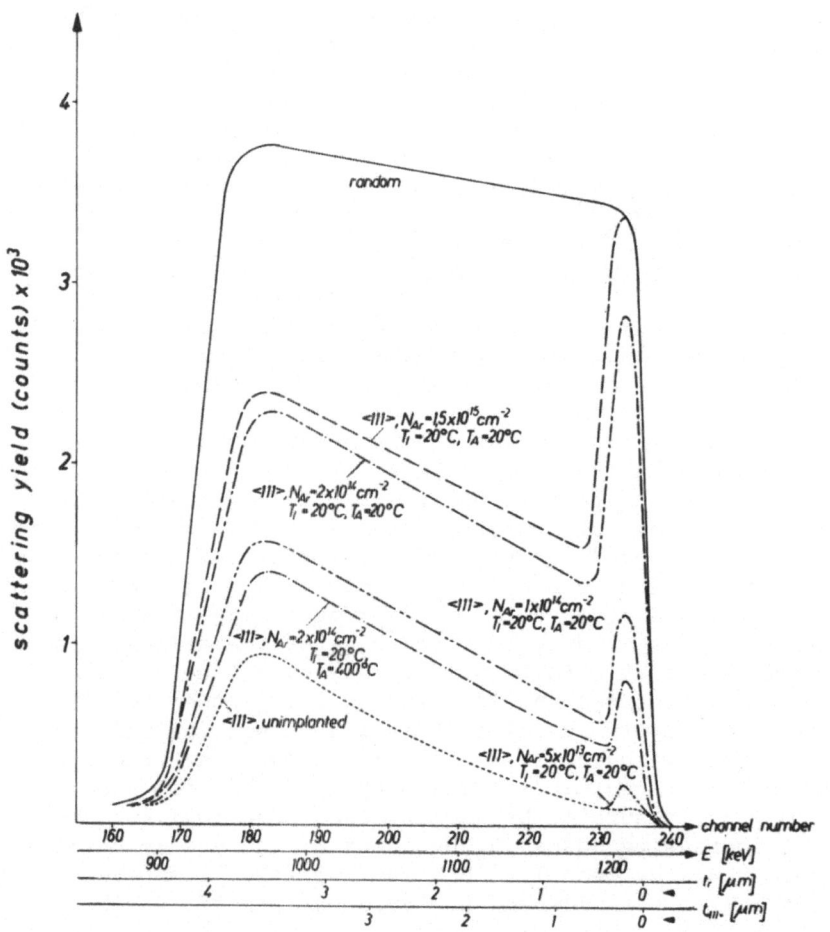

Fig. 5. Energy spectra of backscattered protons (E_o=1.4 MeV) from radiation damaged silicon crystals (impl. temp. T_I = 20°C).
a. Backscattering spectra from the silicon crystal.

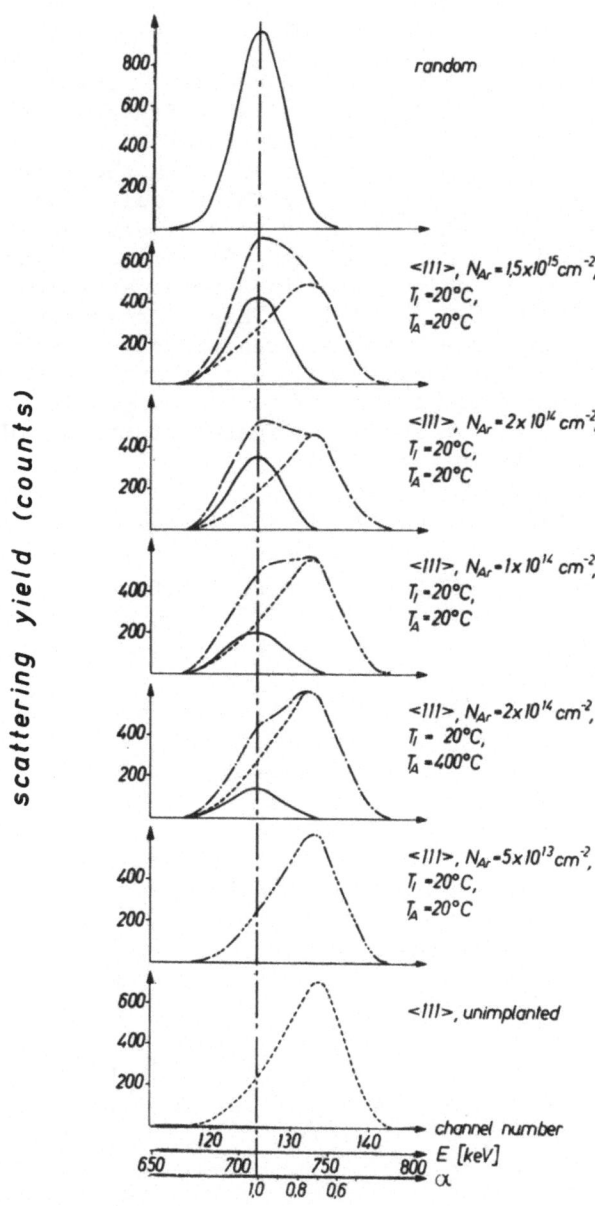

Fig. 5. Energy spectra of backscattered protons (E_o=1.4 MeV) from radiation damaged silicon crystals (impl. temp. T_I = 20°C).
b. Backscattering spectra from the carbon backlayer.

In Figure 6 the backscattering spectra from a silicon crystal implanted at T_I = 200°C (dose 1 · 10^{15} cm^{-2}, t = 2.6 μm) is plotted for different annealing temperatures. In contrast to the room temperature implantation there is no pronounced peak due to direct scattering processes from the damage centers. There is also a considerable annealing during the implantation process, but no annealing can be obtained at temperatures as high as 800°C. The curve denoted by T_A = 800°C is identical with the curve obtained immediately after the implantation without additional temperature process. It is also seen that there is a considerable amount of dechanneling centers in depths behind the damaged layer. The dechanneling of protons is increased in samples annealed at higher temperatures (Fig. 6, T_A = 1030°C). This corresponds to a shift in the mean stopping power towards greater values, seen in the energy spectra from the backlayer. In contrast to the room temperature implantations (Fig. 5a and b) it is not possible to separate the spectra into a random and an "aligned" fraction. With Eq. (7) a mean value for α is obtained:

$$T_A \leq 800°C \; : \; \bar{\alpha} \simeq \alpha_p = 0.75 \pm 0.03$$

$$T_A = 1030°C \; : \; \bar{\alpha} \simeq \alpha_p = 0.80 \pm 0.03$$

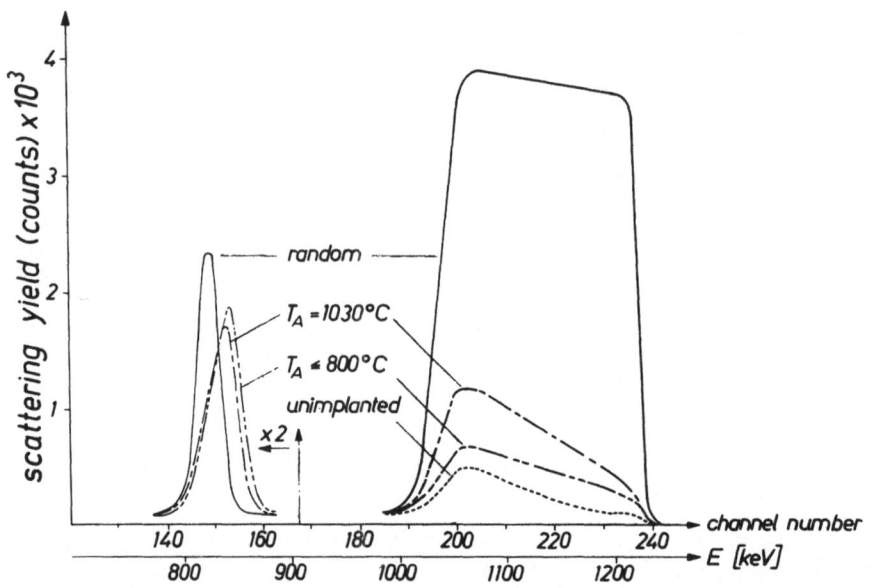

Fig. 6. Energy spectra of backscattered protons (E_0=1.4 MeV) from unimplanted, radiation damaged (T_I = 200°C, N_{Ar} = 1 · 10^{15} ions cm^{-2}) and annealed silicon crystals (t = 2.6 μm).

As shown by optical measurements, the effect cannot be explained
by dimpling of the thin silicon layers. We think that throughout
the crystal there exists a great deal of lattice disorder (strain,
argon bubbles, dislocations, etc.). Electron-microscopic investigations support this assumption.

Discussion

   Dechanneling

      Explanation of the experimental results on dechanneling are
usually based on the theoretical model of Lindhard [1], who calculated the change in the transverse energy distribution by two processes, namely the scattering by nuclei as the result of thermal
vibrations and the scattering by electrons. According to this
theory, several authors [16,17,18,19,20] have calculated the dechanneled fraction of the aligned beam for some crystals (protons
and He^+ ions in Si, W and Ge).
      According to Lindhard's theory the distribution $g(E, E_\perp, t)$
has to be calculated for the channeled fraction of the beam. Neglecting the energy loss of the particles in the case of axial
dechanneling in a perfect crystal, $g(E, E_\perp, t)$ can be described by
a diffusion-like equation:

$$\frac{\partial g(E_\perp, t)}{\partial t} = \frac{\partial}{\partial E_\perp}\left[D(E_\perp)\frac{\partial g(E_\perp, t)}{\partial E_\perp} \right] \tag{8}$$

t: coordinate in the direction of the string.

The diffusion function can be composed additively from a nuclear
and an electronic part:

$$D(E_\perp) = D(E_\perp)_n + D(E_\perp)_{el} \tag{9}$$

Differences between the theoretical and experimental results for the
dechanneling fraction for large traveling paths imply that the nuclear part described theoretically is not exact enough. We tried
an improvement of $D(E_\perp)_n$ in a twofold way:
      1.   Using a more realistic string potential.
      2.   Taking into account the second order contributions.
For calculating the string potential we proceed from the electro-
statical interaction potential between two stiff atoms:

$$V(r) = \frac{Z_1 Z_2 e^2}{r}\, \phi_{12}(r) \tag{10}$$

with the screening function

$$\phi_{12}(r) = \phi_2(r) - \frac{1}{2} \int_0^\infty dr_1 \phi_2(r_1) [\phi'_1(r+r_1) - \phi'_1(|r-r_1|)]$$

(11)

where r is the distance between the atoms (projectile and lattice atom), ϕ_1 and ϕ_2 are the single screening functions of the atoms, calculated by using the Slater model. The string potential is obtained by integration of the atomic potentials along the string:

$$U(r) = \frac{Z_1 Z_2 e^2}{d} \int_{-\infty}^{+\infty} \frac{\phi_{12}(\sqrt{r^2 + t^2})}{\sqrt{r^2 + t^2}} dt$$

(12)

To obtain an analytical expression we try to approximate the string potential, calculated from Eq. (12) using Lindhard's string potential

$$U(r) = \frac{E_o \psi_1^2}{2} \log (1 + \frac{c^2 a^2}{r^2})$$

(13)

ψ_1 is Lindhard's critical angle and a the screening radius. Taking c^2 as a parameter a good agreement between the string potentials after Eq. (12) and Eq. (13) in the range of interest was obtained in the case of H → Si for $c^2 = 1.05$ (see Fig. 7).

If the potential (13) is used and the second order term of the force fluctuation (see Lindhard [1] Eq. 4.10) is not neglected the phase space averaging provides:

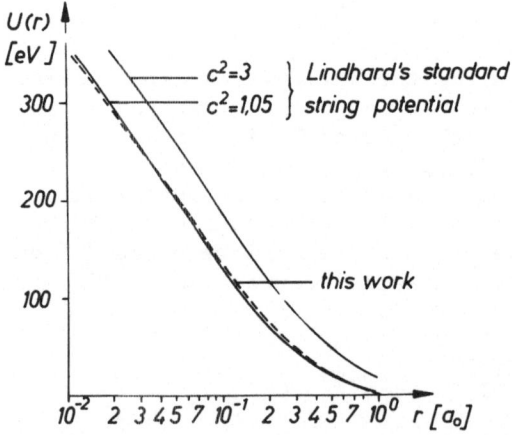

Fig. 7. String potential for protons in <110> direction of silicon.

$$\left(\frac{d\bar{E}_\perp}{dt}\right)_n = \psi_1^2 \pi N d Z_1 Z_2 e^2 \{ \frac{\rho_\perp^2}{c^2 a^2} [(\frac{1}{2} e^{\varepsilon_\perp + \varepsilon_{\perp,o}} + \frac{1}{3})(1 - e^{-(\varepsilon_\perp + \varepsilon_{\perp,o})})^3]$$

$$+ \frac{1}{2}(\frac{\rho_\perp^2}{c^2 a^2})^2 [e^{2(\varepsilon_\perp + \varepsilon_{\perp,o})} - 2 e^{\varepsilon_\perp + \varepsilon_{\perp,o}}$$

$$+ \frac{14}{5} - 4 e^{-(\varepsilon_\perp + \varepsilon_{\perp,o})}$$

$$- 7 e^{-2(\varepsilon_\perp + \varepsilon_{\perp,o})} + 26 e^{-3(\varepsilon_\perp + \varepsilon_{\perp,o})}$$

$$- 24 e^{-4(\varepsilon_\perp + \varepsilon_{\perp,o})} + \frac{36}{5} e^{-5(\varepsilon_\perp + \varepsilon_{\perp,o})}] \} \qquad (14)$$

where $\rho_\perp^2$: mean square amplitude of the thermal vibrations perpendicular to the string;

$$\varepsilon_\perp = 2 E_\perp / (E_o \psi_1^2);$$

$$\varepsilon_{\perp\rho} = 2 U(r_o)/(E_o \psi_1^2), \quad r_o \text{ string radius}).$$

As follows from Eq. (14) $(\frac{d\bar{E}_\perp}{dt})_n$ is not proportional to $\rho_\perp^2$. The diffusion function is obtained by integration of Eq. (14) (and taking also in consideration the electronic part after [16]):

$$D(E_\perp) = \int_o^{E_\perp} \frac{d\bar{E}_\perp}{dt} (E'_\perp) \, dE'_\perp \qquad (15)$$

Introducing the diffusion function in Eq. (8) the distribution of the transverse energy was calculated numerically. With the distribution of the transverse energy the dechanneled fraction of the aligned beam was obtained, following Bonderup and coworkers [16].

The results for the dechanneling of protons along the <110> and <111> axis in silicon crystals are presented in Figs. 8 and 9. Here are plotted the dechanneled fractions calculated with the first and second order contributions of the nuclear part for two string potentials ($c^2 = 3$ and $c^2 = 1.05$) and experimental values from our measurements (see Results), and from Foti and coworkers [20] respectively.

From Fig. 8 and Fig. 9 it is obvious that the terms from the second order ($\sim \rho_\perp^4$) are important especially if a strong screened string potential is used. The best agreement with the experimental values is obtained for the stronger screened potential ($c^2 = 1.05$)

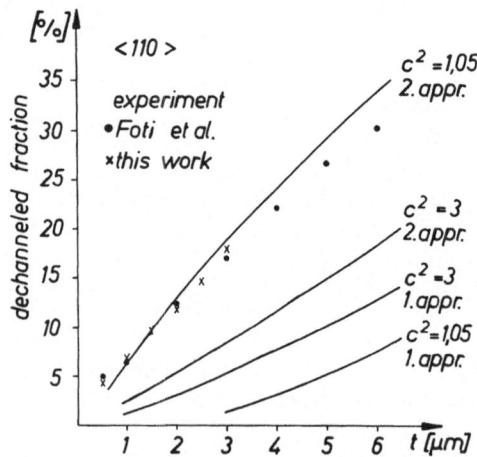

Fig. 8. Dependence of the dechanneled fraction for 1.4 MeV protons along the <110> axis of silicon.

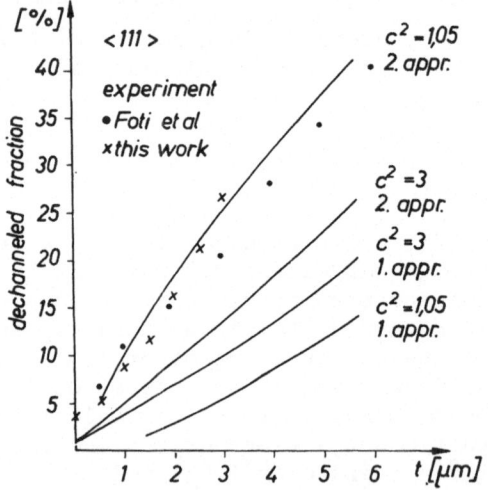

Fig 9. Dependence of the dechanneled fraction for 1.4 MeV protons along the <111> axis of silicon.

taking into consideration nuclear contributions of the second order.
But particularly for larger traveling paths the theoretical values
are too large. Further the curvature of the theoretical curve for
lower depth is contrary to the trend of the experimental values.
It will be necessary to use a more precise string potential which
takes into account the ionization of the projectile. Besides the
assumption of a fixed value for the transverse energy, at which the
transition from the aligned in the random beam occurs, is only an
approximation. Also, we did not take into consideration amorphous
surface layers, the beam divergence, impurities and crystal imper-
fections. All these effects will tend to increase the dechanneling
of the particles and therefore deteriorate the agreement between
our theoretical and experimental results. To attain a more precise
potential experiments will be necessary which are sensitive to the
screening function. As a next step, the theoretical treatment of
the dechanneling as a function of the incidence angle ψ_i (in the
region $\psi_i \leq \psi_1$) would be of advantage. Further it would be important to calculate the influence of disordered atoms on the dechanneling. We will continue our work in this direction.

Energy loss

    The ratio α_p between the most probable stopping powers for
protons incident along the <110> and <111> axes and the random
stopping power are shown as a function of energy in Fig. 10. The
agreement between results taken with samples of different thickness
(1-4 µm) and different backlayer (carbon, LiF, gold) is good. The
scatter in the points is not more than ±5%. Also shown in Fig. 10

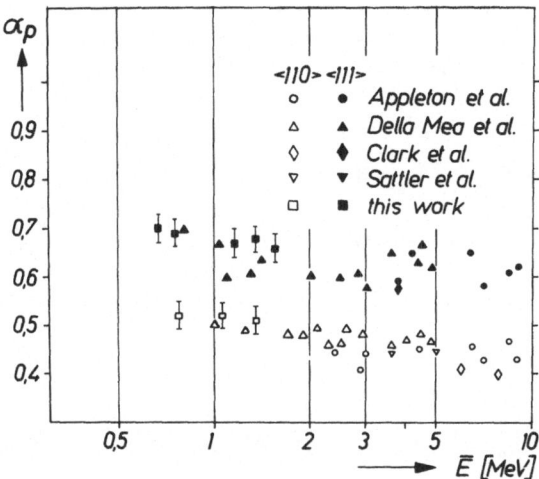

Fig. 10. Ratios between channeled and random stopping powers for
protons in silicon.

are the results of other authors [2,3,4,6]. Contrary to Della Mea
and coworkers [6] we found at the energy region about 1 MeV only a
weak dependence of the α_p values on the energy. On the other hand,
our values are 5% larger on an average.

The dependence of the ratio $\bar{\alpha}$ of the mean stopping powers for
protons incident near the <110> and <111> axis and the random stopping power is plotted as a function of the angle of incidence ψ_i
to this axis in Fig. 11 (E_p = 1.4 MeV, t = 4.0 μm). For both of
the axes $\bar{\alpha}$ increases almost linearly with ψ_i up to $\psi_{1/2}$. For
$\psi_i = \psi_{1/2}$ $\bar{\alpha}$ = 1 is obtained ($\psi_{1/2}$ = 0.35° for the <111> and $\psi_{1/2}$=
0.38° for the <110> axis). For $\psi_i > \psi_{1/2}$ a maximum is obtained which
has a value $\bar{\alpha}$ = 1.03 for both of the axes. Although this lies within the limit of errors this effect was measured for all samples.
By measurements on crystals with different thicknesses we could
establish that the increased stopping power appeared on the surface
of the crystal only. The increased stopping power at the surface
of the crystal for protons with an angle of incidence somewhat
larger than $\psi_{1/2}$ is due to "quasi channeling" effects [21]. After
the entrance in the crystal those protons move in regions near the
string where a higher electron density exists. As there is also
a greater probability of single scattering, at a sufficient depth
the protons are steered into the random beam.

For $0 \leq \psi_i \leq \psi_{1/2}$ the ratio $\bar{\alpha}$ between the mean stopping power and
the random stopping power can be approximated by the following ex-
pression:

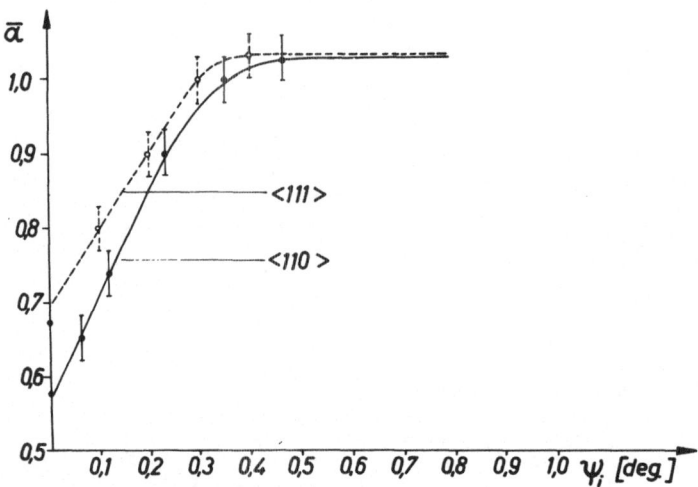

Fig. 11. Ratio between mean stopping power and random stopping power
of 1.4 MeV protons in dependence of angle of incidence to <110> and
<111> axis of 4.0 μm silicon crystals.

$$\bar{\alpha}(t,\psi_i) = \bar{\alpha}(t,0) + [1- \bar{\alpha}(t,0)] \frac{\psi_i}{\psi_{1/2}} \qquad (16)$$

$\bar{\alpha}(t,0)$ can be calculated after Eq. (7) from the results of energy loss measurements.

From the energy spectra of protons scattered by radiation damaged (ionimplanted) crystals (Fig. 5) it can be seen that the beam which has penetrated the damaged layer can be divided in two parts; an aligned component and a random component. For thin damaged layers the mean energy loss ratio $\bar{\alpha}$ behind the layer can be estimated by:

$$\bar{\alpha}(t,N') = \chi_R(t,N') + [1 - \chi_R(t,N')]\alpha_p \qquad (17)$$

$\chi_R(t,N')$ is the normalized random fraction of the beam, which has penetrated the layer containing a defect density N', α_p is the ratio between the most probable stopping power of a crystal without defects and the random stopping power. Eq. (17) is applicable only for the cases in which the total amount of defect centers is small. For heavy damaged crystals the beam spreading due to multiple scattering process must be taken into account.

To determine depth profiles of defect centers in damaged crystals the depth dependence of $\bar{\alpha}$ in the damaged layer must be known. We proceed from the conditions represented in Fig. 12. Plotted are the relative to random spectra normalized spectra. $\chi_1(t)$ is

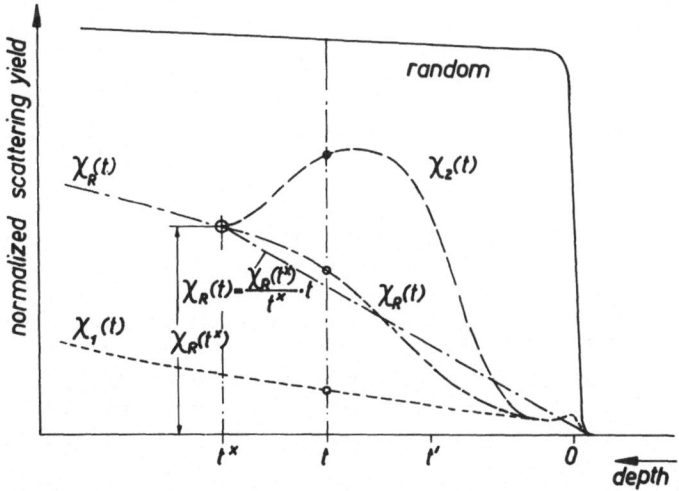

Fig. 12. Schematic diagram of the random and aligned yields of backscattered ions of undamaged and damaged crystals for calculating the mean stopping power.

the aligned spectrum from an undamaged, $\chi_2(t)$ the aligned spectrum from a damaged and $\chi_R(t)$ the dechanneled (random) fraction of the analyzing beam.

We consider the mean stopping power ratio for particles penetrated to the depth t. There are the following contributions:

1. The mean stopping power ratio for particles channeled up to the depth t is $\bar{\alpha}_d(t,N')$, the fraction $(1-\chi_R(t))$. $\bar{\alpha}_d(t,N')$ corresponds in a first approximation to α_p, generally the energy loss due to the defects must be taken into account.

2. The mean stopping power ratio for particles channeled up to a depth t', dechanneled at t' by defect centers, vibrational lattice atoms or impurities and then take a random path to t is

$$\bar{\alpha}_d(t,t',N') = \bar{\alpha}_d(t,N') \cdot \frac{t'}{t} + \frac{t-t'}{t} \tag{18}$$

The fraction dechanneled at t' is determined by $(d\chi_R(t)/dt)_{t'}$. The total of the mean stopping power ratio is obtained by integration over t':

$$\int_0^t (d\chi_R(t)/dt)_{t'} \, \bar{\alpha}_d(t,t',N') \, dt' \ .$$

Thus we obtain for the total mean stopping power ratio:

$$\bar{\alpha}(t,N') = [1-\chi_R(t)] \, \bar{\alpha}_d(t,N') + \frac{1}{t} \int_0^t (d\chi_R(t)/dt)_{t'}$$

$$[t'\bar{\alpha}_d(t,N') + (t-t')]dt' \tag{19}$$

According to Ziegler [22] $\bar{\alpha}_d(t,N')$ can be calculated from the defect density N':

$$\bar{\alpha}_d(t,N') = \alpha_p + (1-\alpha_p) \, \frac{N'(t)}{N_o} \tag{20}$$

N'(t) is the defect density in the range from the surface up to the depth t.

Calculating $\bar{\alpha}(t,N')$ the depth dependence of $\chi_R(t)$ and N'(t) must be known. This can be done by a similar iterative procedure proposed by Westmoreland and coworkers [23] for the analysis of disorder distributions.

For thin damaged layers, often obtained by ion implantation, $\bar{\alpha}(t,N')$ can be calculated for a depth in the range $0 \le t \le t^x$ approximating $\chi_R(t)$ by a straight line (see Fig. 12).

With this Eq. (19) is reduced to:

$$\bar{\alpha}(t,N') = \alpha_p + (1-\alpha_p)\, \frac{N'(t)}{N_o} + \frac{\chi_R(t)}{2}\,(1-\alpha_p)\,(1-\frac{N'(t)}{N_o}) \quad (21)$$

With this formula we have calculated $\bar{\alpha}(t,N')$ using the values $N'(t)$, $\chi_1(t)$ and $\chi_R(t)$ from the energy spectra of the crystals (Fig. 5a and Fig. 6, on the right) and compared it with the measured values (Fig. 5b and Fig. 6, on the left). The results are represented in Table 1. The agreement indicates that reasonable values of $\bar{\alpha}(t,N')$ for implanted crystals may be calculated using Eq. (21).

Conclusion

From the energy of backscattered ions it is possible to obtain the depth t of the scattering centers. Usually the beam is incident in a channeling direction and the scattered ions are measured in a random direction. In this case one starts from a relation like Eq. (3) and makes use of mean random values for the energy of the ingoing and outgoing ions (Eqs. (4) and (6)). For that the depth scale may be calculated from the relation

$$t = \frac{kE_o - E(t)}{N\left[\dfrac{k\bar{\alpha}(t,N')S(\bar{E})}{\cos\phi_1} + \dfrac{S(\bar{E}'')}{\cos\phi_2}\right]} \quad (22)$$

where $\bar{\alpha}(t,N')$ is the ratio between the mean energy loss of the ions scattered at the depth t and the random energy loss.

In good single crystals the energy loss in channeling directions is lower for ions which will penetrate the crystal in comparison with those ions which will be backscattered. This is caused by the fact that primary the most poorly channeled ions will contribute significantly to the backscattered yield. By the simultaneous measurement of $\bar{\alpha}$ for the penetrating ions and for the backscattered ions (from the energy differences of the edges of the silicon spectra) we found $\bar{\alpha}_{<111>} = 0.69 \pm 0.03$; $\bar{\alpha}_{<110>} = 0.52 \pm 0.03$; $\bar{\alpha}_{(110)} = 0.87 \pm 0.03$ for the penetrating ions and $\bar{\alpha}_{<111>} = 0.96 \pm 0.02$; $\bar{\alpha}_{<110>} = 0.95 \pm 0.02$; $\bar{\alpha}_{(110)} = 0.98 \pm 0.02$ for the backscattered ions. This suggests that $\bar{\alpha} = 1$ is a good approximation for the calculation of the depth scale for single crystals containing only a small defect concentration.

The same approximation is also practicable for incidence angles somewhat different to channels. In <111> direction we found in the range $0° \leq \psi_i \leq 0.75°$ for the penetrating ions $\bar{\alpha} = 0.69...1.03$ and for the backscattered ions $\bar{\alpha} = 0.96...1.03$.

The situation is quite different for radiation damaged crystals. The backscattering yield from the depth t consists of two parts, firstly from ions channeled to the depth t and backscattered there directly by lattice defects or impurities and secondly from ions

Table 1

Fig.	Conditions of Implantation	$\bar{\alpha}(t,N')_{meas.}$	$\bar{\alpha}(t,N')_{calc.}$
5	nonimplanted	0.73 ± 0.03	0.73
5	$N_{Ar} = 5 \times 10^{13} cm^{-2}$, $T_I = 20°C$, $T_A = 20°C$	0.75 ± 0.03	0.74
5	$N_{Ar} = 2 \times 10^{14} cm^{-2}$, $T_I = 20°C$, $T_A = 400°C$	0.81 ± 0.03	0.79
5	$N_{Ar} = 1 \times 10^{14} cm^{-2}$, $T_I = 20°C$, $T_A = 20°C$	0.85 ± 0.03	0.84
5	$N_{Ar} = 2 \times 10^{14} cm^{-2}$, $T_I = 20°C$, $T_A = 20°C$	0.88 ± 0.03	0.89
5	$N_{Ar} = 1.5 \times 10^{15} cm^{-2}$, $T_I = 20°C$, $T_A = 20°C$	0.90 ± 0.03	0.89
5	random	1.00 ± 0.03	1.00
6	$N_{Ar} = 1 \times 10^{15} cm^{-2}$, $T_I = 200°C$, $T_A = 20°C$	0.75 ± 0.03	0.74
6	$N_{Ar} = 1 \times 10^{15} cm^{-2}$, $T_I = 200°C$, $T_A = 800°C$	0.75 ± 0.03	0.74
6	$N_{Ar} = 1 \times 10^{15} cm^{-2}$, $T_I = 200°C$, $T_A = 1030°C$	0.80 ± 0.03	0.77

scattered between surface and t which are already in the random
beam. The ratio $\bar{\alpha}(t,N')$ between channeling and random stopping
powers depends strongly from the depth and the defect concentration.
It must be calculated by Eq. (19) or Eq. (21) as an approximation.
For crystals with a small amount of damage $\bar{\alpha}(t,N')$ can be approximated by α_p how it is proposed by Eisen et al. [5]. For crystals
with a very high amount of damage $\bar{\alpha} \simeq 1$ is approached.

Acknowledgments

We wish to thank Dr. Krause for fabrication of epitaxial silicon wafers. We are grateful to Ch. Dierse for careful preparation
of thin silicon crystals. The technical assistance of H. Treff,
W. Ahnefeld, W. Trippensee and G. Lenk are gratefully acknowledged.
We are indebted to K. Hehl and K. Gärtner for useful discussions.

References

[1] J. Lindhard, Mat. Fys. Medd. Dan. Vid. Selsk. 34, No. 14 (1965).
[2] B. R. Appleton, C. Erginsoy, W. M. Gibson, Phys. Rev. 161, 330
     (1967).
[3] A. R. Sattler, G. Dearnalay, Phys. Rev. Letters 15, 59 (1965).
[4] G. J. Clark, D. V. Morgan, J. M. Poate, Atomic Collision
     Phenomena in Solids (eds., D. W. Palmer, M. W. Thompson,
     P. D. Townsend) North-Holland Publ. Co., Amsterdam-London
     1970, p. 388 ff.
[5] F. H. Eisen, G. J. Clark, J. Bøttiger, J. M. Poate, Rad. Eff.
     13, 93 (1972).
[6] G. Della Mea, A. V. Drigo, S. Lo Russo, P. Mazzoldi, G. G.
     Bentini, Rad. Eff. 13, 115 (1972).
[7] G. Götz, K. D. Klinge, Exp. Technik d. Phys. (to be printed).
[8] R. L. Meek, W. M. Gibson, R. H. Brown, Nucl. Instr. Meth. 94,
     435 (1971).
[9] E. Bøgh, Can. J. Phys. 46, 653 (1968).
[10] C. Williamson, J. Pl. Boujot, CEA-2189 (1962).
[11] J. F. Ziegler, M. H. Brodsky, J. Appl. Phys. 44, 188 (1973).
[12] D. A. Thompson, W. D. Mackintosh, J. Appl. Phys. 42, 3969 (1971).
[13] K. Gärtner, Dissertation, Universität Jena (1972).
[14] W. M. Gibson, J. B. Rasmussen, P. Ambrosius-Olesen, C. J.
     Andreen, Can. J. Phys. 46, 551 (1968).
[15] H. Grosser, M. Kolditz, Diplomarbeit, Universität Jena (1973).
[16] E. Bonderup, H. Esbensen, J. U. Andersen, H. E. Schiøtt, Rad.
     Eff. 12, 261 (1972).
[17] K. Björkqvist, B. Cartling, B. Domeij, Rad. Eff. 12, 267 (1972).
[18] A. Fontell, E. Arminen, E. Leminen, Rad. Eff. 12, 255 (1972).
[19] S. U. Campisano, G. Foti, F. Grasso, M. Lo Savio, E. Rimini,
     Rad. Eff. 13, 157 (1972).

[20] G. Foti, F. Grasso, R. Quattrocchi, E. Rimini, Phys. Rev. B3,
     2169 (1971).
[21] L. T. Chadderton, F. G. Krajenbrink, Atomic Collision Phenomena
     in Solids (eds., D. W. Palmer, M. W. Thompson, P. D. Townsend)
     North-Holland Publ. Co., Amsterdam-London (1970) p. 456 ff.
[22] J. F. Ziegler, J. Appl. Phys. 43, 2973 (1972).
[23] J. E. Westmoreland, J. W. Mayer, F. H. Eisen, B. Welch, Rad.
     Eff. 6, 161 (1970).

FLUX PEAKING, DECHANNELING CROSS SECTION, AND DETECTION PROBABILITY IN CHANNELING AND BLOCKING EXPERIMENTS FROM COMPUTER SIMULATIONS AND ANALYTICAL MODELS

H. J. PABST

School of Mathematical and Physical Sciences
University of Sussex
Falmer, Brighton BN1 9QH, Sussex, England

ABSTRACT

Results and interpretations are presented of computer simulation calculations of the channeling and the blocking effect for light energetic ions in the <110> channel of silicon. Oscillations of hyperchanneled ions were detected in the case of channeling. The rule of reversibility is shown to apply for channeling and blocking when equating initial beam divergence and detector acceptance angle. The dechanneling cross section by defects is shown not to be constant over the channel area. Applicability of analytical models and implications for the analysis of experiments are discussed.

Introduction

The effect of correlated glancing collisions of charged particles entering close to low index directions of single crystal materials was suggested by computer calculation simulation [1]. Since then many studies of this kind have been undertaken [2-8], yielding in particular the discoveries of the flux peaking effect [9] (bunching up of ions in channel center) corroborated experimentally [10,13], corroboration of the energy peaking effect [8] (correlation between the ions transverse energy and electronic densities accessible to the ion resulting in variations in energy loss) as can be inferred from transmission experiments (for instance [11]), and variations in nuclear encounter probability [3] as observed in backscattering channeling experiments. Few of these effects are amenable to exact analytical treatment, and in the following the classical channeling theory of Lindhard [12] and the

717

ensuing treatments of flux peaking in statistical equilibrium, as
developed by Andersen et al. [13] and van Vliet [14], will be taken
into account.

The present study looks in more detail into the oscillations
of flux peaking, the applicability of the rule of reversibility in
the case of blocking and presents evidence for scattering center
location dependence of the single scattering dechanneling cross
section, including its application to experiments. The computer
programs employed are adaptions from those developed by Morgan and
van Vliet [2]. They are based on binary collisions in momentum
approximation for the Thomas-Fermi-Moliere potential [2]. Impact
parameter independent electronic stopping and electronic multiple
scattering (the latter proportional to the former) are incorporated
following the suggestion by van Vliet [14]. Thermal vibrations
(triangular distributions as approximations to gaussians [2]) of the
target atoms in two dimensions perpendicular to the beam direction
are taken into account. An isotropic initial beam divergence simu-
lates the effects of geometrical beam collimation and oxide surface
layers.

Channeling

The ions were started just outside the crystal surface distributed over a representative area, comprising one half of the channel
cross section. This area was divided into about 120 squares and
one ion was started from each square choosing the coordinates within the square randomly. Similarly the initial ion directions were
chosen randomly from an isotropic cone of predetermined radial angle (parameter COLL). All channeling simulations were carried out
to 4000 Å penetration depth, and the ions chosen were 300 keV H^+
and 275 keV, 700 keV and 2.3 MeV He^+. Stopping power data of Eisen
and Eisen et al. [11] were employed assuming 50% of the electrons
to contribute to the multiple scattering [2, 12]. The thermal vibrations of the target atoms corresponded to a temperature of 150°
K.

A selection of graphs of the ion flux at channel center (CC,
solid lines) and the tetrahedral interstitial position (T, dotted
lines; for position of T see Fig. 6) as function of penetration
depth are collected in Fig. 1. Graphs 1a and 1b show the ion flux
at CC and T for 300 keV H^+ and 275 keV He^+ for an initial beam divergence of 0.1 ψ_1 (incidence angle causing additional transverse
energy of up to 1% of Lindhard row potential), where ψ_1 is the
Lindhard critical angle according to Morgan and van Vliet [15] of
1.0° and 1.6°, respectively. Two systems of peaks can be recognized, one denoted by capital letters (both CC and T exhibit maxima,
oscillation in (001) plane), the other denoted by small letters
(maximum at CC coincides with minimum at T, oscillation in (1$\bar{1}$0)
plane). The corresponding wavelengths are sampled in Table 1,
which also contains the two cases 0.7 MeV and 2.3 MeV He^+, Figs. 1e

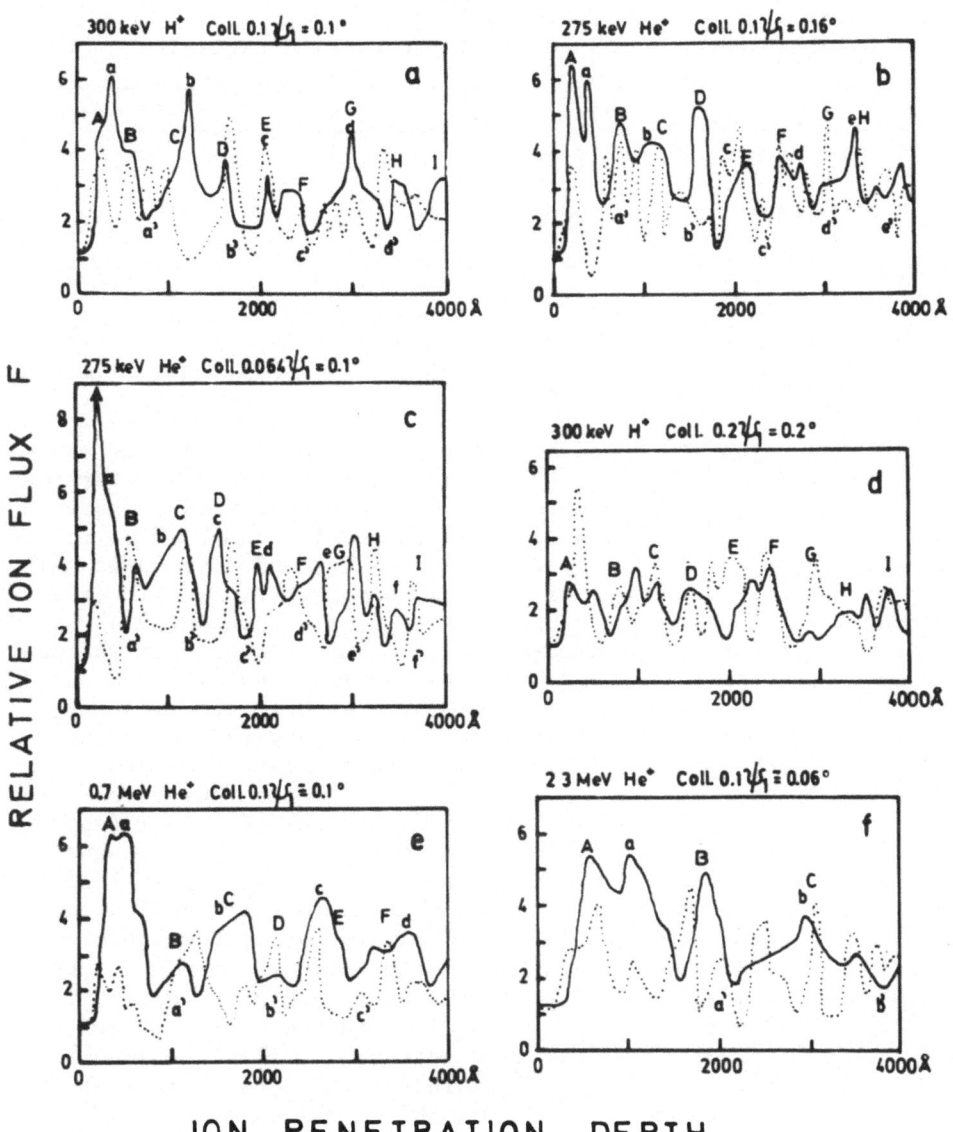

RELATIVE ION FLUX F

ION PENETRATION DEPTH

Fig. 1. Channeling ion flux relative to uniform random in the <110>
channel of silicon at the channel center or potential minimum CC
(solid lines) and at the tetrahedral interstitial position T (dotted
lines) as function of ion penetration.

a)   300 keV H$^+$, COLL = $0.10\psi_1$ (= 0.10°)
b)   275 keV He$^+$, COLL = $0.10\psi_1$ (= 0.16°)
c)   275 keV He$^+$, COLL = $0.06\psi_1$ (= 0.10°)
d)   300 keV H$^+$, COLL = $0.20\psi_1$ (= 0.20°)
e)   700 keV He$^+$, COLL = $0.10\psi_1$ (= 0.10°)
f)   2.3 MeV He$^+$, COLL = $0.10\psi_1$ (= 0.06°)

Table 1
Oscillation wavelengths of best channeled particles

	300 keV H$^+$	275 keV He$^+$	700 keV He$^+$	2.3 MeV He$^+$
λ_A/Å	960	850	1230	2310
λ_a/Å	1650	1500	2200	3850
$K_A \cdot Å^2$/eV	12.9	15.1	18.3	17.0
$K_a \cdot Å^2$/eV	4.35	4.83	5.71	6.12

and 1f. The height and especially the separation of peaks as shown in Fig. 1 are fairly independent of changes in the initial starting conditions brought about by variation of the random number sequence employed, that is without changing the external parameters of initial beam direction and collimation, ion energy and electronic multiple scattering. Reduction of the beam divergence, Fig. 1c (maximum additional transverse energy from angle of incidence 0.4% of Lindhard row potential), increases the initial peak height, whereas an increase in beam divergence, Fig. 1d (maximum additional transverse energy 4% of Lindhard row potential), reduced it. This suggests the conclusion that the initial peak is largely due to the presence of hyperchanneled ions of sufficiently low transverse energy to participate in the harmonic bowl oscillation described above: the area surrounded by the potential line through T (approx. 2% of Lindhard row potential) is approximately 1/8 of the channel area, the presently used sampling area in channel center is 1/35 of the channel area, hence the bunching up of the hyperchanneled ions for which the potential line through T limits the accessible area will produce initial and subsequent peaks of flux 35/8 = 4.4 superimposed on the fairly even (with depth) flux peaking effect of ions with higher transverse energy, not subject to the harmonic bowl approximation, hence of varying wavelengths. Similarly channeling in the <111> direction in silicon exhibits two distinct initial flux peaks for the potential minimum [16]. This case and the general effects of impact parameter dependent electronic stopping power will be dealt with in a subsequent publication.

From the above it is deduced that a good proportion of hyperchanneled ions in wide open channels undergo quite regular oscillations and transmission experiments on channels deviating from circular symmetry may show effects due to predilected planes of oscillation, as here for silicon <110> for the planes (001), shorter wavelength, and (1$\bar{1}$0), longer wavelength.

In another series of calculations concerning 300 keV H$^+$ in Si <110> the initial beam divergence was varied between 0.1 ψ_1 and 1.0 ψ_1 (in steps of 0.1 ψ_1) radial angle of the isotropic cone to test

the rule of reversibility and will therefore be presented and discussed in conjunction with the blocking simulations.

Blocking

     For the blocking computer simulations the ions were started
inside the crystal from several points in the channel area <110> of
silicon, subject to similar random thermal displacements as the
crystal atoms. The initial ion directions were distributed isotrop-
ically in a cone of radial angle 1.0 ψ_1 centered on the <110> direction, compare Fig. 2 (CC 0-11 Å). The starting points chosen were
8 points at distances 0.00, 0.05, 0.10, 0.15, 0.20, 0.25 (labelled
R00 to R25), 0.50 (T) and 0.75 (CC) L.U. (1.0 L.U. $\hat{=}$ 2.714 Å) from
the <110> row in the <001> direction, i.e. in the ($1\bar{1}0$) plane, and
the last two positions correspond to the tetrahedral interstitial
position, and the channel center or potential minimum, respectively
(see also Fig. 6). In addition the two bond centered positions
(BC1 and BC2, centers of sides of hexagon, see also Fig. 6) were
employed as starting points. All other parameters are as in the
channeling simulations of the preceding section.
     The ion emergence pattern was obtained from the recording of
two orthogonal components of the transverse angle every 3.8 Å for
each of 200 ion trajectories, averaged over trajectory distances
from 2000 - 3000 Å (equivalent to 260 identical crystal surfaces).
This procedure is thought to be permissible because of the lack of
defined trajectories in axial channeling (but for hyperchanneled
ions as described in the previous section) and the random distur-
bances from thermal vibrations and electronic multiple scattering.
     Figure 2 shows these ion emergence patterns for 300 keV H$^+$
through equiflux lines (arbitrary units) in the sequence CC, T, R00
to R25, BC1, BC2 and finally the compound bond center BC (numerically BC = (BC1 + BC2)/2). It can be seen from Fig. 2 by comparing
the position of the dark shaded areas with the ψ_1 circle that the
highest probability of emergence angle correlates well with the
initial average transverse energy. In addition the effect of the
planes intersecting in the <110> channel is clearly discernible,
in particular the "wide" plane (111) (diagonal line): for ions
started with low initial transverse energy through high flux in
the planes (CC, T, BC2), and for ions started in areas of high initial transverse energy (R00 to R15) through high flux in the
shoulder of the planes.
     These emergence patterns were also averaged for circular sym-
metry and normalized to the number of ions originally started into
those circular areas, i.e. to the number of ions started into equivalent solid angles, simulating the case of no scattering effect
by the crystal structure. By definition this yields the relative
detection probabilities F' for scattering centers in the blocking
case for detectors of various radial acceptance angles (in absence

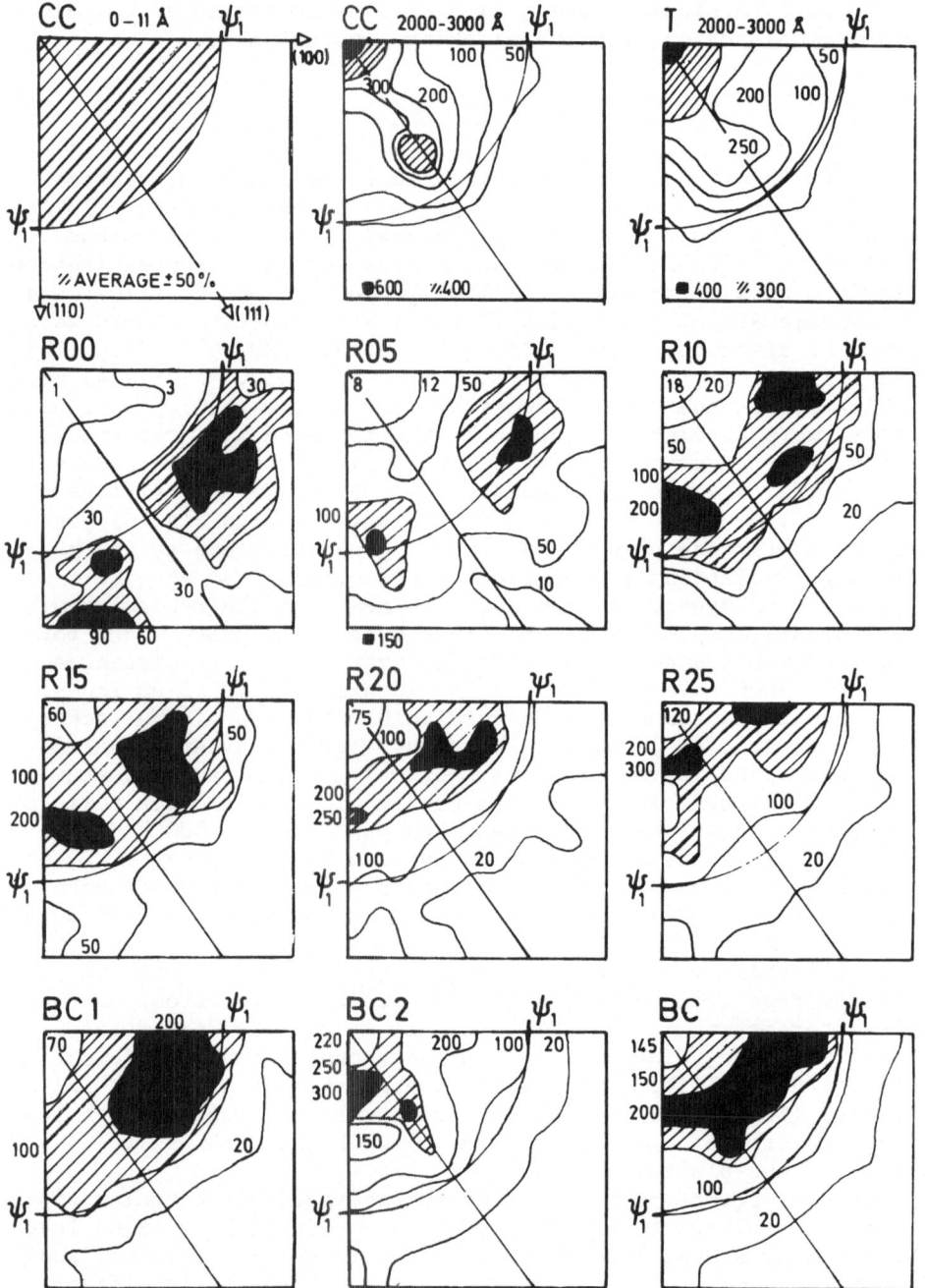

Fig. 2. Ion emergence patterns for blocking simulation of 300 keV
H$^+$ ions started from different positions in the <110> channel of
silicon; ion flux in arbitrary units.

CC 0-11	Isotropic starting conditions exemplified for CC, small penetration (11Å), the arrows point in directions of major planes, ψ_1 is the Lindhard critical angle of channeling.
CC 2000-3000Å	Channel center
T	Tetrahedral interstitial position
R00	Atomic row
R05	0.05 L.U. (= 0.1357Å) from atomic row in (1$\bar{1}$0)
R10	0.10 L.U. (= 0.2714Å) from atomic row in (1$\bar{1}$0)
R15	0.15 L.U. (= 0.4071Å) from atomic row in (1$\bar{1}$0)
R20	0.20 L.U. (= 0.5438Å) from atomic row in (1$\bar{1}$0)
R25	0.25 L.U. (= 0.6795Å) from atomic row in (1$\bar{1}$0)
BC1	Bond centered position in {100} plane
BC2	Bond centered position in {111} plane
BC	(BC1 + BC2)/2

of a crystal F' ≡ 1). These results are presented in Fig. 3, the parameter being the scattering center location.

Similarly Fig. 4a shows the detection probability F' as function of the distance of the scattering center from the atomic row, the parameter being the detector acceptance angle. Fig. 4b which shows ion flux profiles in the same plane for the channeling computer calculations described in the preceding section, the parameter being the radial collimation angle. The rule of reversibility is seen to be well fulfilled within statistical accuracy (± 10%), when equating F' and F for equal detector acceptance angle and initial beam divergence, respectively. This fact is of considerable advantage as the channeling case can adequately be described by the formulas deduced by van Vliet [14]: near atomic rows at distances ρ, the ion flux is given by

$$F(\rho) = \left[1 + \frac{2\delta E_\perp}{E\psi_1^2} \left(1 + \frac{\rho^2}{C^2 a^2}\right) \right] \log \frac{1}{1 - Nd\pi\rho^2}$$

where $\delta E_\perp$ is the mean additional transverse energy due to the initial beam divergence, $E\psi_1^2$ is the maximum transverse energy compatible with channeling, (Ca) stems from Lindhard's potential, N is the atomic density, and d is the interatomic spacing of the row

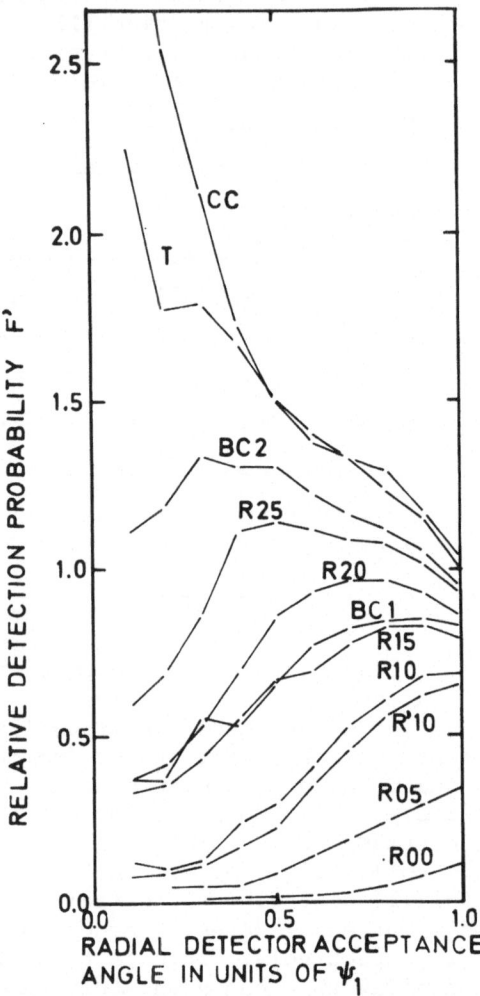

Fig. 3. 300 keV H$^+$ in silicon <110> ion detection probabilities F'
as function of the radial acceptance angle of the detector relative
to uniform random case for different scattering center positions,
as in Fig. 2. R'10 is 0.10 L.U. from an atomic row in (001).

considered. This formula is valid as long as $F(\rho) \leq 2\rho/\rho_0^2$ (random
case), where ρ_0 is the channel radius.
    For channel center a good estimate may be obtained from van
Vliet's formula [14]

$$F_{cc} = 1 + \log \left(\frac{A_o k}{\pi \delta E_\perp}\right) ,$$

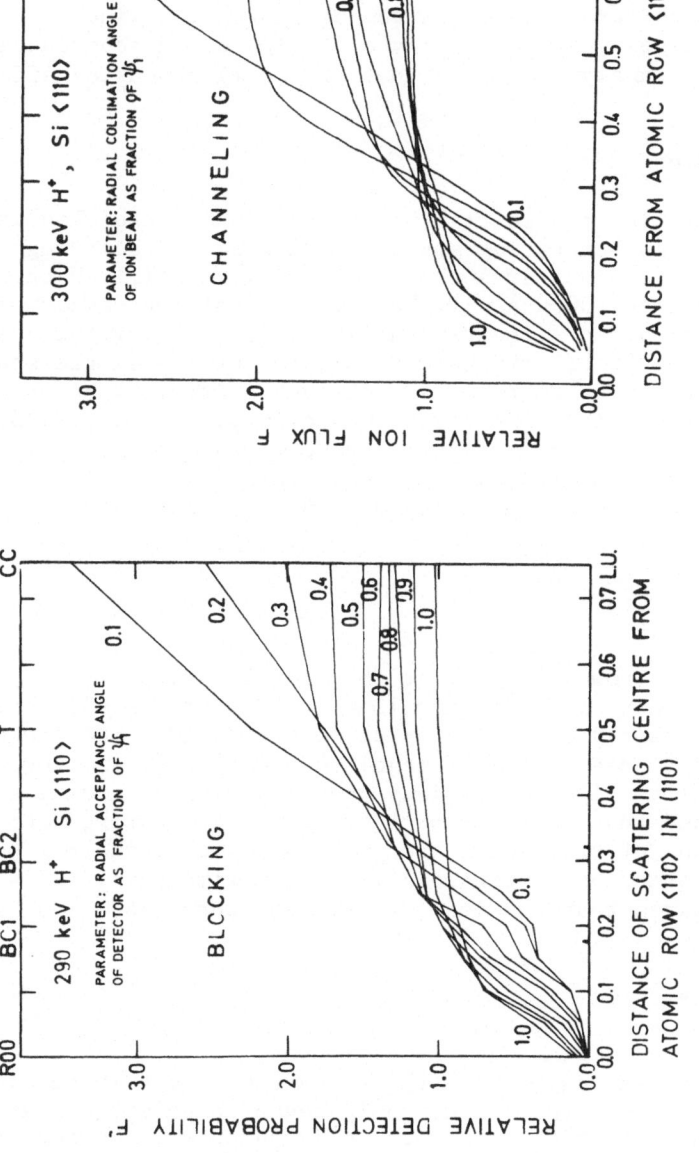

Fig. 4. Comparison of relative detection probability F', blocking (a), and relative ion flux F, channeling (b), for corresponding detector acceptance angles and initial beam collimations; computer simulation for 300 keV H⁺ in silicon <110>; statistical error ± 10%.

notation as above and A_0 the channel area and k the harmonic bowl
parameter, given as averages by van Vliet [14] to k_{H+} = 4.3 eV/Å^2
and k_{He+} = 8.5 eV/Å^2 for the <110> channel in silicon. Both these
formulas were found to be reasonably accurate for $\delta E_{\perp} \leq 0.3\ E\psi_1^2$.

The complete picture of distributions of ion flux or detection probability may then be obtained from graphs of the potential shells
of the channel concerned assuming the equipotential lines to represent equiflux lines, as also employed in Fig. 6, in accordance with
van Vliet's [14] analytical treatment of the flux peaking effect.

Dechanneling Cross Section

By interpreting the initial ion directions of the blocking
simulations as ion scattering angles at different scattering center
locations in the channel area the number of ions still channeled
after, for instance, 3000 Å will be a measure of the dechanneling
cross section of defects located in different areas of the channel
potential [16]. The traditional calculation of the single scattering dechanneling cross section assumes that scattering by an angle
larger than the critical angle for channeling, i.e. ψ_1 [12,15],
will transfer a channeled ion into a random trajectory [17,18,19,20].
This corresponds to a gain in transverse energy from 0 to $\geq E\psi_1^2$.
The momentary transverse energy of an ion is defined by

$$E_{\perp} = E\psi^2 + U(r)$$

where U(r) is the row potential at the ions location.
As single scattering dechanneling cross sections of energetic
light ions are fairly small (order of magnitude 10^{-18} cm^2) compared
with the channel area (order of magnitude 10^{-15} cm^2) and the channeled ion density falls off rapidly when approaching atomic strings
it can be argued that the main part of single scattering dechanneling affects ions of transverse energy comparable to the row potential at the scattering center location. For simplicity it is,
therefore, assumed that the ions transverse energy is

$$E_{\perp} = U(r)$$

i.e. we consider the case where the transverse angle is zero. For
dechanneling to occur the ion now only needs to gain the energy
$E\psi_1^2 - E_{\perp}$ in a scattering event, to be dechanneled. Obviously,
scattering by an angle smaller than ψ_1 will suffice, and indeed
$\psi_D(r)$ defined by

$$E\psi_D^2(r) = E\psi_1^2 - E_\perp = E\psi_1^2 - U(r)$$

is the position dependent dechanneling critical angle. For the Lindhard potential

$$U(r) = \frac{E\psi_1^2}{2} \log \left(\left(\frac{Ca}{r}\right)^2 + 1 \right),$$

hence

$$\frac{\psi_D^2(r)}{\psi_1^2} = 1 - \tfrac{1}{2} \cdot \log \left(\left(\frac{Ca}{r}\right)^2 + 1 \right).$$

The position dependent dechanneling cross section is, therefore, given (integration of the Coulomb differential cross section from ψ_D to π) by

$$\sigma_D(r) = \frac{\pi}{4} d^2 \psi_1^2 \times \frac{1}{1 - \tfrac{1}{2} \log \left(\left(\frac{Ca}{r}\right)^2 + 1 \right)},$$

$$\sigma_D(r) = \sigma_D(r_o) \times \frac{1}{1 - \tfrac{1}{2} \log \left(\left(\frac{Ca}{r}\right)^2 + 1 \right)},$$

where r_o is the channel radius, hence $\sigma_D(r_o)$ is the traditional single scattering dechanneling cross section. The ratio $\sigma_D(r)/\sigma_D(r_o)$ as given by the above equation was calculated ($C = 3^{\tfrac{1}{2}}$) and the results are compared in Fig. 5 with the ratios of number of ions still channeled after 3000 Å for ions started in channel center to the corresponding number of ions still channeled for other starting points as described previously. It is seen that near atomic rows the dechanneling cross section can be up to about twice the traditional value and the bond centered positions average a value of $\sigma_D(r) = 1.2 \, \sigma_D(r_o)$. The implications of this finding for the analysis of damage by means of channeling are outlined in the following.

According to single scattering theory, disregarding the flux peaking effect, i.e. assuming uniformly $F \equiv 1$, the concentration of displaced atoms at depth t, $\frac{N(t)}{N_o}$, is given by [17,18,19,20]

$$\frac{N(t)}{N_o} = 1 - \{\gamma(t) \, [1 - P(t)]\}^{-1},$$

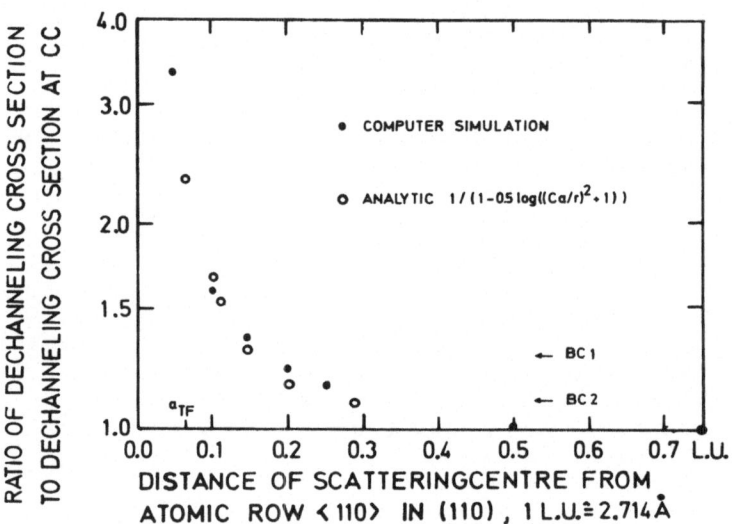

Fig. 5. Ratio of dechanneling cross section to dechanneling cross
section for potential minimum as obtained from computer simulations
(solid circles) and an analytical model (open circles) as function
of the distance of the scattering center from the atomic row <110>
in the (1$\bar{1}$0) plane.

where N_0 is the atomic density, P(t) is the dechanneling probability
before depth t, i.e. in single scattering theory

$$P(t) = 1 - \exp[- \sigma_D \int_0^t N(t') \, dt'] \quad .$$

The experimental parameter $\gamma(t)$ is defined by [20]

$$\gamma(t) = [y_r(t) - y_a(t)]/[y_r(t) - y_a'(t)]$$

and

$y_r(t) \equiv$ random backscatter yield

$y_a(t) \equiv$ aligned backscattered yield from unimplanted
         sample

$y_a'(t) \equiv$ aligned backscattered yield from implanted
         sample .

For non uniform ion flux F, only the fraction F of the number of
scattering centers will contribute, hence

$$F \frac{N(t)}{N_0} = 1 - \{\gamma(t)[1 - B(t)]\}^{-1}$$

where $B(t) = 1 - \exp [- \sigma_D(r) \int_0^t F \, N(t') \, dt']$.

Two cases may now be envisaged: (i) a disorder peak is being
analyzed [18,19,20], (ii) uniform low concentration damage [21].
     In the first case (i) there is a depth t = T which is greater
than the maximum depth of disorder, hence N(T) = 0. Then the total
number of displaced atoms per unit area, D_1, is given by

$$D_1 = \int_0^T N(t') \, dt' = (F \, \sigma_D(r))^{-1} \ln \gamma \, (T)$$

(but for F and the r dependence of $\sigma_D(r)$ this formula was derived
by Hart [20]). The total number of displaced atoms may also be obtained from the direct scattering contribution from the damage peak
in the way described by Hart [20], which quantity may be termed D_2
(D in Ref. 20). If the direct scattering and the dechanneling are
caused by the same displaced atoms in single scattering, non uniform
flux F will cause D_2 to be inversely proportional to F, hence when
comparing D_1 and D_2, which should be equal in this theory, the ratio
of D_1 and D_2 is only susceptible to the magnitude of $\sigma_D(r)$, which
in the light of the above may be regarded as a fitting parameter
yielding information on the scattering center location. Hence, in
Hart's study [20] a value of the dechanneling cross section slightly
above $\sigma_D(r_0)$ points to a quite random arrangement of scattering centers or alternatively to displaced atoms not in channel center and
also not very near atomic rows (see Fig. 5). By comparison in the
studies of Feldman and Rodgers [18] and of Westmoreland et al. [19]
a "fitting" dechanneling cross section of nearly twice the value of
$\sigma_D(r_0)$ may be interpreted as due to scattering centers in the vicinity of atomic rows.
     In the second case (ii) of uniformly distributed low concentra-
tions of defects approximations can be derived for B(t) and F N(t)/N_0:

$$B(t) \approx F \, \sigma_D(r) \int_0^t N(t') \, dt'$$

and

$$F \frac{N(t)}{N_o} \approx \frac{y'_a(t) - y_a(t)}{y_r(t)} - B(t)$$

Hence by defining the normalized yield changes by

$$\Delta Y_{SA}(t) = \frac{y'_a(t) - y_a(t)}{y_r(t)}$$

and assuming a uniform defect distribution, i.e.

$$\int_o^t N(t') \, dt' = N.t$$

one finally obtains for single alignment (SA)

$$\Delta Y_{SA}(t) = F.N/N_o + F.\sigma_D(r).N.t$$

as derived previously [16,21].

Morita et al. [23] have derived that an experimental dechannel-ing cross section can be obtained from a combination of the changes in the dechanneling parameter $\frac{1}{\Lambda}$ and the changes in the normalized minimum yield Y_{SA} due to introduction of defects:

$$\sigma_D = \frac{\Delta \frac{1}{\Lambda}}{N_o \Delta Y_{SA}} .$$

This can be shown in the following manner, assuming d to be the irradiation dose. For a dechanneling parameter to be defined, an equation of the following form must hold:

$$1 - (Y_{SA}(t,0) + \Delta Y_{SA}(t,d)) = (1 - (Y_{SA}(0,0)$$
$$+ \Delta Y_{SA}(0,d))) \, e^{-\frac{1}{\Lambda}}$$

where $\Lambda = \Lambda(N,t)$. Then for $\Delta Y_{SA}(t,d) = F.c + F.c.N_o.\sigma_D.t$ (i.e. $N/N_o = c$), with application of the rule of l'Hôpital, the dechannel-ing parameter extrapolated to the sample surface is

$$\frac{1}{\Lambda}\Big|_{t=0} = \frac{\frac{\partial}{\partial t} Y_{SA}(t,0)\big|_{t=0} + F c N_o \sigma_D}{1 - (Y_{SA}(0,0) + cF)}$$

Here $\frac{\partial}{\partial t} Y_{SA}(t,0)$ is the intrinsic, undamaged crystal dechanneling parameter $\frac{1}{\Lambda_a}$ assumed to be constant. The changes in $\frac{1}{\Lambda}\big|_{t=0}$ with changes in the defect concentration c due to irradiation can be obtained by calculating $\frac{\partial \frac{1}{\Lambda}\big|_{t=0}}{\partial c}$, which yields approximately (omitting a term $(F/\Lambda_a).\Delta c$)

$$\Delta \frac{1}{\Lambda}\Big|_{t=0} = \Delta c. \frac{1 - Y_{SA}(0,0)}{(1 - Y_{SA}(0,0) - \Delta Y_{SA}(0,d))^2} \cdot N_o.\sigma_D$$

which again can be approximated, because of $Y_{SA}(0,0) << 1$ and $\Delta Y_{SA}(0,d) < Y_{SA}(0,0)$ (assumption), by

$$\Delta \frac{1}{\Lambda}\Big|_{t=0} = \Delta c.F.N_o.\sigma_D \quad .$$

On the other hand we have

$$\Delta Y_{SA}(0,d) = \Delta c.F \quad ,$$

from which follows

$$\sigma_D = \frac{\Delta \frac{1}{\Lambda}\big|_{t=0}}{N_o \Delta Y_{SA}(0,d)} \quad ,$$

independent of the ion flux at the scattering center. This means that the experimental dechanneling cross section can be obtained from the extrapolation of bulk properties as expressed in magnitude and slope of channeling spectra. Due to this possibility of obtaining experimental dechanneling cross sections from single alignment of undamaged and damaged crystals, estimates of the scattering center location are possible, which then yield estimates of the ion flux at the scattering center location, hence again a good estimate

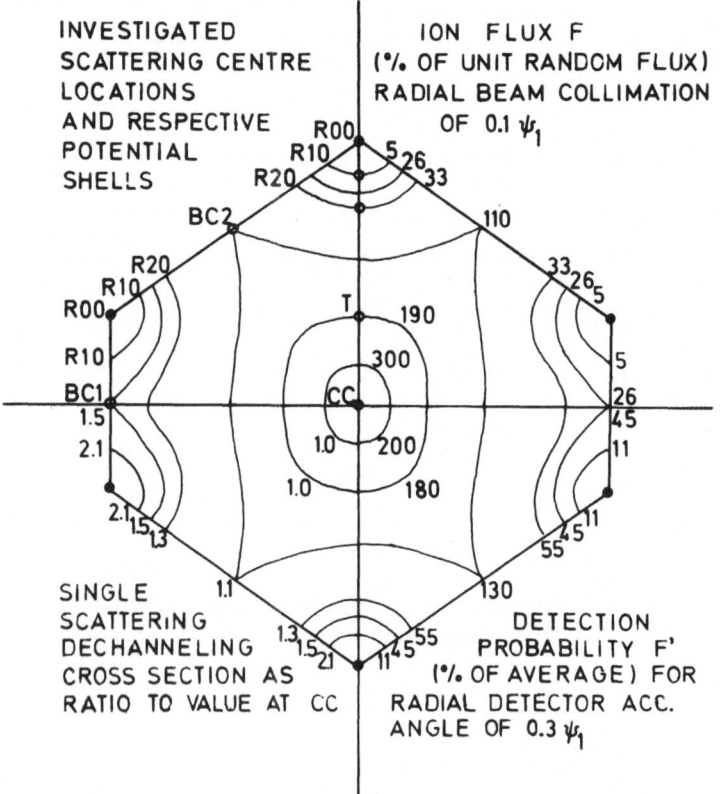

Fig. 6. Location dependent quantities in the channel area of silicon <110> for light energetic ions: scattering potential (TFM),
ion flux F, detection probability F', and relative single scattering
dechanneling cross section.

of the actual number of scattering centers is possible. For the
case of the <110> channel in silicon, Fig. 6 gives the three scattering center position dependent quantities in a semianalytic con-
text, as ion flux F, detection probability F' and dechanneling cross
section ratio $\sigma_D(r)/\sigma_D(r_o)$ are linked to the respective potential
lines. The figures given are independent of ion energy and valid
for both H^+ and (in good approximation) for He^+ for the beam divergences and detector acceptance angles specified. It may be seen
that for experimental determinations of the dechanneling cross sections with a relative error of 25% the ion flux F (channeling) or
detection probability F' (blocking) can be obtained to within a
factor of two, which accuracy then also applies to scattering center
concentrations deduced through the equations given above.

     Finally, it is pointed out that changes in dechanneling cross
section have been observed [16,22], in particular attention is drawn
to Fig. 1 in Ref. 22. The n-type low temperature damage curve

exhibits a change in slope separated by some "annealing". This can be explained quantitatively (based on experimental values of $\sigma_D(r)$ by a change-over from scattering centers near channel center ($F \sim 1$) to scattering centers near atomic rows ($F \sim 0.05$) assuming one scattering center per defect in the first case and 6 scattering centers per defect after the "anneal", corresponding to a threefold reduction in scattering yield after the anneal from the extrapolation of the first section of the curve.

Acknowledgments

My sincere thanks go to Dirck van Vliet for letting me have a complete listing of the <110> channeling computer program and for encouragement to use it. Financial assistance was provided first by the Ministry of Defence, Procurement Establishment, and later by the Science Research Council.

References

[1] M. T. Robinson, D. K. Holmes and O. S. Oen, "Proc. Int. Colloq. Ion Bombardment," Paris, December 1961; M. T. Robinson and O. S. Oen, Phys. Rev. 132, 2385 (1963).

[2] D. V. Morgan and D. van Vliet, Can. J. Phys. 46, 503 (1968); D. V. Morgan and D. van Vliet, "Atomic Collision Phenomena in Solids," Ed. by D. W. Palmer, M. W. Thompson and P. D. Townsend, Horth Holland, 1970, p. 476.

[3] J. H. Barrett, Phys. Rev. B 3, 1527 (1971).

[4] I. G. Massa and G. J. Clark, "Proceedings of the European Conference on Ion Implantation," Reading, 1970, Peter Peregrinns Ltd., 1970, p. 207.

[5] R. B. Alexander and J. M. Poate, rad. eff. 12, 211 (1972).

[6] A. Desalvo, R. Rosa and F. Zignani, Lettere al Nuovo Cimento 2, 8, 390 (1971).

[7] V. A. Eltekov, D. S. Karpuzov, Yu. V. Martynenko and V. E. Yurasowa, "Proceedings of the International Conference on Atomic Collision Phenomena in Solids," Brighton 1969, Ed. by D. W. Palmer, M. W. Thompson and P. D. Townsend, North Holland, 1970, p. 657.

[8] H. D. Carstanjen and R. Sizmann, rad. eff. 12, 141 (1972).

[9] D. V. Morgan and D. van Vliet, rad. eff. 5, 157 (1970).

[10] F. H. Eisen and E. Uggerhøj, rad. eff. 12, 233 (1972).

[11] F. H. Eisen, G. J. Clark, J. Bøttiger and J. M. Poate, "Atomic Collisions in Solids IV," Ed. by S. Anderson, K. Björkquist, K. Domeij and N. G. E. Johánsson, Gordon and Breach 1973, p. 377.

[12] J. Lindhard, Kgl. Dans. Vid. Sels. Mat Fys. Medd. 34, No. 14 (1965).

[13] J. N. Andersen, O. Andreasen, J. A. Davies and E. Uggerhøj, rad. eff. 7, 25 (1971).

[14] D. van Vliet, rad. eff. 10, 137 (1971).

[15] D. V. Morgan and D. van Vliet rad. eff. 8, 51 (1971).

[16] H. J. Pabst, Thesis, University of Sussex, 1973, unpublished.

[17] E. Bøgh, Can. J. Phys. 46, 653 (1968).

[18] L. C. Feldman and J. W. Rodgers, "Proceedings Conf. Atomic Collision Phenomena in Solids," Brighton 1969, Ed. by D. W. Palmer, M. W. Thompson and P. D. Townsend, 1970, p. 128.

[19] J. E. Westmoreland, J. W. Mayer, F. H. Eisen and B. Welch, rad. eff. 6, 161 (1970).

[20] R. R. Hart, rad. eff. 6, 51 (1970).

[21] H. J. Pabst and D. W. Palmer, "Radiation Damage and Defects in Semiconductors," Ed. by J. E. Whitehouse, Conference Series No. 16, IOP, 1973, p. 438.

[22] H. J. Pabst and D. W. Palmer, this conference.

[23] S. Miyagawa, K. Morita, N. Matsunami, K. Tachibana and N. Itoh, rad. eff. 13, 271 (1972).

ANALYSIS OF BLOCKING LIFETIME EXPERIMENTS*

Y. HASHIMOTO[†]
*Rutgers University, New Brunswick, New Jersey 08903
and Bell Laboratories, Murray Hill, New Jersey 07974*

J. H. BARRETT
*Solid State Division, Oak Ridge National Laboratory
Oak Ridge, Tennessee 37830*

W. M. GIBSON
*Bell Laboratories, Murray Hill, New Jersey 07974
U.S.A.*

ABSTRACT

Analysis of blocking lifetime experiments depends
on the best information available about particle trajectory distributions in crystals and therefore relates to
a wide variety of channeling effect studies such as flux
peaking, impurity ion location, and dechanneling. Such
experiments and analysis also offer the possibility to
gain additional useful insight into particle motion in
crystals. The compound nuclear recoil direction and lifetime determine the displacement distribution of the emitting atoms, giving a convenient and well defined starting
condition for trajectory distribution (flux peaking)
studies. Since only one displacement direction relative
to a particular lattice row direction is involved, the
approach of particle trajectories towards statistical
equilibrium can be explored through displacement and
distortion effects in the emitted particle angular distribution.

We have carried out a detailed study of a particular
case: The blocking of 4.0 MeV protons by <110> rows in
germanium crystals. Extensive Monte Carlo computer simulation calculations [1], multistring continuum model
calculations [2] and single string continuum calculations
have been made for this case. For values of $v_\perp \tau$ up to
about 0.3 Å, the multiple string and Monte Carlo calcula-
tions have the form $\chi = \chi_0 + C'Nd\pi 2(v_\perp \tau)^2$ where C' is
about half as large as the coefficient C in the expression
$\chi = CNd\pi\rho^2$ that has been found [3] for the minimum yield

just inside the surface of the crystal. For $v_\perp \tau > \sim 0.3$ Å appreciable dependence on recoil direction in the lattice appears and structure in the emission angular distribution curves is observed for some recoil directions. The recoil direction effects are correlated with flux peaking in channeled particle trajectory distributions. Comparison of multistring continuum calculations with Monte Carlo results gives a useful view of the influence of multiple scattering on different parts of the emission distribution as a function of emission position and emission depth in the crystal. The emission depth dependence also shows the emission pattern to change from shifted assymetric distributions to symmetric distributions centered on the axial direction as the emission depth is increased beyond about 2000 Å.

Simple approaches to inclusion of multiple scattering effects into continuum calculations are investigated as well as semiempirical approaches to determination of mean lifetime values from measured angular distributions without extensive computer computations. Finally, deviations from the situation of a single exponential decay function are examined with emphasis on the possibility of determining the form of the decay function by measurements of the emission distribution at different recoil angles and directions. As a test of accuracy and applicability, the computed emission distributions are compared to an extensive series of experimental measurements [4].

References

*Work supported in part by the National Science Foundation, the U. S. Atomic Energy Commission (under contract with Union Carbide Corporation) and Bell Laboratories.

†On leave from the University of Tokyo.

[1] Y. Hashimoto, J. H. Barrett and W. M. Gibson, Phys. Rev. Lett. 30, 995 (1973).
[2] Y. Hashimoto, Bull. Am. Phys. Soc. 17, 560 (1972); Y. Hashimoto and J. H. Barrett, Bull. Am. Phys. Soc. 18, 119 (1973).
[3] J. H. Barrett, Phys. Rev. B3, 1527 (1971).
[4] W. M. Gibson, Y. Hashimoto, R. J. Keddy, M. Maruyama and G. M. Temmer, Phys. Rev. Lett. 29, 74 (1972); Bull. Am. Phys. Soc. 16, 557 (1971); 17, 560 (1972); 18, 119 (1973); and 18, 668 (1973).

CHANNELING, BLOCKING, AND RANGE MEASUREMENTS USING THERMAL NEUTRON INDUCED REACTIONS

J. P. BIERSACK and D. FINK
Hahn-Meitner Institute
Berlin, Germany

ABSTRACT

The emission of light particles from exo-energetic (n,p), (n,t) or (n,α) reactions can be used for detecting depth distributions and lattice positions of trace amounts of foreign atoms in crystalline samples. The advantages over p backscattering experiments are: (i) there occurs no background from the host lattice, (ii) thermal neutrons cause almost no radiation damage to the crystal, (iii) the method is particularly suitable for light atoms which are difficult to detect by other means. The method is demonstrated for low concentrations of ^{6}Li and ^{10}B in ionic crystals, semiconductors and metals. In some cases narrower blocking dips and lower minimum yields are observed than in other comparable blocking experiments. Range profiles of lithium incident on niobium and silver with an energy of 220 keV are measured, and are found to agree with the theoretically predicted Edgeworth expansion.

Introduction

The present paper describes a method suitable for detecting depth distributions and lattice locations of light atoms in crystalline environments. The method utilizes MeV protons, tritons or α particles emitted after the capture of thermal neutrons. Suitable candidates undergoing (n_{th},p) or (n_{th},α) reactions with cross sections of the order of barns to kilobarns can be found among the isotopes of light atoms, e.g., ^{3}He, ^{6}Li, ^{7}Be, ^{10}B, ^{14}N, ^{17}O. Heavy atoms, however, do not have sufficiently large cross sections. This may be considered as an advantage, as heavy host lattices do

not cause much of a background. Therefore, in contrast to other
methods--particularly to backscatter techniques--preferably light
atoms in heavy substances can be studied.

Further advantage over other methods are: neutrons cause less
radiation damage than beams of charged particles, because of smaller
scattering cross sections and less momentum transfer.[†] Also, the
concentration profile of light atoms is not changed or removed
during the measurement, as in secondary ion mass spectrometry
(SIMS) or with radioactive tracer methods, where layers of material
are removed by sputter techniques.

The feasibility of the proposed method will be demonstrated
for a range measurement and for channeling/blocking experiments.

Experiments on Depth Distributions

The depth distribution of 220 keV $^6Li^+$ ions, implanted at a
random direction into niobium samples is measured through the energy
spectrum of α particles emitted after $^6Li(n,\alpha)T$ reactions. The
sample is placed in a thermal neutron beam of about $10^4 n/cm^2 sec$,
extracted from the "thermal column" of the FMRB reactor facility in
Braunschweig, W.-Germany. The α spectrum, as shown in Figure 1,
is well resolved at the chosen setting of the multichannel analyser.
It exhibits a small peak at 2.0 MeV, i.e., the full α energy of
the reactions, resulting from Li accumulated at the surface of the
sample, and a non-Gaussian depth distribution (gradual rise from
the surface, steep decline beyond the maximum).

At the far right of the energy spectrum, the triton peak
appears, reflecting the same depth profile of Li, but less well
resolved at the setting of 16 keV/channel. Beyond this peak, i.e.,
above 2.7 MeV, the number of counts drops to zero, indicating that
there exists no background. From this it is concluded that the
long tail extending from the triton peak to lower energies (corresponding to about 4 μm depth) is real, and must be attributed to a
fraction of implanted lithium which becomes mobile at low temperatures. The same conclusion is drawn from the occurence of a surface peak (resolved in the α spectrum) which builds up at room
temperature within a few weeks, or at 500°C within a few hours.
This peak also must be due to the fraction of initially mobile
Li atoms which become trapped at the surface,[††] while the rest of
the Li remains in the original profile up to about 900°C. More
details on the diffusion of Li in Nb are being published in the
subsequent paper [1].

[†]In the case of well thermalized neutrons, the (n,α) or (n,p)
reaction of the impurity atoms under investigation remains the
only source of damage.

[††]In addition to the effect of the mere surface, also impurity
layers may enhance the trapping at the surface, as the experiment
was not performed in ultra high vacuum.

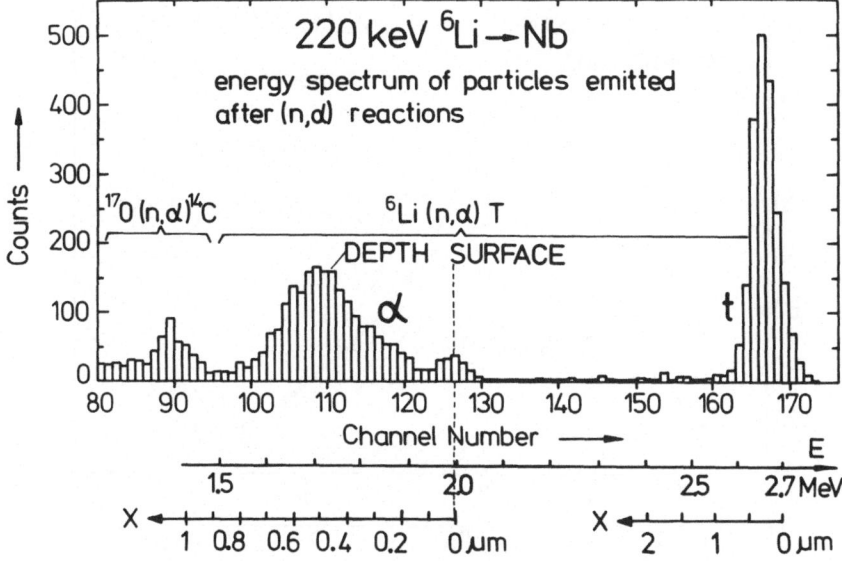

Fig. 1. Energy spectrum of particles emitted after [6]Li(n,α)T reactions; [6]Li was implanted at 220 keV into niobium.

The surface peak of Li is a suitable means for calibration of energy and for testing the resolution of the experimental equipment. In the present case, a least square fit with a Gaussian applied to the α surface peak (Fig. 2) yields the location of the surface to be in channel 125.5 (corresponding to 2.0 MeV), and the depth resolution is found to be σ = 0.053 μm (corresponding to 29.7 keV).

The peak at the far left of Fig. 1 may be due to surface layers of oxygen--on the sample and on structural materials in the vacuum chamber--which become visible through the [17]O(n,α)[14]C reaction, identified by the α energy of 1.4 MeV.

The conversion of the energy spectrum into a Li depth profile, as indicated in Fig. 2, requires the knowledge of the stopping power of α particles in niobium in the energy range of 1.5 to 2 MeV, i.e., just beyond the stopping power maximum.

As there are no experimental data known on MeV alphas incident on Nb, the interpolation formula of Biersack [2,3] is used which includes both the Lindhard-Scharff [4] and the Bethe [5] stopping power relations, and gives good agreement with otherwise available experimental data in the transition region, where S_e reaches its maximum. This interpolation formula is composed of

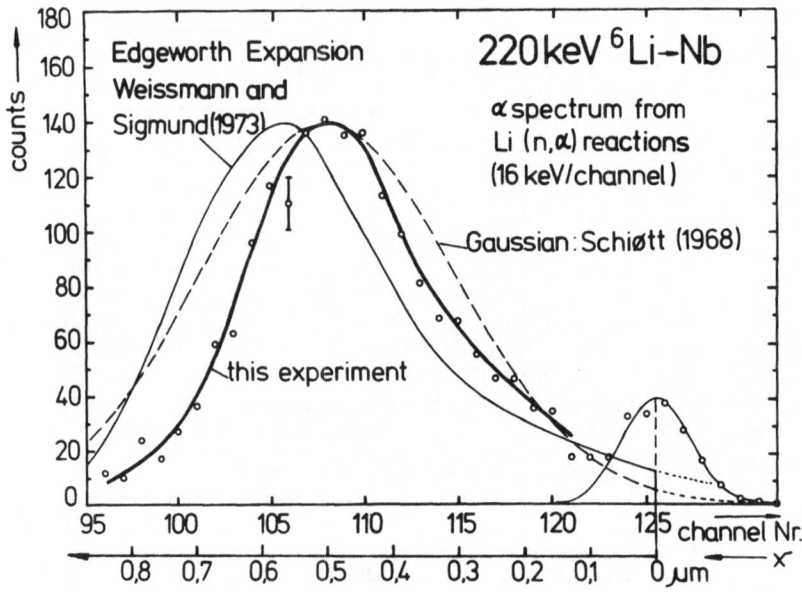

Fig. 2. Range profile of implanted ^{6}Li (derived from α energy spectrum) in comparison with theoretical predictions. Gaussian fitted at right, indicates location of surface and resolution of apparatus.

$$S_L = \xi_e \frac{8\pi Z_1 Z_2 e^2 a_o N}{(Z_1^{2/3} + Z_2^{2/3})^{3/2}} \frac{v}{v_o} \ , \ \xi_e \simeq Z_1^{1/6} \tag{1}$$

(Lindhard et al. [4])

and

$$S_B = \frac{8\pi Z_1^2 e^4 N}{I_o} \cdot \frac{\ln(\varepsilon + 1 + b/\varepsilon)}{\varepsilon} \ , \ \varepsilon = \frac{2m_o v^2}{Z_2 I_o} \ , \ b \simeq 5 \tag{2}$$

(low energy modification [2] of Bethe's formula)

by simply adding the reciprocals:

$$\frac{1}{S_e} = \frac{1}{S_L} + \frac{1}{S_B} \ . \tag{3}$$

Using I_0 = 10.47 eV as the best adjusted value of the Bloch constant for niobium [6], one obtains S_e = 396 keV/μm for the initial α energy of 2.05 MeV, and S_e = 412 keV/μm for E_α = 1.73 MeV which corresponds to an α particle emerging from a depth of 0.5μm (about the mean depth of Li deposition) at an angle of 45° to the surface (direction of observation by the particle detector). As the uncertainty of the formula in this energy region is estimated to be 5 to 10% the change in S_e is neglected, and the constant value of 396 keV/μm is used throughout, leading to the depth scale depicted in Fig. 2.

The analysis of the [6]Li range profile--after subtracting the surface peak--yields the experimental data of Table 1 in comparison with the theoretical predictions of H. Schiött [7], and R. Weissmann and P. Sigmund [8]. Obviously all theoretical values are larger than could be explained by experimental errors, or uncertainty of the stopping power.

If the experimental curve is deconvoluted with the resolution function (obtained from the surface peak), the experimental curve becomes still narrower, $<(x-\bar{x})^2>^{1/2}$ = 0.144μm. The narrowness and the skewness of the curve show very drastically that the fit with the Gaussian of theoretical width is rather poor, see dashed line in Fig. 2. By adding the next higher (third order) term in an Edgeworth expansion and readjusting the height, one obtains a much better qualitative agreement between experiment and theory, as judged by the shape of the distributions. Visual inspection of the profiles, solid lines of Fig. 2, reveals that in this case the difference is reduced to a mere scale factor of 1.14 in the depth scale.

A similar experiment on 220 keV Li implanted into silver is conducted and evaluated along the same lines as described above, except for the stopping power, where experimental data are applied. This experiment leads to the following results: (i) All of the Li atoms implanted into silver are highly mobile at low temperatures (25 - 600 C). (ii) The surface also acts as a trap with a binding energy of the order of 1 eV. (iii) The mean projected range of (0.45±0.05)μm agrees well with the theoretical predictions [7,8];

Table 1

Depth distribution of 220 keV [6]Li implanted into niobium

			This Experiment	Theory
Most probable range	$\hat{x}$	=	0.50μm	0.56μm, ref.[8]
Mean range	$\bar{x}$	=	0.46μm	0.50μm, ref.[7]
Standard deviation	$<(x-\bar{x})^2>^{1/2}$	=	0.16μm	0.20μm, ref.[7]
"Skewness"	$<(x-\bar{x})^3>^{1/3}$	=	-0.11μm	-0.21μm, ref.[8]

higher order moments of the range distribution can not be presented
accurately, as the profiles were already affected by room temperature diffusion.

Channeling and blocking measurements

The experiments on channeling and blocking effects are performed inside the reactor, as the high neutron flux requirements
are not met in the extracted n beams of the FMRB reactor facility.
Crystals of about 0.3 cm diameter are placed close to the graphite
block of the thermal column at a flux between $5 \cdot 10^9$ and $5 \cdot 10^{10}$
neutrons/cm^{2}sec. The detecting device for the emitted α particles
is an acetate cellulose foil located at a distance of 80 cm from
the crystal, subtending a solid angle of about 5° x 5° with an
angular resolution of 0.1°. Cellulose acetate foils detect
charged heavy particles above 0.5 MeV, but are not sensitive to
the radiation background of β's, γ's, or thermal neutrons. Irradiation times of 1 to 10 days are sufficient for producing a density
of 100 - 1000 traces per mm^2 foil. After etching, the traces reach
a size which can easily be observed under microscopes with electronic
counting facilities. The results are graphically displayed
in the form of isohypses, as shown in Figures 3 through 6. The
lines connect points of equal trace densities, and are labeled
with the number of counts (traces) per 0.236 mm^2 (field of view of
the microscope).
The first experiments were carried out with lithium fluoride
[9] crystals containing 3.7 atom % ^{6}Li in regular lattice positions.
The observed blocking patterns, Fig. 3, exhibit relative minimum
yields χ_{min}, of 10 to 14% in the main axial dips, and 30 to 50% in
the prominent planar dips (measured outside the axial shoulders).
The results of several experiments on LiF are compiled in Table 2.
The width of axial dips is obtained by averaging over concentric

Table 2

Results of blocking measurements on LiF crystals

Crystallographic Direction	Blocking Angle (HWHM)	Relative Yield χ_{min}
<100>	0.7°	10%
<110>	0.7°	14%
<111>	0.6°	10%
{100}	0.2°	30%
{110}	0.2°	40%
{111}	0.2°	50%

rings around the axis (slightly narrower dips would be obtained in scanning through the axis along a "random" direction, i.e. towards a shoulder).

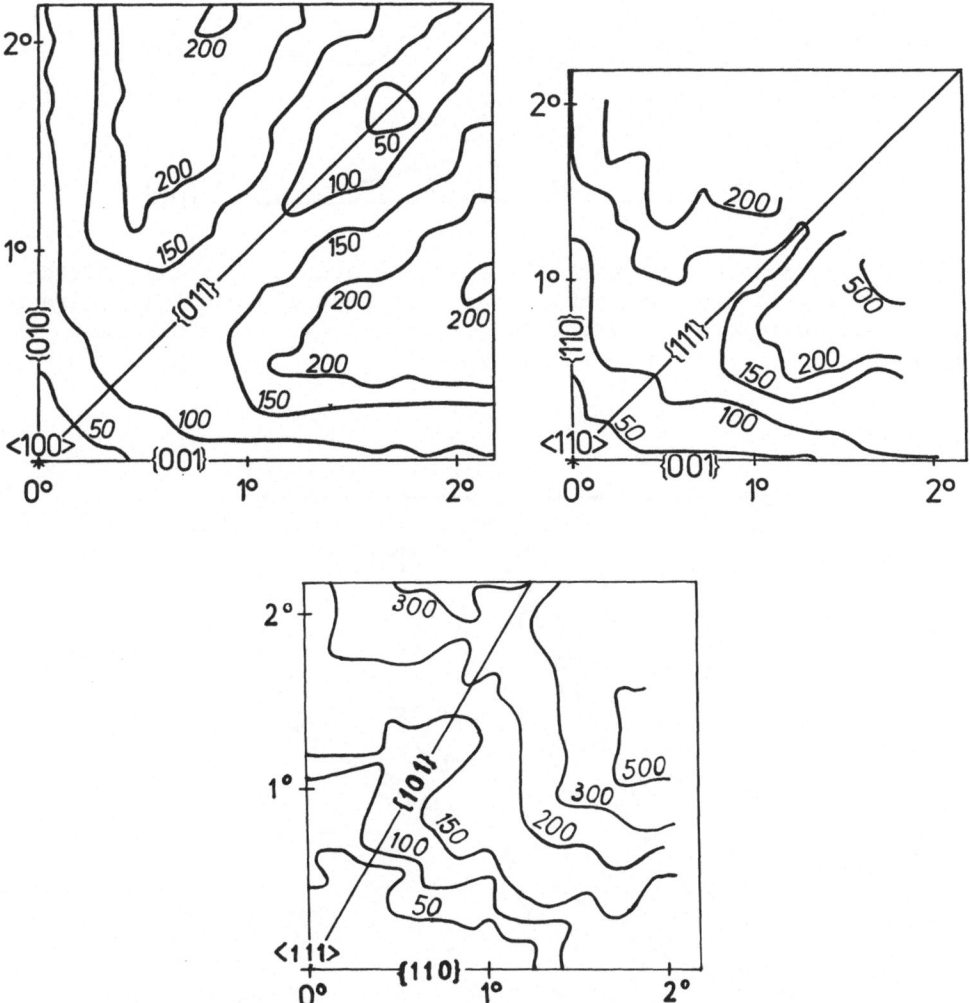

Fig. 3. Contour maps (isohypses) of particle emission in directions near the <100>, <110>, <111> axis of LiF crystals (3.7% ^{6}Li) measured as track density on cellulose acetate foils. (Average of 4 to 8 sectors, according to symmetry).

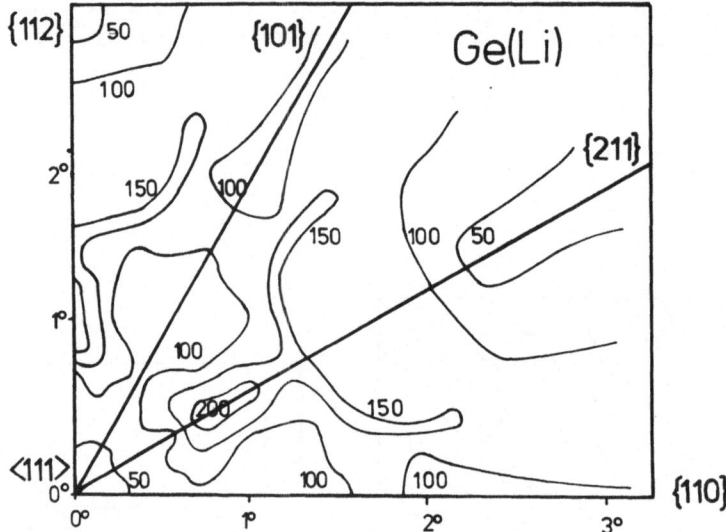

Fig. 4. Contour maps of particle emission around the <111> direction of a Ge(Li) crystal, containing less than 100 ppm [6]Li. (Trace
density in cellulose acetate foils, averaged over 12 equivalent
sectors).

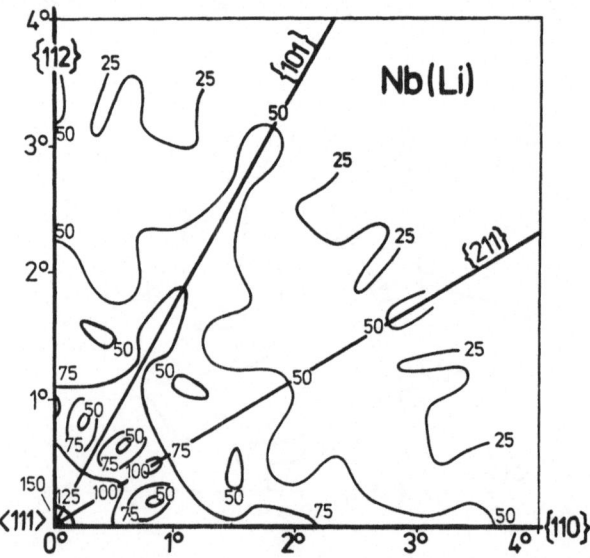

Fig. 5. Channeling pattern of Nb(Li) around <111> direction
(average trace density of 12 equivalent sectors). [6]Li was implanted
to a concentration of a few percent atomic fraction.

The results agree reasonably well with theoretical predictions of Lindhard [10], Varelas and Biersack [2], or Barrett [11], if the calculated channeling angles are converted to blocking angles by applying a factor $\sqrt{2}$. The observation of axial channeling in the <100> direction of LiF by Amberger [12], who used an 8.8 MeV α particle source, yields a larger angle, when reduced to the presently studied energies. The experimental results of Price and Kelly [13], obtained with a 1.3 MeV He$^+$ beam on LiF, are consistent with the present measurements.

A second experiment with a much lower ^{6}Li concentration (atomic fraction $\lesssim 10^{-4}$) was performed with a germanium single crystal into which lithium of natural abundances has been introduced by diffusion. For obtaining ideal channeling patterns, this method of doping the crystal is preferable to ion implantation, as radiation damage and local oversaturation of Li is avoided. With the diffusional doping, all Li is expected to be on interstitial sites which is consistent with the blocking pattern observed around the <111> axis. This emission pattern, as depicted in Fig. 4, exhibits clear axial and planar effects, of halfwidth somewhat smaller than predicted by theory [10,11]. The measured data are compiled in Table 3.

In the yet unknown system niobium-lithium a study was carried out, in order to find the lattice location of Li which was implanted at 220 keV into a Nb crystal (bcc). Pure ^{6}Li was implanted at room temperature to maximum concentrations of a few percent. The subsequent reactor experiment revealed a channeling pattern around the <111> axis, as shown in Fig. 5. This is consistent with Li atoms occupying interstitial (octahedral or tetrahedral) lattice sites. As the channeling effect is less pronounced than in the system Ge(Li), it may be concluded that a considerable fraction of lithium inhabits random positions, e.g. in vacancy clusters or minute precipitations. This may also explain the reduced mobility of a large fraction of implanted Li (cf. last chapter).

An example of using the ^{10}B$(n,\alpha)^7$Li reaction is given in Fig. 5 which depicts the blocking pattern of the mineral kernite.

Table 3

Results of a blocking experiment with Ge(Li)

Crystallographic Direction	Blocking Angle (HWHM)	Relative Yield χ_{min}
<111>	0.5°	26%
{110}	0.2°	66%
{211}	0.2°	51%

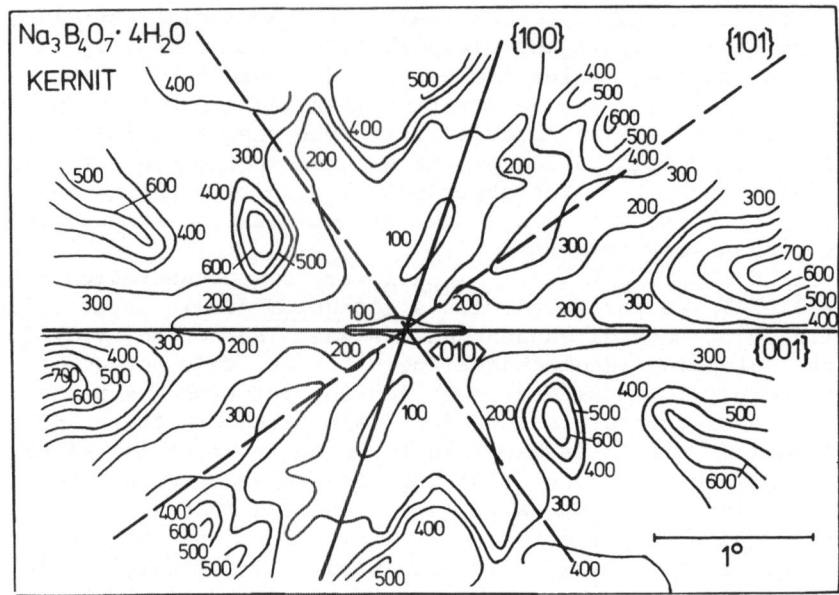

Fig. 6. Blocking pattern of $Na_3B_4O_7 \cdot 4H_2O$ (kernite) in <010>
direction. Particles are emitted after $^{10}B(n,\alpha)^7Li$ reactions
(content of ^{10}B is 3%).

Here, a crystal of low symmetry (monoclin-prismatic) and complicated
basis (104 atoms in a large unit cell of 15 Å length) is studied in
the hope of detecting shadow effects originating in the basis rather
than in the crystal structure. Kernite has the chemical composition
$Na_3B_4O_7 \cdot 4H_2O$ and is available in well-grown and stable crystals.
The content of ^{10}B is 3%. The $^{10}B(n,\alpha)$ reaction of thermal neutron
cross section 3837 barns releases α particles of 1.78 MeV (6%),
1.48 MeV (94%), and ^7Li recoils of 1.02 and 0.84 MeV, corresponding-
ly. All emitted particles are detected with the acetate cellulose
foil. The <010> blocking dip as well as the major crystal planes
show up clearly, cf. Fig. 6, while other structures are not yet
explained in full.
      Some other blocking experiments with an improved angular reso-
lution of 0.06° were performed by placing the crystal on a minia-
turized goniometer inside the reactor beam tube, and using a surface barrier detector near the outer end of the beam tube (at a
distance of 2.3 m from the sample). In cases of planar blocking,
somewhat narrower and deeper dips were obtained than in foil
experiments corresponding to the improved resolution in angle and
depth (energy).

Resume

The method of using thermal neutron induced nuclear reactions leads to the following results:

(i) Range distributions of 220 keV ^{6}Li in Nb and Ag are easily measured without background problems; they agree reasonably with third order Edgeworth expansions predicted by theory.

(ii) Channeling and blocking effects are clearly observed in the systems LiF (3.7% ^{6}Li), Ge–Li ($\lesssim$100 ppm ^{6}Li), Nb–Li (1...10% ^{6}Li), and a kernite mineral (3% ^{10}B), and partly can be interpreted in terms of foreign atom lattice locations.

It is believed that the method can be developed into a suitable tool for studying lattice locations, range distributions and transport processes of trace amounts of light atoms in solids.

Acknowledgment

We are indebted to the reactor division of the Physikalisch-Technische Bundesanstalt, Braunschweig, Germany, for making available to us the FMRB reactor facility. We are very much obliged to Prof. Adam, Technische Universität Berlin, for his kind help in preparing the single crystal samples of niobium. We are also grateful to our colleagues Dr. S. Roth, E. Santner, U. Morfeld for discussions and practical support.

References

[1] J. P. Biersack, D. Fink, Proceedings of the Intern. Conf. on Ion Beams to Metals, Albuquerque, Oct. 2-5, 1973.

[2] C. Varelas, J. P. Biersack, Nucl. Instr. and Meth. 79, 213 (1970).

[3] J. P. Biersack, to be published.

[4] J. Lindhard, M. Scharff, Phys. Rev. 124, 128 (1961).

[5] H. Bethe, Ann. Phys. 5, 325 (1930).

[6] R. M. Sternheimer, Phys. Rev. 145, 247 (1966).

[7] H. Schiött, Mat. Fys. Medd. Dan. Vid. Selsk. 35, No. 9 (1966).

[8] R. Weissmann, P. Sigmund, Rad. Effects 19, 7 (1973).

[9] J. P. Biersack and D. Fink, Nucl. Instr. and Meth. 108, 397 (1973).

[10] J. Lindhard, Mat. Fys. Medd. Dan. Vid. Selsk. 34, No. 14 (1965).

[11] J. H. Barrett, Phys. Rev. B3, 1527 (1971).

[12] R. W. Amberger, Z. Physik 236, 352 (1970).

[13] P. B. Price and J. C. Kelly, proceedings of this conference.

EXPERIMENTAL INVESTIGATION OF THE REVERSIBILITY RULE AT NON ZERO DEPTH*

S. U. CAMPISANO, G. FOTI, F. GRASSO, and E. RIMINI
Istituto di Struttura della Materia dell'Università di Catania
Corso Italia, 57 - I 95129, Catania, Italy

ABSTRACT

The validity of the reversibility rule between source and detector has been tested by experiments on channeling and blocking performed just beyond the crystal surface. In the present work the measurements have been extended to finite depth inside the crystal by MeV proton backscattering from main axes of Si and Ge. The aligned yield in blocking is higher than that corresponding in channeling for all the investigated beam energies, crystal axes and target temperatures. No direct comparison can be made in this case between channeling and blocking because in channeling experiments all the detected dechanneled particles come from a monoenergetic source, while in blocking the detected channeled particles originate from sources of different energies because of the energy losses experienced by the incident particles along the incoming random trajectories. For a comparison between channeling and blocking all the blocking sources should be then reported to the same energy of the channeling source. The simple scaling parameter z/E has been used for this reduction on the basis of experimental dechanneling results and for the lack of a quantitative theoretical prediction. The corrected blocking aligned yield remains slightly higher (10-20%) than the corresponding channeling one, and the reversibility rule seems then to be valid even in the occurence of multiple scattering within 20% in the framework of this simple data treatment.

749

Introduction

The duality between channeling and blocking phenomena was stated by Lindhard [1] on the basis of general arguments. This reversibility between source and detector has been experimentally verified [2] just near the crystal surface. As pointed out by Lindhard [1] this rule should be valid even in occurence of multiple scattering, i.e., blocking and channeling measurements should be identical at finite depth for the same experimental conditions. A large number of experiments to investigate blocking have been performed near the surface or by transmission measurements through thin crystals [3], i.e., for very small energy loss; on the other hand the blocking phenomenon has been widely applied for nuclear lifetime measurements [4] and for lattice damage studies [5,6]. In these cases, where the effects of energy losses cannot be neglected for the finite depths involved, there is a lack of systematic experimental information. The aim of this work is then to give a detailed description of the depth dependence of blocking effects in comparison with channeling.

The aligned yield at finite depth in the blocking condition is usually higher than the corresponding one in channeling for the same incident beam energy and for the same crystallographic axis, as recently shown [7]. No direct comparison can be made, however, between the two results because in channeling experiments all the detected dechanneled particles come from a monoenergetic source while in the blocking experiments the detected channeled particles originate from sources having different energies because of the energy losses experienced by incident particles along the incoming random trajectories. This consideration has been utilized to compare channeling and blocking results obtained by measurements with protons of energies ranging between 0.3 and 1.5 MeV on different axes of Si and Ge single crystals.

Experimental

Channeling measurements have been performed by the backscattering technique using a proton beam of energy ranging between 0.3 and 1.5 MeV, and collimated within 1 mrad by means of a telescopic arrangement of 2 mm diameter circular apertures 2 m apart. The crystal target was mounted in a goniometer which allows orientation of the crystal within 0.5 mrad in any space direction. The temperature of the crystal in thermal contact with a thermostatic bath can be set at 80°K to 300°K. Scattered protons to 155° were detected by a 25 mm^2 surface barrier detector at a 8 cm distance. Standard electronics were used to feed pulses to a 4096 channel pulse height analyzer. The energy resolution of the system was 10 keV FWHM for 1.0 MeV protons.

Blocking measurements have been carried out in the same experimental condition. The target was rotated to 25° to align the

major axis of the crystal to the direction of a detector set 1.5 m
from the crystal. The detector was mounted in an x-y translation
system and the sensitive area was reduced in such a way as to produce the same collimation as in channeling experiments. The blocking alignment has been made either by small angular deflections of
the goniometer stage or by translation of the detector. In any
case the random entrance of the beam has been checked by another
detector set near the crystal target.

In the experimental set up attention has been paid to assure
complete symmetry between blocking and channeling. Several circular
apertures have been located along the detection line to reduce the
background due to particles scattered from the walls. This contribution has been measured by masking the center of the detector leg
and it amounted to 0.5% of the unmasked random yield. The used
crystals were cut normal to the investigated major axis. Because
the lower yield in channeling conditions could be attributed to
better collimation of the incident beam as due to the accelerator
focus condition, a thin nickel foil has been put in the first aperture of the collimating telescope to spread out the beam. In these
conditions the beam collimation is determined only by the telescope
geometry. A negligible difference has been observed between the
aligned yields measured with and without the nickel foil.

Statistically significant yield in blocking requires a bombardment dose two orders of magnitude higher than in channeling. This
dose produces, especially at low temperature, a detectable damage,
so that the aligned yield in blocking has been recorded every 10μC;
as soon as damage effect appeared the beam was translated to a
virgin spot.

Results and Discussion

The validity of the reversibility rule at non zero depth can
be tested by a suitable analysis of the energy spectra obtained in
channeling and blocking conditions respectively. Some typical
backscattered energy spectra are reported in Fig. 1 for 0.6 MeV H^+
entering or revealed along different axes of Si and Ge single crystals. The aligned yield in blocking is, except at high energies,
a factor two larger than the corresponding channeling yield. This
trend is independent of the crystal target and of the crystal
directions. Moreover, this same behaviour is also seen by varying
the beam energy as shown in Fig. 2 and the crystal temperature as
reported in Fig. 3. The relative changes of the minimum yield in
blocking at finite depth with decreasing crystal temperature from
300° to 80°K are one half of those corresponding to channeling
measurements. The absolute variation with temperature is then the
same in both experiments.

For a more detailed investigation angular scans have been also
performed in channeling and blocking conditions. The results are

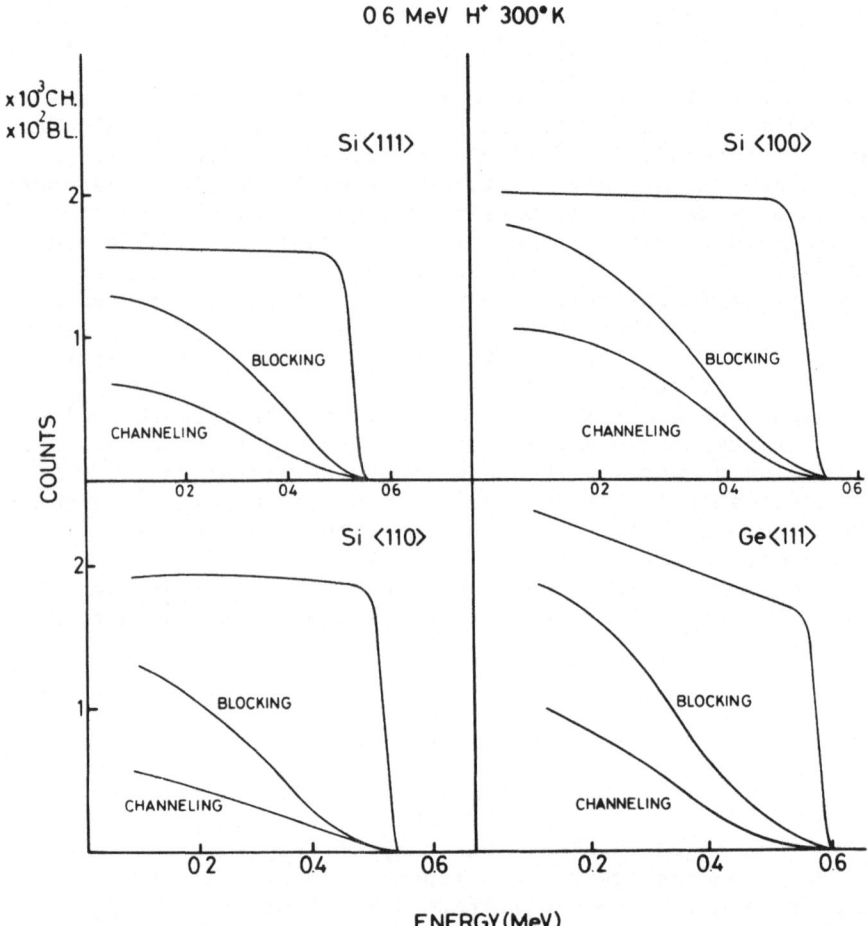

Fig. 1. Energy spectra of 0.6 MeV H$^+$ backscattered from a Si and Ge crystal at 300°K and for different conditions of beam incidence. The aligned channeling yields are measured for beam incidence along the <110>, <111>, <100> directions of Si and the <111> of Ge, the aligned blocking yields are measured for random incidence and for emergence along the same axes.

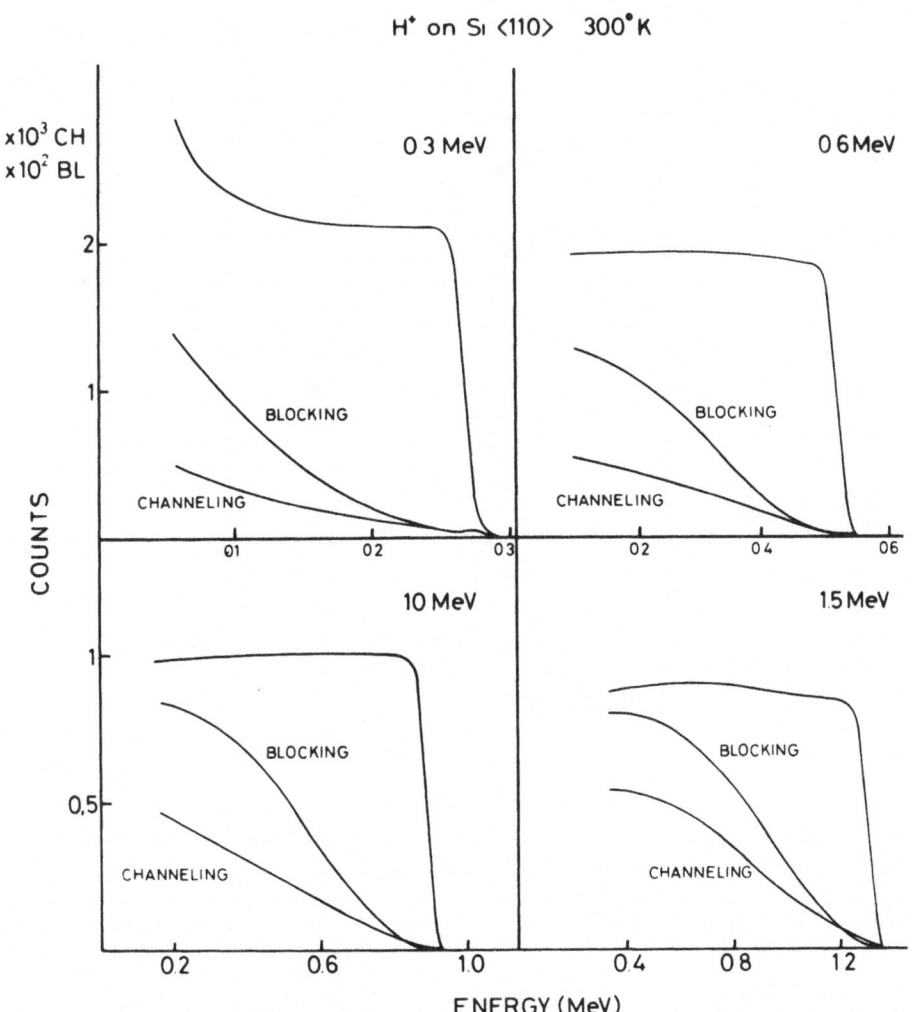

Fig. 2. Energy spectra for proton of incident energy 0.3, 0.6, 1.0 and 1.5 MeV and backscattered from a Si crystal at 300°K. The aligned channeling and blocking yields are recorded for beam incidence and for beam emergence along the <111> direction respectively.

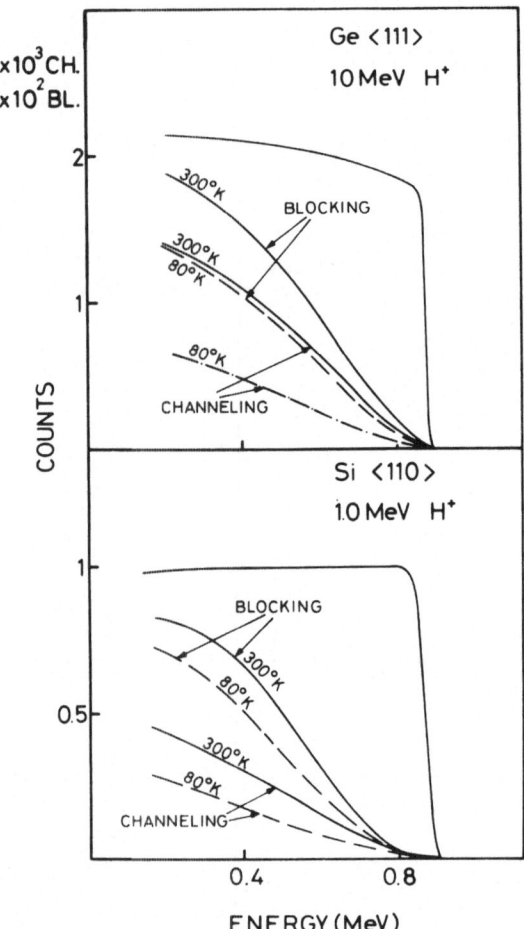

Fig. 3. Backscattered energy spectra for 1.0 MeV H$^+$ entering Si and Ge crystals. The aligned channeling and blocking yields are recorded for beam incidence and for beam emergence along the <111> directions respectively and for 80° and 300°K target temperatures.

shown in Fig. 4 for 1.0 MeV H$^+$ in the <110> Si crystal and for two different revealed energy intervals; one near the maximum energy, i.e., near the surface, and the other at 2/3 of this energy. The angular yield profiles measured near the surface coincide within experimental uncertainties, while going deeper in to the crystal there is a difference in the minimum yield which causes a change in the half width at half minimum $\psi_{1/2}(z)$ which is higher for blocking than for channeling.

All these experimental results verify the reversibility rule only for negligible penetration depths. In fact, blocking and channeling measurements should be identical, when no slowing down

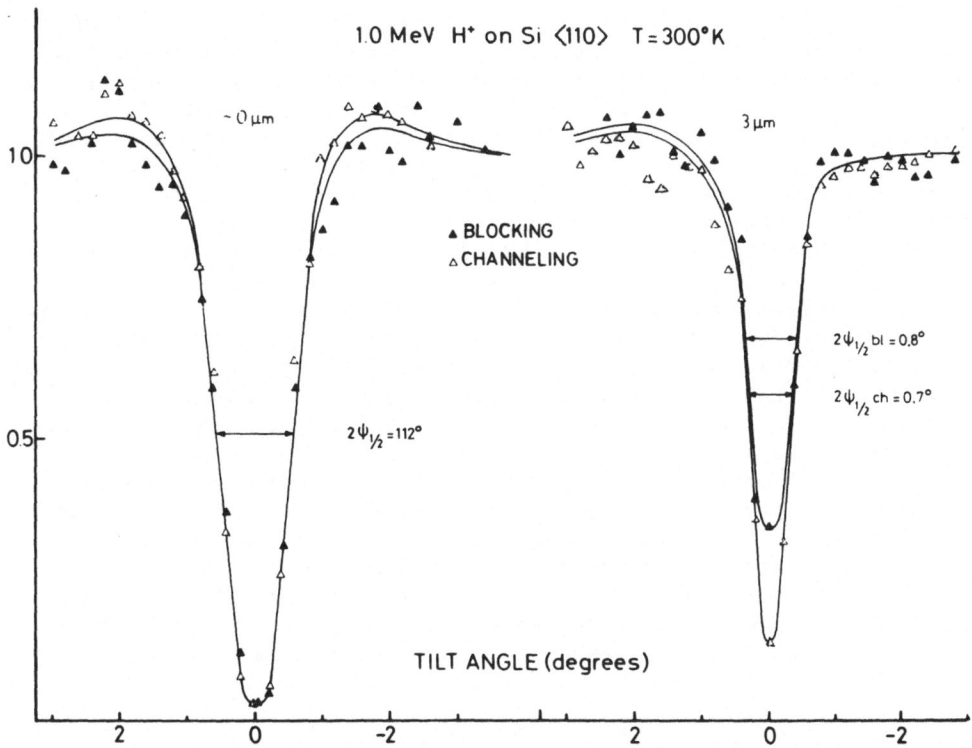

Fig. 4. Angular yield profiles for 1.0 MeV H⁺ in a <110> Si, for blocking (▲) and channeling (Δ) measured near the crystal surface (left hand part) and at a depth of 3 μm inside the crystal (right hand part).

processes are involved, if they are compared for the same particles, energies, crystals, crystal directions, etc. In the present experiments (see Figs. 1, 2, 3 and 4) differences between blocking and channeling measurements performed with the same incident beam energy and with the same geometry arise at finite depth inside the crystal and can be attributed to slowing down processes. As pointed out above, in the blocking case there are several souces emitting backscattered particles whose energy decreases with increasing distance from the surface because of the experienced energy losses. It is then important to see if, after accounting for these energy losses, blocking and channeling measurements maintain their duality even in occurence of multiple scattering.

Consider now in detail how the blocking and channeling aligned yields arise, and assume for simplicity the same stopping power independent of the particle trajectories. In a channeling experiment particles enter the crystal channel with an energy E_o, and at a path length z_1 from the surface some of these leave the channel

going in the random component of the beam. From this exit depth
they interact with all the crystal atoms thus contributing to the
backscattered energy spectrum with a step whose leading edge E_1
corresponds to the outcoming trajectories of length $z_1/\cos\theta$
starting from z_1. Other particles which are dechanneled at another
depth $z_2(z_2>z_1)$ produce another step whose leading edge $E_2(E_2<E_1)$
is shifted towards low energy values. The overlap of all these
steps produces the aligned yield.

In a blocking experiment particles with an energy E_0 enter the
crystal in a random direction. After a path length $z_1/\cos\theta$ some
of these suffer a wide-angle scattering in the detection direction.
In the aligned blocking yield all the particles channeled between
z_1 and the surface are detected at the same energy E_1. In the same
way, from all the particles scattered at a depth $z_2/\cos\theta$ in the
detection direction only those channeled between z_2 and the surface
are revealed as a peak at a lower energy E_2. The overlap of all
these peaks produces the aligned yield in blocking experiments.
A schematic representation of the processes from which channeling
and blocking aligned yields arise is shown in Fig. 5.

To summarize, the measured aligned yield in channeling at a
given backscattered energy E results from particles of incident
energy E_0 dechanneled between the crystal surface and the depth z.
The relationship between z and E is the following

$$E = k^2 \left[E_0 - \int_0^z \frac{dE}{dx} \, dx \right] - \int_0^{z/\cos\theta} \frac{dE}{dx} \, dx$$

where k^2 is the kinematic factor which depends on the mass M of the
crystal atoms, on the mass m of the projectile and on the scattering
angle θ.

The measured aligned yield in blocking at the same detected
energy E results from particles emitted at a depth z from the sur-
face with an energy $k^2 (E_0 - \int_0^{z/\cos\theta} \frac{dE}{dx} \, dx)$, and channeled between z
and the surface, the blocking energy spectrum comes out then from
sources emitting with different energies; for instance for 1.0 MeV
H^+ incident beam the source located at 3 μm from the crystal surface emits with an energy 15% lower than that located at the crystal
surface. The implications of the previous considerations must be
then taken into account for a meaningful comparison between channeling and blocking experiments. The first step involves then the
reduction to the same energy of all the blocking sources.

Previous measurements [8,9] performed in Si and Ge in the same
energy range here adopted have shown that the simple z/E parameter
includes both the energy and the depth dependence of dechanneled

CHANNELING

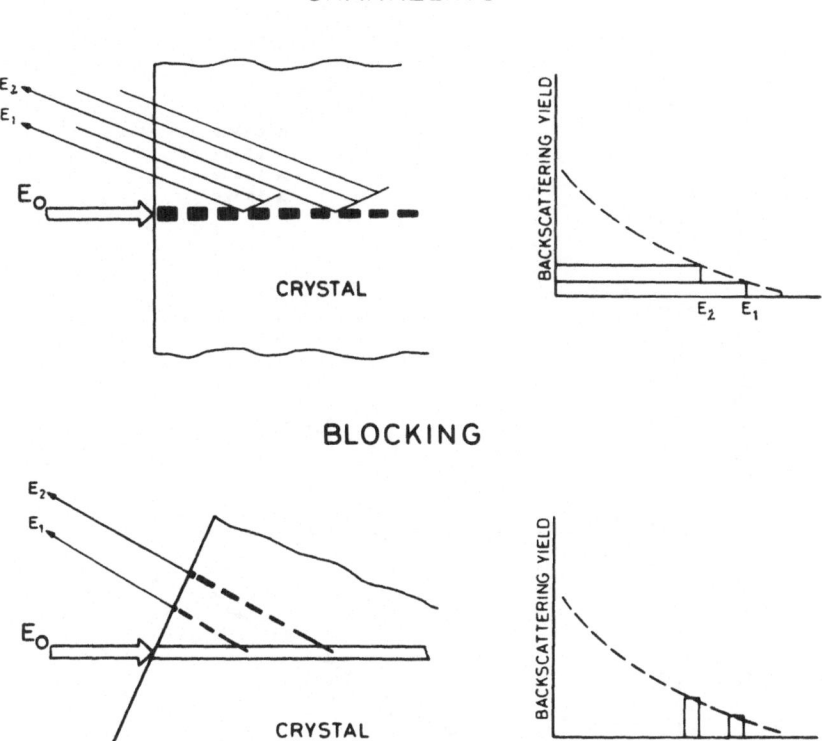

BLOCKING

Fig. 5. Schematic representation of the scattering processes from which the aligned yields in channeling (upper part) and in blocking (lower part) arise respectively.

fractions. For a wider energy range (0.5–12 MeV) of protons in tungsten a better fit of the experimental dechanneled fractions has been found [10] by using $z\,E^{-\alpha}$ with $\alpha = 1/4$. This marked contrast with the silicon and germanium behaviour is still now a puzzling feature because on the basis of nuclear and electronic contribution to dechanneling one might expect to observe an energy dependence that is intermediate between $E^{-1/2}$ and E^{-1}. Because this important energy dependence could be determined in part by the crystal work-manship, the depth and energy dependence of dechanneled fractions has been again measured on the same crystal samples used for block-ing experiments. Fig. 6 shows the obtained z/E dependence of the measured dechanneled fractions for different axes of Si and Ge.

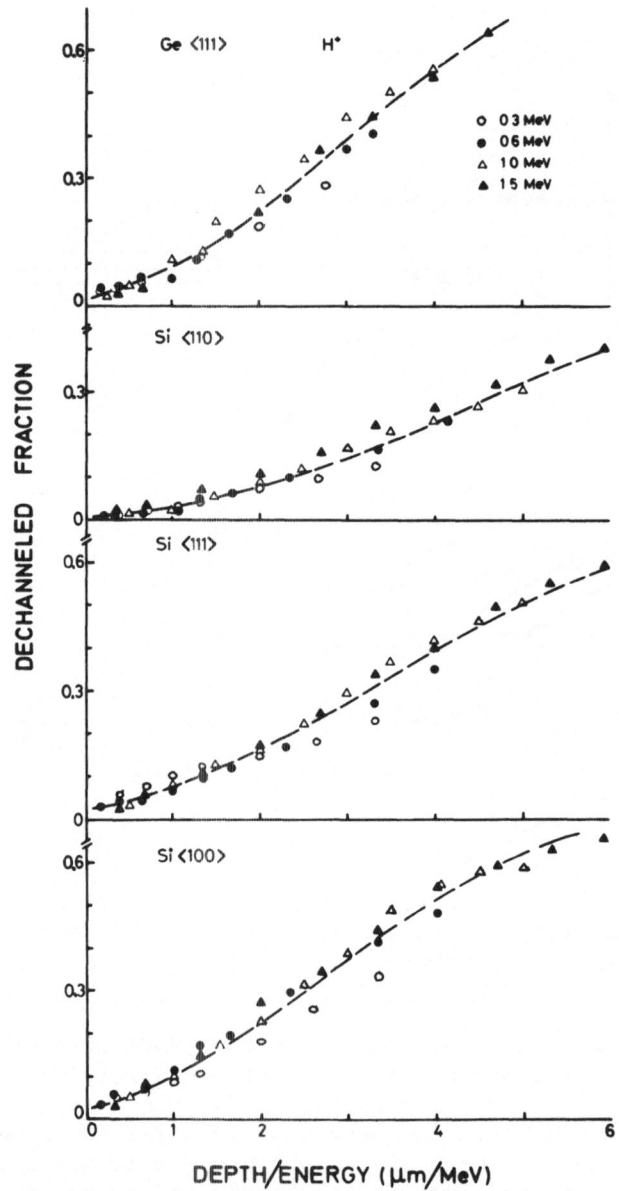

Fig. 6. Dependence of the proton dechanneled fraction on z/E
along the <111> axis of Ge and the <110>, <111>, <100> axes of
Si respectively. The proton beam energies are 0.3(o), 0.6(●),
1.0(Δ), 1.5(▲) MeV.

The energy to depth conversion scale has been obtained following
the usual procedure from reported stopping powers [11] and from
the experimental geometry. The same stopping power has been assumed
for random and channeled trajectories.

As one can see the different dechanneled curves collapse for
all the investigated crystal and crystal axes, thus supporting the
use of the z/E as a scaling parameter at least for Si and Ge in
this energy range. To reduce the blocking sources at the same
energy as the corresponding channeling source the z/E scaling parameter has been used; this choice seems reasonable on the basis of
the experimental results shown in Fig. 6 and for the lack of a
detailed quantitative theory on the energy dependence of dechanneling. The effect of this scaling procedure is shown in Fig. 7. In
the upper part the aligned yield for blocking experiments is reported as given directly from the experimental data, the scaled
values according to the z/E dependence are reported instead in the
lower part and all the emitting sources have been reduced to the
same energy of the corresponding channeling experiments [by taking:
z(scaled) = z(measured) $E_O/E(z)$]. For comparison the aligned
yield for channeling are also shown as dashed lines. This procedure
allows then a meaningful comparison between channeling and blocking.
In all the reported curves aligned yield in blocking is higher
than the corresponding to channeling by a factor which ranges from
10% to 25%. To avoid the scatter between curve and curve and to
see if this trend is a common feature of the measurements, all the
channeling and the corrected blocking data have been plotted in
Fig. 8 as a function of z/E. The full lines delimit the region
in which the experimental points lie; the width of this region is
nearly the same for blocking and channeling. In all the investigated cases the blocking curves are located higher than the channeling ones.

The differences decrease with increasing Si channel width
(i.e., 10% for <100>, 20% for <111>, 25% for <110>) and are nearly
zero for the same axes of Si and Ge. They seem then related to
effects depending on the geometry of the channel.

The still remaining difference between channeling and blocking
measured aligned yield could be attributed to different effects.
A small effect is related to the difference between the angular
distribution of the incident beam particles (channeling) and of
the emitted beam particles (blocking). In channeling, beam-focus
conditions can produce a too-well collimated beam while in blocking
the angular distribution of the emitted beam is spread out. This
effect has been measured by means of a nickel foil set in the first
aperture of the channeling leg and negligible differences have been
observed in the channeling aligned yields so measured. A more
important correction arises if one accounts for the different
energy losses experienced by particles along random or channeled
trajectories. In this case one obtains three different depth
scales, one for random yield, another for channeling aligned yield,
and the last for blocking aligned yield. The two aligned depth

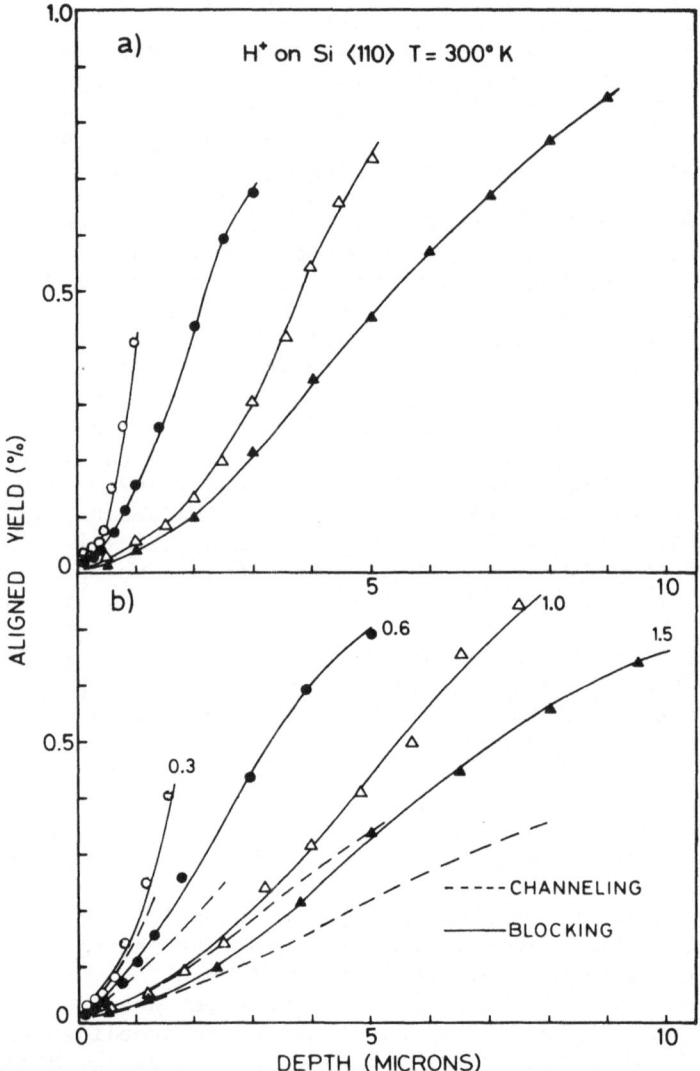

Fig. 7. Normalized aligned yields in blocking condition for 0.3(o), 0.6(•), 1.0(Δ), 1.5(▲) MeV incident proton and revealed along the <110> direction of Si are reported in the upper part as raw data; the same data scaled through the z/E parameter are reported in the lower part. The dashed lines refer to the corresponding yields for channeling.

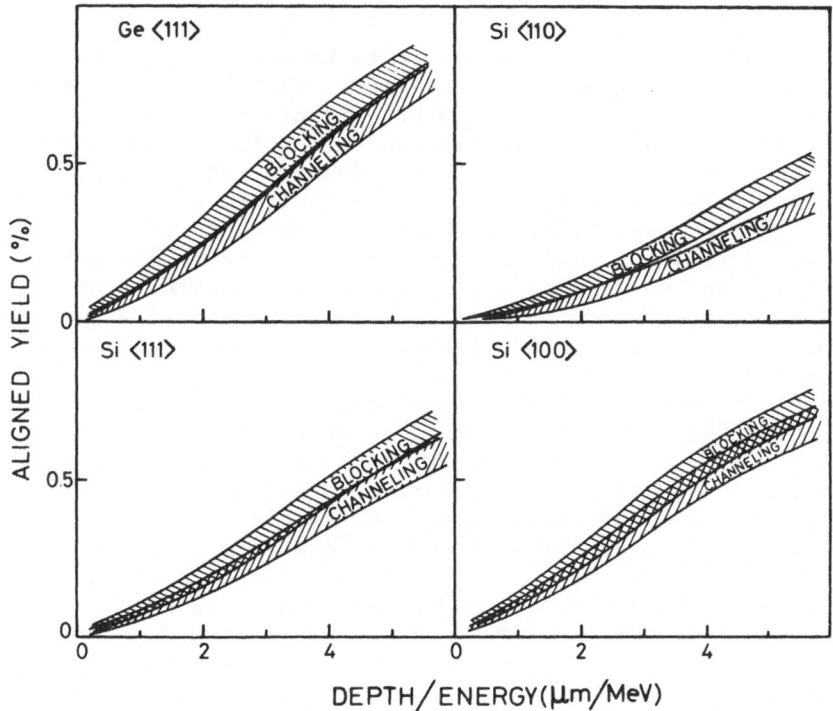

Fig. 8. Dependence of blocking and channeling normalized aligned yields on z/E parameter for proton energy ranging between 0.3 and 1.5 MeV and for the <111> axis for Ge and for the <110>, <111>, <100> axes of Si.

scales are not equal because in channeling the low stopping power is experienced by particles during their incoming trajectories; in blocking instead it is experienced by particles during their out-coming trajectories at a lower energy than that for channeled particles. Due to this difference a path length z traversed by a particle in a channeling experiment produces a larger energy loss than that required for the same length z traversed by a particle in a blocking experiment. With these energy-to-depth conversion scales, the aligned yields (in % of the random yield) decrease both in channeling and in blocking with respect to those computed using the same random scale for all the spectra. The decrease in the blocking aligned yield is moreover larger than that in the channeling aligned yield. Assuming for instance a ratio of 0.5 : 1 between channeling and random stopping powers the difference between blocking and channeling aligned yield is reduced 50% with respect to the raw data.

These corrections should reduce then the small disagreement between channeling and blocking aligned yields found with the present simple data analysis and treatment.

Summary and Conclusion

Channeling and blocking measurements have been performed in Si and Ge crystals at 80° and 300°K temperatures and for proton energies ranging between 0.3 and 1.5 MeV. Attention has been paid to obtaining symmetrical experimental conditions for channeling and blocking measurements. The depth dependence of the aligned yields has been investigated to test the validity of the reversibility rule even in the case of reduced multiple scattering.

The raw data taken for the same energy of the incident beam show in all cases an aligned yield higher for blocking than for channeling. The experimental critical angle depends less on depth in blocking than in channeling, while both minimum yield and critical angle are equal near the crystal surface.

It must be pointed out that in blocking the aligned yield arises from sources which emit particles whose energy decreases with increasing distance from the surface. This fact must be then taken into account for a meaningful comparison between channeling and blocking; i.e., all the blocking sources should be reported to the same energy of the channeling source. The scaling parameter z/E has been used for this reduction. This choice has been made on the basis of the experimental channeling results and because of the lack of a detailed quantitative theory on the depth dependence of dechanneling. After this correction the blocking data are slightly higher than the channeled ones; the difference ranges between 10% and 25% and it seems correlated to the channel width. This disagreement could be reduced further on using a more sophisticated data treatment accounting for the different energy losses between random and channeled trajectories and for the difference between the angular distribution of the incident beam in channeling and of the emergent beam in blocking.

The reversibility rule has been tested to be valid even in the presence of multiple scattering within 20% in the framework of this simple data treatment. This interpretation of the blocking spectra is of interest for a correct use of the blocking phenomenon in its applications as for instance in nuclear lifetime measurements, in lattice location of impurities and in damage profiles obtained by means of uni-axial or double alignment techniques.

It will be of interest to investigate the depth dependence of blocking also in the planar case and in the axial case for other elements, for instance W where a quite different energy dependence behavior has been found for the dechanneled fraction.

Acknowledgments

The help of G. Caruso and V. Scuderi in performing measurements is gratefully acknowledged.

References

*Supported in part by Gruppo Nazionale di Struttura della
Materia del Consiglio Nazionale delle Ricerche and by Centro
Siciliano di Fisica Nucleare e di Struttura della Materia.

[1] J. Lindhard, Kgl. Danske Videnskab Selskab Mat. Fys. Medd.
No. 14 (1965).

[2] E. Bøgh and J. L. Whitton, Phys. Rev. Letters 19, 553 (1967).

[3] For a review see D. S. Gemmel, "Channeling and Related Effects
in the Motion of Charged Particles Through Crystals" (Rev.
Mod. Phys., to be published).

[4] W. M. Gibson and K. O. Nielsen, Phys. Rev. Letters 24, 114
(1970).

[5] V. S. Kulikanskas, M. M. Malov and A. F. Tulinov, Sov. Phys.
JETP 26, 321 (1968).

[6] B. R. Appleton and L. C. Feldman, in Atomic Collision Phenomena
in Solids (North-Holland, Amsterdam, 1970) p. 417.

[7] V. S. Andreev, V. N. Gagaev, I. K. Goliukov, A. A. Puzanov,
A. F. Tulinov, Proc. of the III All Union Conference on the
Physics of Charged Particle Interaction with Single Crystals,
Moscow, 1972, p. 140.

[8] G. Foti, F. Grasso, R. Quattrocchi, and E. Rimini, Phys. Rev.
B3, 2169 (1971).

[9] S. V. Campisano, F. Grasso and E. Rimini, Rad. Effects 9, 153
(1971).

[10] J. A. Davies, L. M. Howe, D. A. Marsden, and J. L. Whitton,
Rad. Effects 12, 247 (1972).

[11] C. F. Williamson, J. P. Boujot, and J. Picard, Saclay Report
No. R 3042 (1966) (unpublished).

EFFECT OF REACTION TIME ON THE MINIMUM YIELDS OF AXIAL AND PLANAR BLOCKINGS

F. FUJIMOTO, K. KOMAKI, H. NAKAYAMA,* M. ISHII,*
and K. HASATAKE*
College of General Education
University of Tokyo
Tokyo, Japan

ABSTRACT

The effect of reaction time on the minimum
yield of the axial and planar blockings is studied
by use of the $^{27}Al(p,\alpha)^{24}Mg$ resonance reactions
at various energies. The result shows that the
minimum yield of the axial blocking and mean
square recoil distance have a linear relation.

Introduction

Several groups have investigated intensively the reaction
times ($10^{-15} \sim 10^{-18}$ sec) for the inelastic processes [1-3],
single resonance reactions [4,5] and fission processes [6,7] by
utilizing the blocking effects, and different techniques and
procedures of analysis were used.

The present authors obtained the reaction times for the
$^{27}Al(p,\alpha)^{24}Mg$ resonance reactions at various proton energies from
the ratio of dip areas in the planar blockings with and without
the lifetime effect [4]. Furthermore, we studied the relations
between the lifetimes and the minimum yields in the axial and
planar blockings [8]. In the present work, we report the last
study more in detail and discuss the analysis methods to determine
the lifetimes from the planar and axial blocking effects.

Experimental Arrangements

The experimental arrangements were reported in the previous
paper [4]. Here we explain the outline.

A thick aluminum crystal was mounted on a goniometer so that the (111), (11$\bar{1}$) planes and the <1$\bar{1}$0> axis are inclined with about 1°, 71.5°, and 84.5° against the incident beam direction, where it is expected that the (111) planar blocking has no lifetime effect. The α-particles from nuclear reaction were detected by celluloid films which were located at the distance of 35 cm from the target. The angular resolution at the films was about 10'.

After the bombardment, the films were developed and the tracks of the α-particles were observed. The distributions of track density around the (111) and (11$\bar{1}$) planar blocking dips were obtained in the region between 1 and 2 cm from the center of the <1$\bar{1}$0> axial blocking dip by counting the tracks for each rectangular area with 0.25 mm width and 10 mm length. For the <1$\bar{1}$0> axial blocking, the distribution along the direction normal to the incident beam was measured by counting the tracks for each small rectangular area with 0.25 mm width and 1.0 mm length. The experiment was carried out once or twice for each resonance reaction at the proton energies E_p = 1183, 937, 731, and 633 keV.

Experimental Results

The angular distributions of the planar blocking dips observed at the 937 and 633 keV resonances are shown in Fig. 1a and b. In Fig. 1a we can see that the minimum yield for the (11$\bar{1}$)

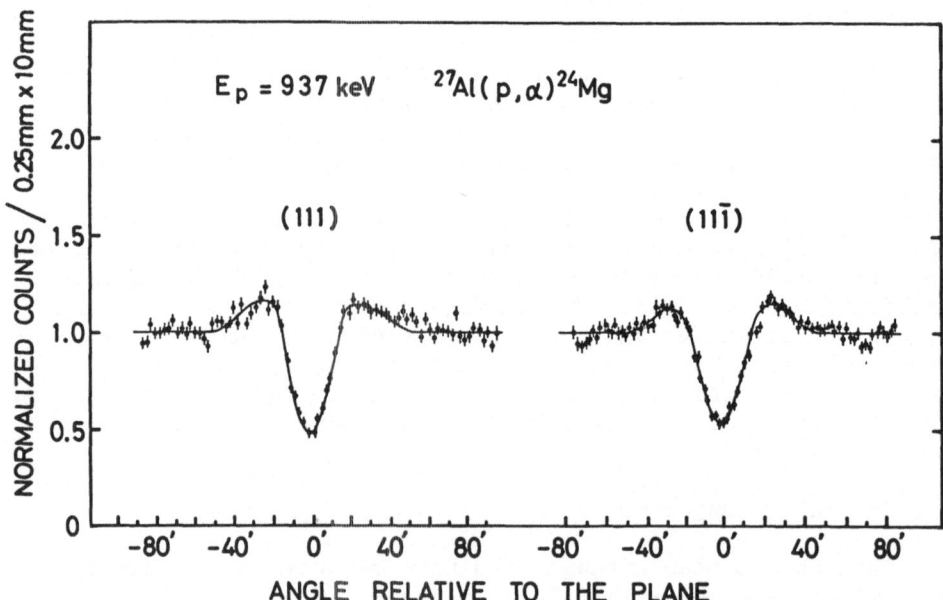

Fig. 1a. The (111) and (11$\bar{1}$) planar blocking dips observed at E_p = 937 keV.

plane is slightly higher than that of the (111) plane. For the
633 keV resonance, the blocking dip of the (11$\bar{1}$) plane is much
shallower than that of (111), though the statistics is poor due
to the low yield. On the other hand, the minimum yields of the
(11$\bar{1}$) and (111) blocking dips at E_p = 1183 keV agreed within the
statistical error. The average minimum yields of the (111) blocking dips at each resonance were nearly equal.

The ratios of the dip areas of the (111) and (11$\bar{1}$) blocking
were obtained from the observed angular distributions. The dip
areas depend on the upper limit of the yield integration over
scattering angle; however, the ratios were almost constant at each
resonance in the angular range shown in Fig. 2. Further, the
ratios obtained from different experiments at the same energy agreed
within the statistical errors, even if the minimum yields vary
slightly. The reaction times, τ, were obtained from the ratios of
dip areas, applying the analytical formula given by Komaki and
Fujimoto [9]. The obtained reaction times at the energies of
E_p = 633, 731, 937, and 1183 keV were τ = $(1.35 \pm 0.19) \cdot 10^{-16}$,
$(5.91 \pm 0.75) \cdot 10^{-17}$, $(1.13 \pm 0.75) \cdot 10^{-17}$, and $\leq 1.4 \cdot 10^{-17}$
sec, respectively.

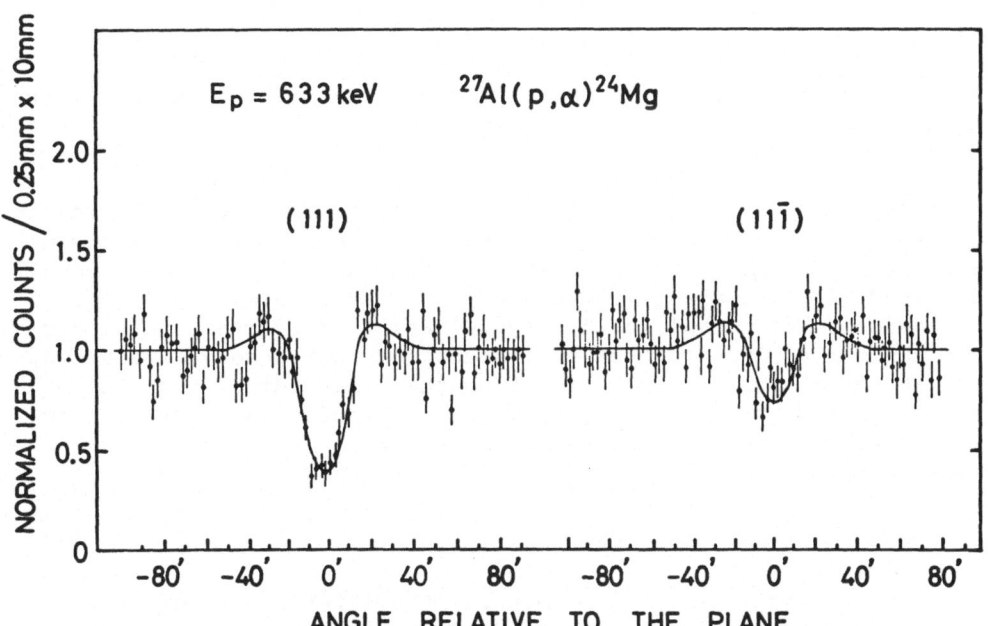

Fig. 1b. The (111) and (11$\bar{1}$) planar blocking dips observed at
E_p = 633 keV.

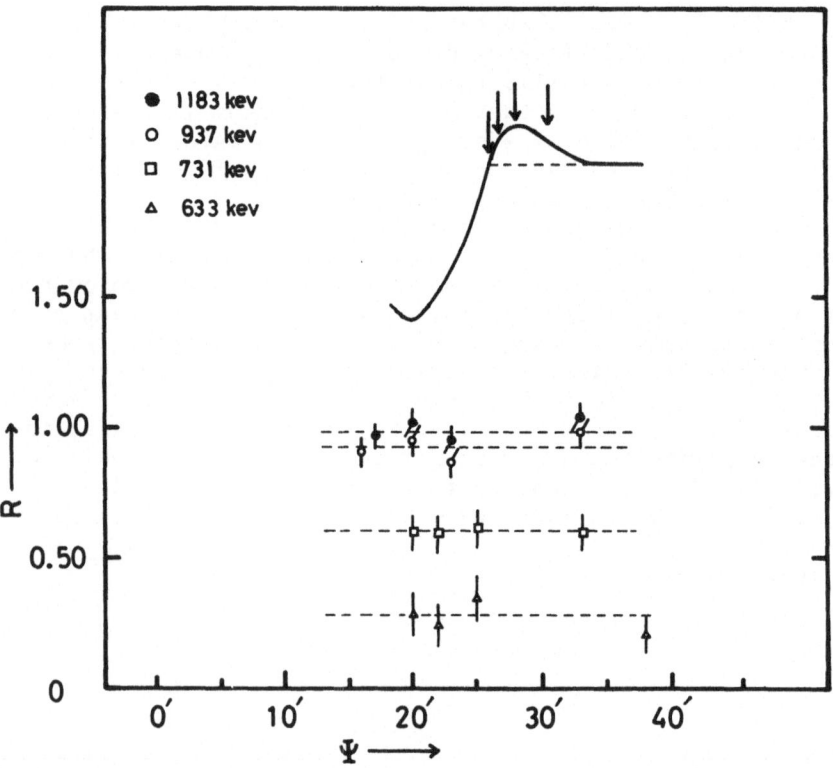

Fig. 2. Variations of the ratio of dip areas with the upper limit of the yield integration over scattering angle.

The <110> axial blocking dips at various resonance energies are shown in Fig. 3. In Fig. 4, the relation between the minimum yield of the <110> axial blocking dip and the square of mean recoil distance, $(v_\perp \tau)^2$, are given, where $v_\perp$ is the recoil velocity of compound nuclei normal to the crystal axis. The error limits for the minimum yield have been estimated from four observed values at the bottom of the dip, that is, from the track number for 1x1 mm^2 rectangular region at the bottom. Fig. 4 indicates that the lifetime effect of the minimum yield of the axial blocking dip can be represented by

$$\chi = \chi_0 + 2CNd\pi(v_\perp \tau)^2, \tag{1}$$

where χ_0 is the minimum yield without the lifetime effect, N the atomic density and d the atomic distance in the string direction.

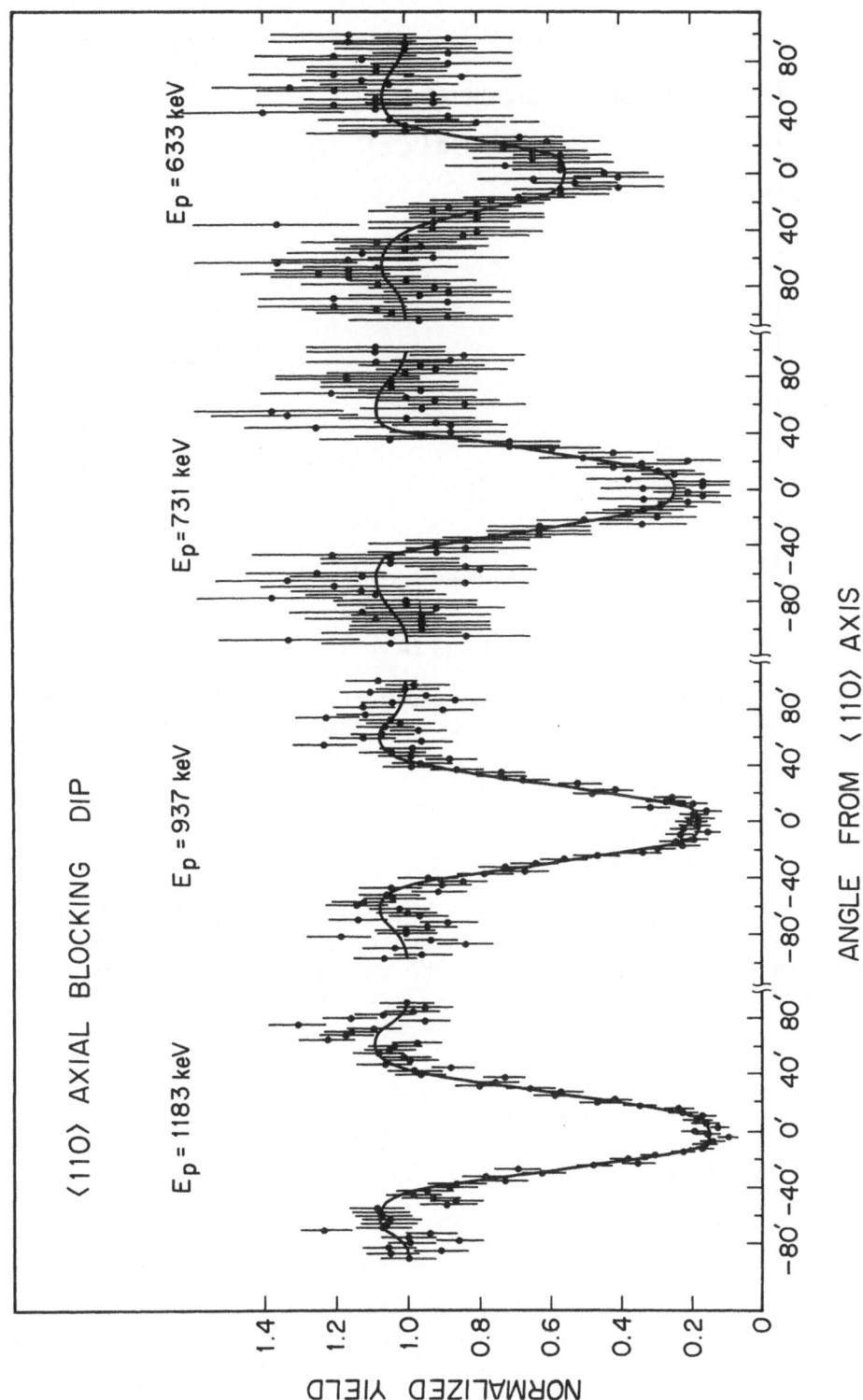

Fig. 3. The <110> axial blocking dips observed at various resonance energy.

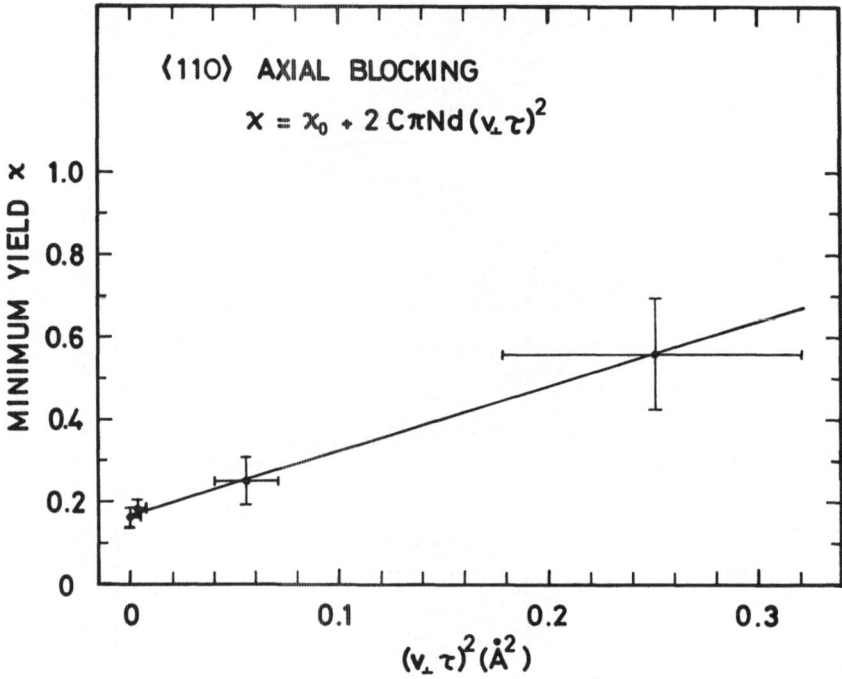

Fig. 4. Variations of the minimum yield of the <110> axial blocking with the mean square recoil distance from the atomic string.

C is given by

$$C = 1.44 \pm 0.31 . \tag{2}$$

We tried to get a similar relation on the minimum yield of the $(11\bar{1})$
planar blocking, χ'. The result is shown in Fig. 5, where the
abscissa is $v'_\perp \tau$, the prime meaning the quantities for the planar
case. From Fig. 5, we can get the relation:

$$\chi' = \chi'_o + C' \, v'_\perp \tau / d_p , \tag{3}$$

where d_p is the lattice spacing and

$$C' = 1.17 \pm 0.37 . \tag{4}$$

Discussions

Gibson and Nielsen [3] have proposed the formula (1) with the
cut-off distance r_c under the assumption that the particles emitted

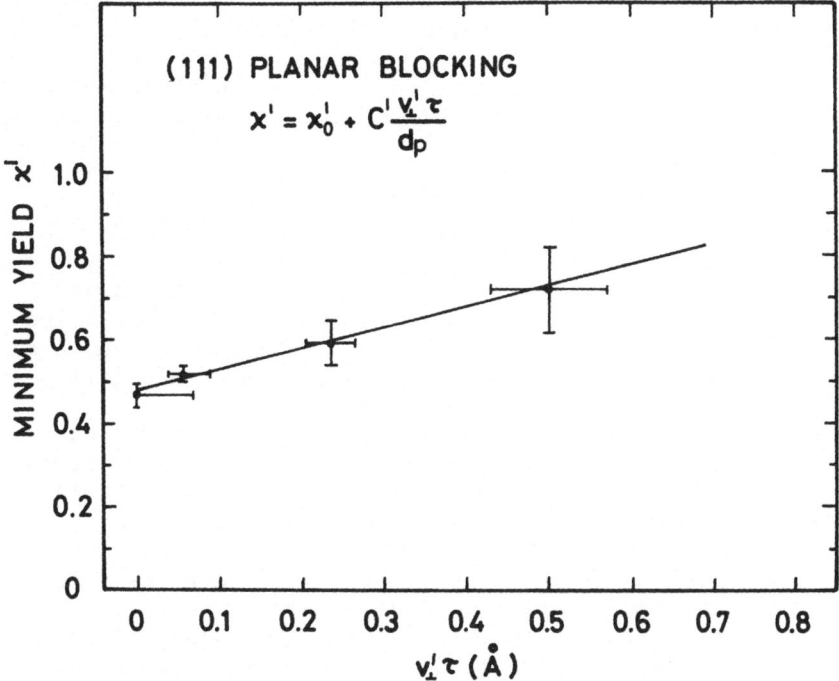

Fig. 5. Variation of the minimum yield of the (11$\bar{1}$) planar blocking with the mean recoil distance from the lattice plane.

outside r_c from the atomic strings do not show the blocking effect
and r_c is 4 $\sim$ 3 times of a_{TF}, the Thomas-Fermi radius. The parameter C = 2.5 was adopted under the assumption that the effect of
displacement of emitting atom on the minimum yield is equivalent
for the case of the thermal vibrations. The latter effect was calculated by Barrett [10] using computer simulation and C was estimated
to be C $\simeq$ 3.

Recently, Sharma et al. [5] investigated the reaction time of
the $^{27}Al(p,\alpha)^{24}Mg$ reaction at E_p = 633 keV by use of the <110> axial
blocking. The minimum yield observed under the same crystal arrangement as ours shows a good agreement. They confirmed also the linear
relation between $(v_\perp\tau)^2$ and χ as well as the present work, changing
$v_\perp$ at the same resonance energy. The reaction time was decided by
putting C = 1.5 $\sim$ 4 in (1). The lower limit C = 1.5 has been
estimated because C may approach to 1.5 for large displacements of
emitting atoms. The obtained reaction time, τ = (1.0 ± 0.3) · 10^{-16}
sec is consistent with our result within the error limit including
the variation in the choice of C.

Barrett [10] has calculated the temperature dependence of the
planar minimum yield and shown that the minimum yield is represented
by a linear relation with the thermal vibration amplitude. The

corresponding value of C' in (3) is estimated as C' = 4.5 which is much larger than that for the lifetime effect given by (4), as well as the axial case.

Hashimoto et al. [11] have carried out a large scale Monte Carlo calculation on the variation of the axial minimum yield with mean recoil distance for the Ge(p,p')Ge reaction. The results show a strong dependence on recoil direction and indicate that the analytical and computer calculations based on the string theory give large error in the axial blocking case. On the other hand, the results suggest that the minimum yield can be represented by the formula (1) in the small recoil distance. However, the value of C is much smaller than that for the thermal vibrations. The value given by (2) has the same tendency as their results, though we cannot directly compare, since the parameters in their calculation are quite different from the present case.

The formula given by Komaki and Fujimoto [9] is obtained by considering the orbits of particles in the periodic planar potential. We can easily calculate the formula numerically for any crystal. Another simple point for the planar case is no effect of recoil direction.

The axial blocking effect is more sensitive than the planar one. However, the accuracy of the lifetime determined from experiment does not depend only on the sensitivity, but also on the statistical error. The minimum yield of the axial blocking dip can be obtained by counting the particles in the limited angular region, while the statistical error for the planar case can be made small by counting in the long region along a plane. Fig. 4 indicates that the errors induced by the statistics in both cases are the same magnitude, where the region in the planar case is ten times longer than that in the axial one. It is not so difficult to get better statistics by taking longer region.

From the above consideration, the lifetime measurement by use of the planar blocking may be more useful than that by axial one, unless we analyze by means of the result of a large scale Monte Carlo calculation.

References

*Department of Applied Physics,Tokyo Institute of Technology,Tokyo.
[1] M. Maruyama, K. Tsukada, K. Ozawa, F. Fujimoto, K. Komaki, M. Mannami, and T. Sakurai, Phys. Lett. 29B, 414 (1969); Nucl. Phys. A145, 581 (1970).
[2] G. J. Clark, J. M. Poate, F. Fuschini, C. Maroni, I. G. Massa, U. Uguzzoni, and E. Verondini, Nucl. Phys. A173, 73 (1971).
[3] W. M. Gibson, Y. Hashimoto, R. J. Keddy, M. Maruyama, and G. M. Temmer, Phys. Rev. Lett. 29, 74 (1972).
[4] K. Komaki, F. Fujimoto, H. Nakayama, M. Ishii, and K. Hisatake, Phys. Lett. 38B, 218 (1972); Nucl. Phys. A204, 545 (1973).
[5] R. P. Sharma, J. V. Anderson, and K. O. Nielsen, Nucl. Phys. A204, 371 (1973).

[6] W. M. Gibson and K. O. Nielsen, Phys. Rev. Lett. 24, 114 (1970).

[7] Yu. V. Mekikov, Yu. D. Otstavnov, A. F. Tulinov, and N. G. Chetchenin, Nucl. Phys. A180, 241 (1972).

[8] F. Fujimoto, K. Komaki, H. Nakayama, M. Ishii, and K. Hisatake, Rad. Effects 20, 141 (1973).

[9] K. Komaki and F. Fujimoto, Phys. Lett. 29A, 544 (1969); Phys. Stat. Sol. (a) 2, 875 (1970).

[10] J. H. Barrett, Phys. Rev. B 3, 1527 (1971).

[11] Y. Hashimoto, J. H. Barrett, and W. M. Gibson, Phys. Rev. Lett. 30, 995 (1973).

HIGH INDEX PLANAR CHANNELING IN SILICON

H. E. ROOSENDAAL, W. H. KOOL, F. W. SARIS,
*FOM-Instituut voor Atoom- & Molecuulfysica
Kruislaan 407, Amsterdam, The Netherlands*

and

W. F. VAN DER WEG[*]
*Philips' Natuurkundig Laboratorium
Amsterdam, The Netherlands*

ABSTRACT

The backscattering yields for several planar
directions have been measured for 120, 160 and 190
keV protons, incident on Si crystals. The measured
half-angles at half-minimum $\psi_{1/2}$ and minimum yields
χ_{min} are discussed in terms of an analytical model,
in which the effect of surface transmission is included.

Comparison between the experimental values and
the model gives good agreement, when taking as the
minimum distance of approach for the channeling
protons the sum of the Thomas-Fermi screening radius
a and the mean square thermal displacement of the
lattice atoms perpendicular to the plane $\rho_\perp$. In
addition the stopping power near the planes is discussed yielding values of 3 to 5 times higher than
for random incidence.

---

[*] On leave at California Institute of Technology,
Pasadena, California, U.S.A.

SECTION IX
CHANNELING

DOUBLE PLANAR ALIGNMENT SCATTERING WITH A VERY THIN CRYSTAL

M. J. GAILLARD, J.-C. POIZAT and J. REMILLIEUX
*Institut de Physique Nucléaire, Université Claude Bernard Lyon-I and
Institut National de Physique Nucléaire et de Physique des Particules
43, Bd du 11 novembre 1918, 69621 Villeurbanne, France*

and

F. ABEL, M. BRUNEAUX and C. COHEN
*Groupe de Physique des Solides de l'Ecole Normale Supérieure
9, quai Saint Bernard, 75005 Paris, France*

ABSTRACT

We have observed variations with incident
energy of the scattering yield of H^+ and He^+ ions
measured in a particular configuration of planar
double alignment. The structure of these variations is related to the oscillating nature of the
trajectories of the ions in planar channeling.
Experimental oscillation lengths are deduced for
H^+ and He^+ ions in the (111) direction, and for
H^+ ions in the (110) direction.
We have performed some calculations, in which
the stochastic model of Abel et al., describing the
scattering probability in single alignment conditions,
has been extended to the double alignment geometry.

Introduction

It is well known [1] that charged particles channeled along a
planar direction follow trajectories that oscillate between two
individual atomic planes. It has been shown also that the large
angle scattering probability of the unchanneled fraction exhibits
depth dependent oscillations [2-4]. Although they are rapidly
damped with increasing depth, they indicate that unchanneled particles feel the planar potential and undergo some kind of periodic
collective motion.

In a preliminary paper [5] we have shown that the scattering
probability from a thin crystal in a particular geometry of double

779

planar alignment exhibits an energy dependence related to the particular behaviour of unchanneled particles.

We present here a more extensive study of the double alignment
scattering yield of H^+ and He^+ ions incident along (111) and (110)
planar directions in thin gold crystals of various thicknesses,
in the 0.4 to 2 MeV energy range. Variations of this yield as a
function of the incident energy can be observed if the number of
interactions suffered by particles with atomic planes is small.
Then, the knowledge of the crystal thickness enables us to calculate
the oscillation lengths of the trajectories. This is of particular
interest if the energy resolution of the detector makes it impossible to observe any structure in the backscattering spectrum.
The oscillation lengths obtained by these two methods (backscattering spectra and double planar alignment scattering) have been compared when it was possible to use simultaneously both of them (for
He^+ ions). Furthermore, the energy dependence of the double alignment scattering yield has been calculated by using the reversibility
principle between channeling and blocking, and the stochastic model
of Abel et al. [6].

Experimental

The particles of the incident beam are injected along the
direction of a low index planar channel of a thin crystal. Among
the wide angle scattered particles, the particles detected are those
emerging from the crystal along the same planar channel (or along
a planar channel of the same family).

The wavelengths of the trajectories inside the planar channels
of these particles are nearly the same before and after the scattering event as long as the energy change due to the wide angle scattering can be neglected, just as the slowing down of the ions inside
the crystal.

At a given incident energy E, the probability for scattered
particles to emerge from the crystal along the blocking direction
depends on the ratio of their pathlength P inside the crystal to
the oscillation length L(E) (half-wavelength). Let n be this ratio,
such as

$$P = n\, L(E)$$

We can expect that this probability reaches, at least for
small values of n, a maximum (or a minimum) when the incident energy
E is such that this number n is an integer (or half-integer). But
we can also expect to observe collective effects if all the detected
particles have had the same pathlength inside the crystal wherever
the scattering event has occurred; it will be so if the foil is in
the bisecting plane of the incident beam and the blocking directions.

Any fluctuation of the double alignment scattering yield with
energy will be related to the quasi-periodic nature of unchanneled

particle trajectories. Such measurement will enable us to deduce n, the number of interactions of unchanneled particles with atomic planes, and the oscillation lengths associated with such trajectories.

The experimental apparatus is shown in Figure 1. A beam of H^+ or He^+ ions from a 2 MeV Van de Graaff accelerator is collimated to a half-width angle $\leq 0.03°$. The beam spot size on the crystal is 0.5 mm. The self-supporting [111] orientated gold crystals are obtained by the technique described in ref. [3]. The sample thickness is measured by comparing the widths of energy spectra of α particles backscattered respectively from the crystal (for a random incidence) and from amorphous gold foils of known thicknesses. The accuracy of such measurements is about 4% in the considered thickness range (300 – 600 Å).

The crystal is tilted into planar channeling conditions by means of a single alignment detector sustending a large solid angle. The double alignment detector can be moved in a X-Y box. It has a middle position (X = Y = 0) lying in the horizontal plane of the incident beam, at $\Theta_S = 46°$ off it, and is collimated by a slit parallel to the blocking planar direction (angular acceptance <0.08°). The foil is kept in a vertical plane so the tilt angle θ required to meet the conditions on pathlengths is determined by Θ_S. In fact the value of θ is 24°. Although the foil is not exactly in the bisecting plane of the incident beam and blocking directions,

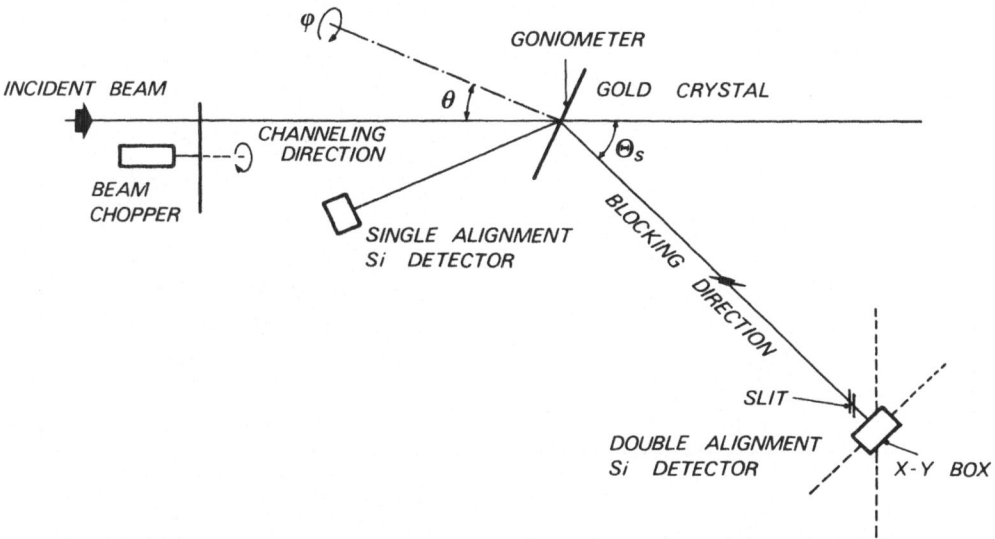

Fig. 1. Double alignment configuration. The double alignment Si detector (blocking direction) is collimated by a slit orientated in a direction parallel with the studied planar channel (angular aperture <0.08°). The angular spread of the incident beam is <0.03°.

the dispersion of the pathlengths of the scattered particles is
quite small and the geometry is such that open axial directions
are avoided.

In the (110) case, the channeling and blocking directions lie
in the same plane, but in the (111) case, two different (111) directions are used.

The incident beam current was measured with a beam-chopper.

Results

For each incident energy of the ion beam, the normalized single
alignment scattering yield χ_{SA} and the normalized double alignment
scattering yield χ_{DA} were simultaneously registered. The random
scattering yields were measured while the gold crystal was regularly rotated around the surface normal (the total rotation being

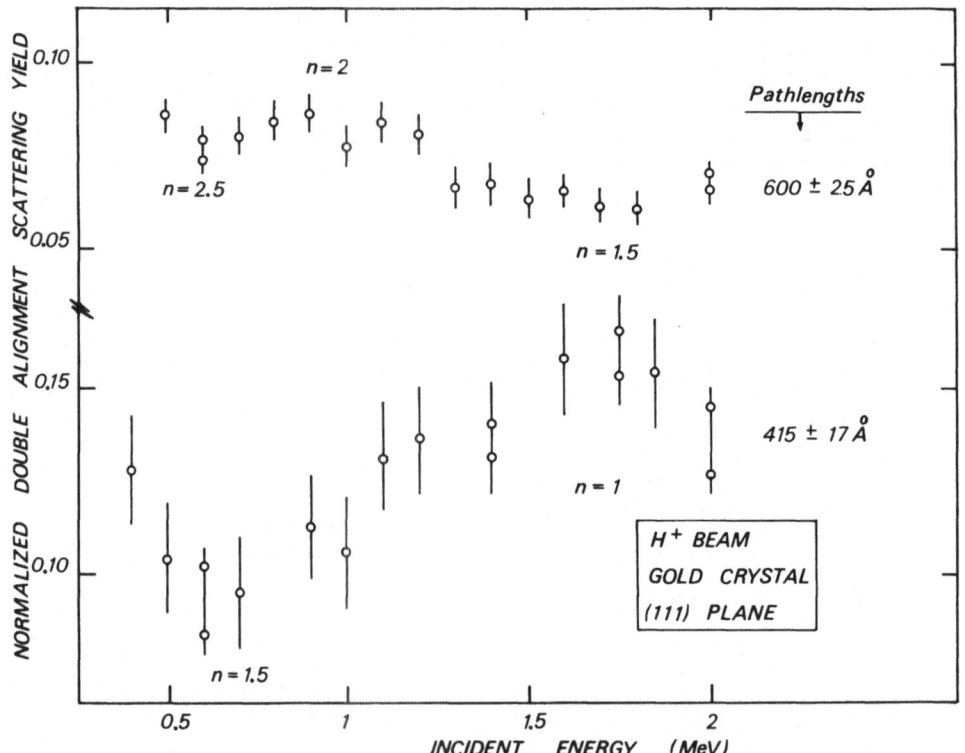

Fig. 2. Variation of the normalized double alignment scattering
yield versus the H^+ incident energy measured along (111) planes
for two gold crystals. Channeling direction in the (11$\bar{1}$) plane,
blocking direction in the (1$\bar{1}$1) plane. The number of oscillations
n is indicated near the corresponding extremum.

$\Delta\phi = 120°$). The ion dose needed to measure χ_{DA} at a given energy was in some cases large enough (typically 1 $\mu C/mm^2$ for 1 MeV H^+ ions) to produce damages affecting the measurements; it was reasonable to correct the measured double alignment yield χ_{DA} by the factor $(\chi_{SA}(0)/\chi_{SA})^2$, $\chi_{SA}(0)$ being the single alignment scattering yield of the non-damaged crystal. This factor could often be neglected and was generally lower than 10%. Besides, the initial qualities of these thin crystals estimated by measuring $\chi_{SA}(0)$ are slightly different. But, even if the double alignment scattering yield is weakly dependent on the initial crystal quality, its variations with energy are very little affected.

Figure 2 shows the normalized double alignment scattering yield versus the H^+ incident energy (from 0.4 to 2 MeV) measured along (111) planes of two gold crystals in which pathlengths were respectively 415 and 600 Å . Strong variations are observed for the first crystal. The extrema unambiguously correspond to a

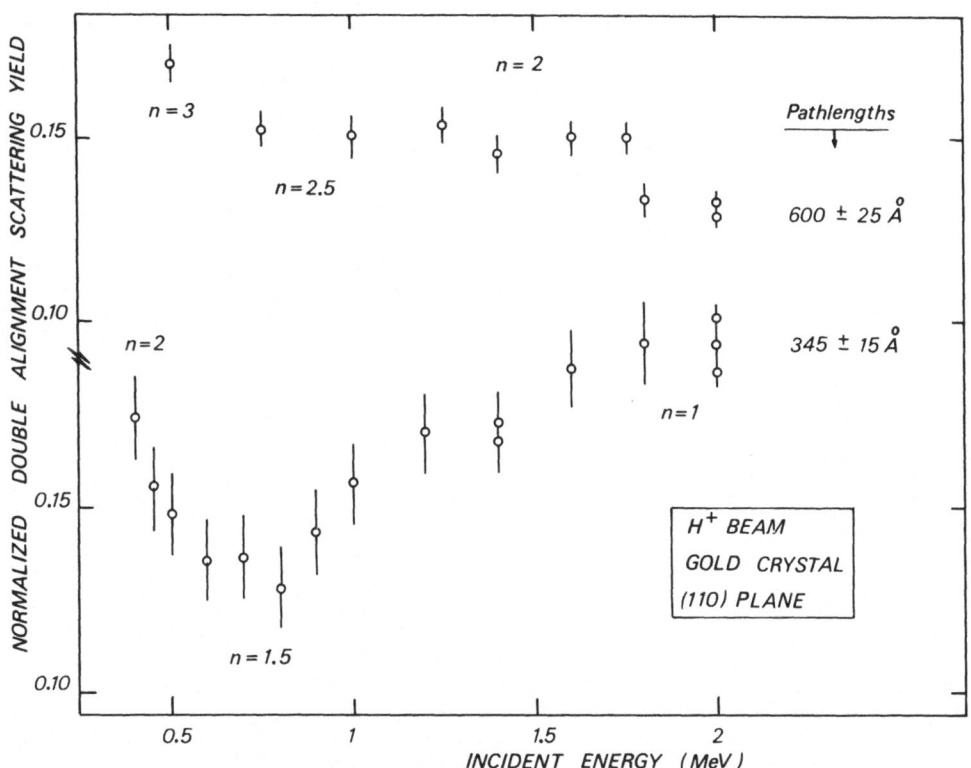

Fig. 3. Variation of the normalized double alignment scattering yield versus the H^+ incident energy measured along (110) plane of gold crystals. Channeling and blocking are in the same (110) planar direction. The number of oscillations n is indicated near the corresponding extremum.

number of oscillations n = 1 and n = 1.5 of the H⁺ ions inside the
crystal. For the second one in which the number of oscillations is
larger, only an attenuated structure is observed.

Figure 3 shows the same data for the (110) plane. In this
case, a stronger damping of the structure when the pathlength
increases is observed.

Figure 4 shows that, if He⁺ ions are used, fluctuations are
almost invisible for the (111) direction and quite invisible for
the (110) direction, even with a very thin crystal (pathlength
360 Å).

Discussion

These results provide first an experimental determination of
the oscillation length L(E) of unchanneled particles (having a
transverse energy equal to the height of the potential barrier) in
planar channeling.

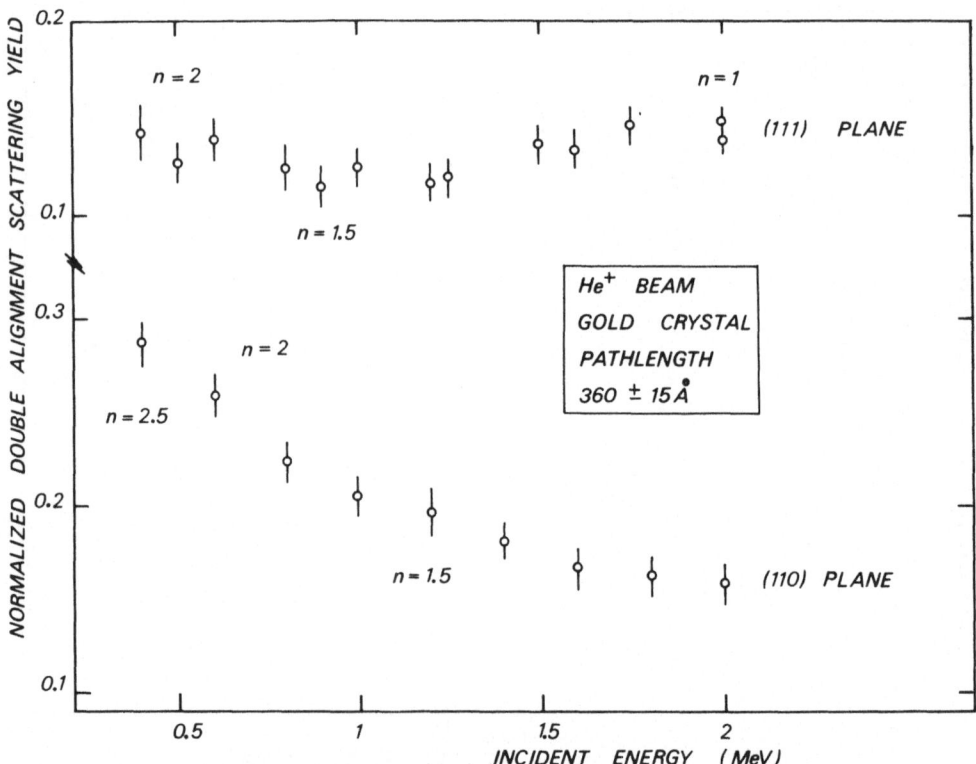

Fig. 4. Variation of the normalized double alignment scattering
yield versus the He⁺ incident energy measured along (110) and (111)
planes of a same thin gold crystal. The n values corresponding to
the number of oscillations in the planar directions concerned are
indicated near the corresponding extrema.

   If the crystal thickness is known, one can determine the oscillation lengths at the energy corresponding to the extrema observed
on the Figures 2-4.
   Two uncertainties are associated with this technique, one about
the position of the extrema (energy losses in the crystal are negligible), the other about the crystal thickness.
   The oscillation lengths so determined for H^+ and He^+ ions in
the (111) directions are reported on the Figure 5.
   In the case of He^+ ions in the (111) plane, it is possible
to determine the oscillation lengths by this double alignment
technique and also by the backscattering method. Our previous

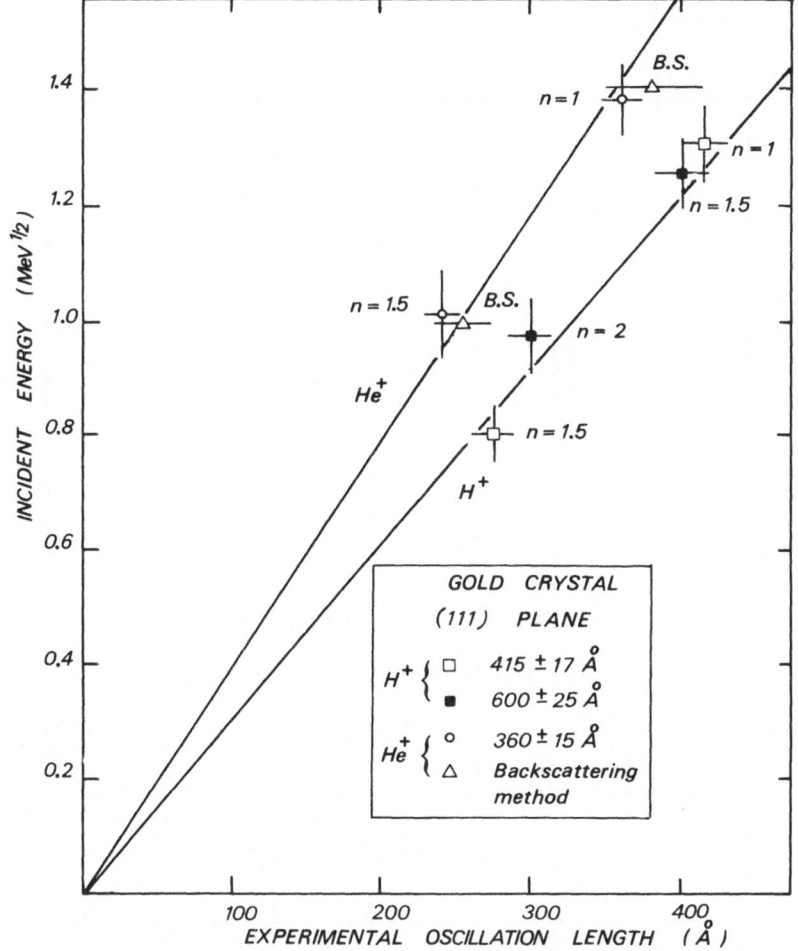

Fig. 5. Oscillation length for H^+ and He^+ ions in the (111) planes
of gold crystals: a) measured on the Figures 2-4; b) obtained by
the backscattering method (for He^+ ions).

results obtained by this last method [3] for 1 MeV and 2 MeV He$^+$ ions are also reported on Figure 5. The comparison shows a good agreement between the two methods.

The variation of the experimental oscillation lengths L with the energy E is compatible with the expression L = K$\sqrt{E}$. The straight lines reported on Figure 5 are such as:

$$L_{H^+}(111) = 325 \sqrt{E} \; (\overset{\circ}{A}, \; MeV)$$

$$L_{He^+}(111) = 255 \sqrt{E} \; (\overset{\circ}{A}, \; MeV).$$

In the same way, the best fit in (110) plane for H$^+$ ions is:

$$L_{H^+}(110) = 270 \sqrt{E} \; (\overset{\circ}{A}, \; MeV).$$

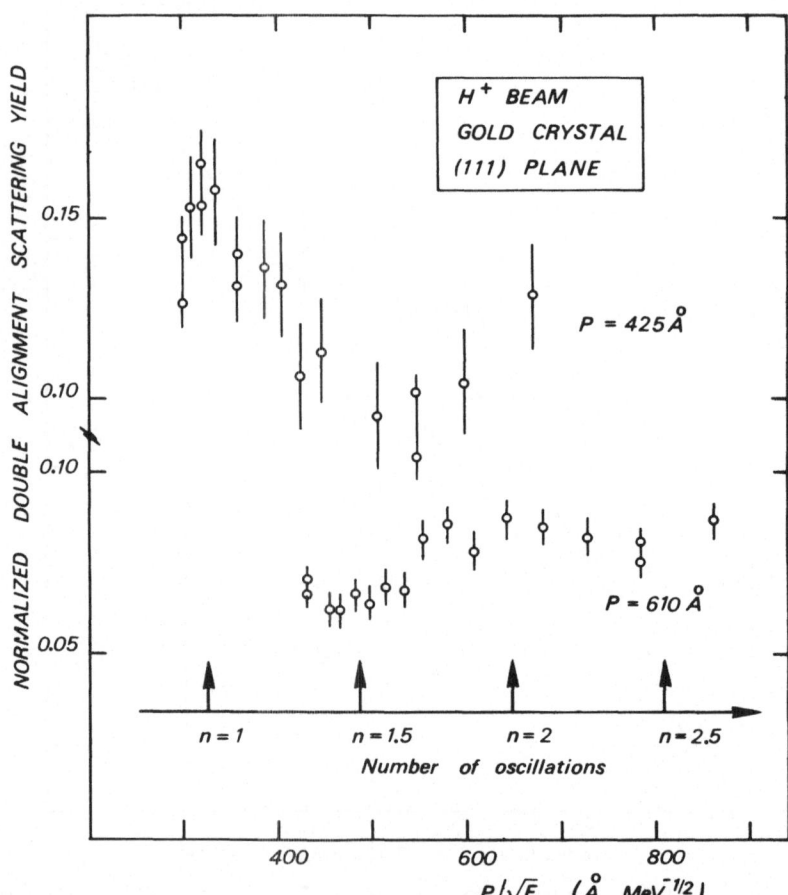

Fig. 6. Normalized yield for a H$^+$ incident beam in the (111) double alignment configuration as a function of P/$\sqrt{E}$ ($\overset{\circ}{A} \cdot$MeV$^{-1/2}$) and also of the number n of oscillations.

It can be noticed that these expressions indicate a dependence on the planar spacing d_p and the charge of incident ions Z through the factor $(d_p/Z)^{0.37}$. It is interesting to compare it with the $(d_p/Z)^{0.5}$ dependence which could be anticipated by assuming that the oscillation length depends on the critical angle ψ_c, in the first approximation, simply as d_p/ψ_c.

Then we can express all the experimental data obtained with various targets for a given planar direction in terms of $P/\sqrt{E}$ or the number of oscillations n. For each crystal, we have to determine a pathlength P compatible with the expressions of L vs E and the incident energy at which the extrema of χ_{DA} vs E are observed. The P values so obtained have always been found compatible with the measured thicknesses.

Figures 6 and 7 show the normalized double alignment scattering yield measured with H^+ ions in (111) and (110) planes versus

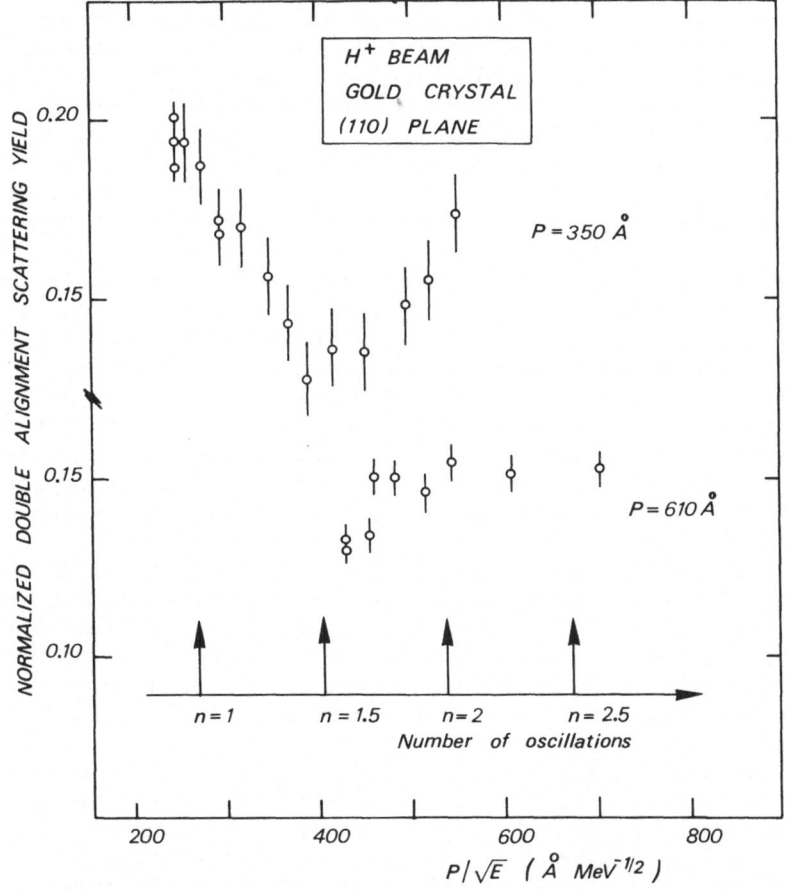

Fig. 7. Normalized yield for a H^+ incident beam in the (110) double alignment configuration as a function of $P/\sqrt{E}$ ($\text{Å} \cdot \text{MeV}^{-1/2}$) and also of the number n of oscillations.

$P/\sqrt{E}$; the P values used for each crystal and the n scale are also
reported on the figures. In this representation, the strong damping observed when the number of the oscillations increases indicates
that the statistic equilibrium is reached after a few oscillations
along the planes; about 2.5 oscillations for the (111) planes and
2 for the (110) planes.

It can be noticed that double planar alignment experiments
in which particles scattered at a small depth are detected emerging
from the front surface at 180° or at 90° (with a crystal tilted at
≃45°) would give χ_{DA} values independent on energy and close to the
maxima observed here, the number of oscillations in the crystal
would always be an integer (if the kinematic energy loss due to the
scattering event is small). It is possible that discrepancies
between experiments [7,8] and Barrett calculations [9] about double
axial alignment scattering yield could be explained by such effects,
although the transverse motion of particles is much more complex
in axial directions.

Let us mention also that the geometrical differences in the
(111) and (110) planes measurements are qualitatively similar to
the differences in axial double alignment measurements performed
respectively at 90 and 180° in terms of the spatial distribution of
the struck atoms in the blocking transverse plane [9]. However,
this effect is thought to be much smaller here (thermal motions
are less effective in planar than in axial blocking or channeling)
and would affect only the amplitude of the χ_{DA} variation with
energy.

Some calculations of the double alignment scattering yield
versus incident energy have been performed. We use a stochastic
model [6] which calculates the backscattering probability as a
function of depth in planar channeling and we apply the rule of
reversibility between channeling and blocking: the probability
for a particle scattered at a certain distance x of the exit surface
to emerge from the crystal in the blocking direction is assumed to
be equal to the probability for a particle entering the crystal
in the channeling direction to be scattered at a penetration depth
x.

The probability for a particle entering the crystal in the
channeling direction to emerge in the blocking direction is the
sum over all depths x of the product of two probabilities: probability for that particle to be scattered at a depth x and probability to emergy after the scattering event in the blocking direction.

Such calculations have been performed for (111) planes with
H^+ and He^+ incident ions.

The values of the parameters used in this calculation have
been extracted from theoretical adjustments of backscattering
experiments with He^+ ions in a gold crystal (pathlength 980 ± 40 Å).
One of these fits is shown on Figure 8 in the case of 1.5 MeV He^+
ions. Figure 9 shows the comparison between the calculated and
the experimental results for the double alignment scattering yield
of He^+ incident beam in (111) planes.

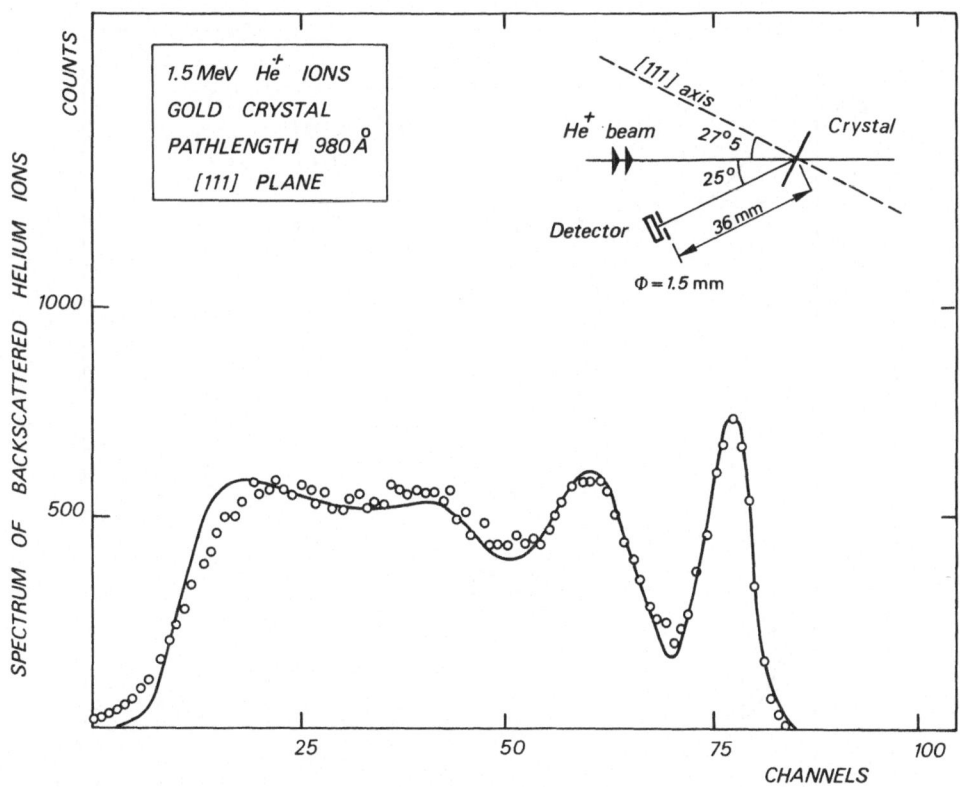

Fig. 8. Single alignment backscattering energy spectrum of 1.5 MeV
He[+] ions incident along the (111) direction. Experimental points
and spectrum calculated by using the stochastic model (solid curve).

    We have also evaluated the parameters for protons in the same
(111) case. The calculated curve and the experimental points are
shown on Fig. 9.
    Rather good fits are obtained. Nevertheless, further calcula-
tions are necessary to improve and test the values of these parameters and their variations with the incident energy, the nature of
the incident ions and the nature of the target.

References

[1] B. R. Appleton, S. Datz, C. D. Moak, and M. T. Robinson, Phys.
    Rev. B4, 1452 (1971).
[2] E. Bøgh, Rad. Effects 12, 13 (1972).
[3] J. C. Poizat and J. Remillieux, Le Journal de Physique 33,
    1013 (1972).
[4] F. Abel, G. Amsel, M. Bruneaux and C. Cohen, Phys. Letters
    42A, 165 (1972).

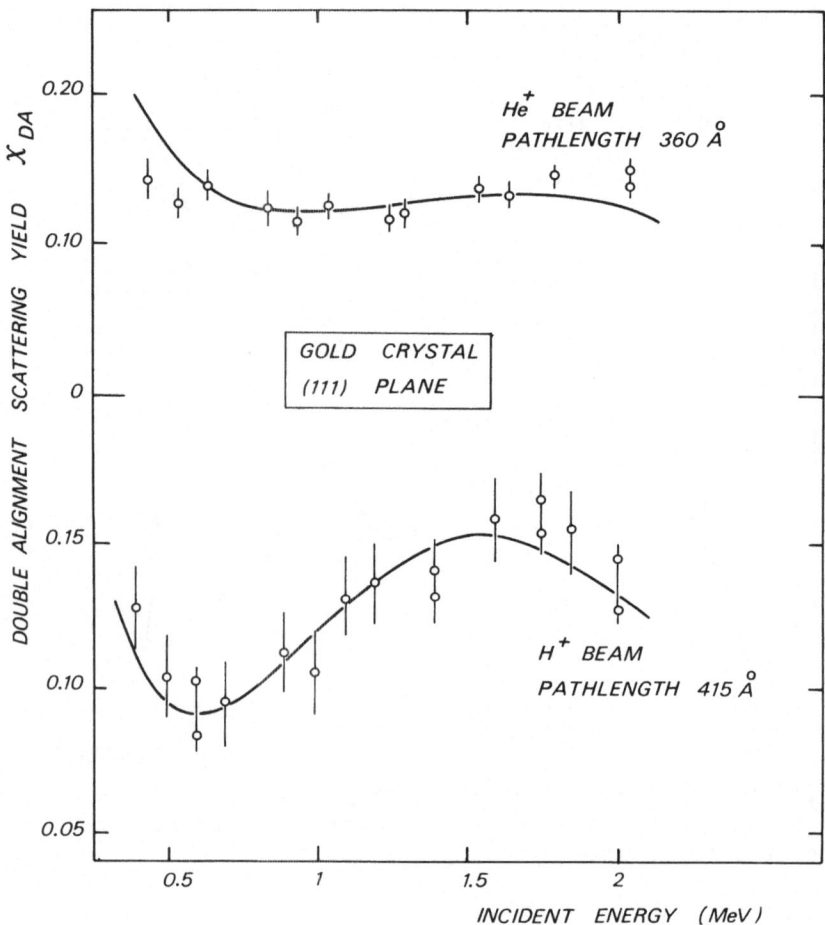

Fig. 9. Double alignment scattering yield as a function of energy of H^+ and He^+ ions incident along the (111) direction. Experimental points and curve calculated by using an extension of the stochastic model to the double alignment geometry (solid curve).

[5] M. J. Gaillard, J. C. Poizat and J. Remillieux, Phys. Letters 45A, 325 (1973).
[6] F. Abel, G. Amsel, M. Bruneaux and C. Cohen, C. R. Acad. Sci. Paris 276, 267 (1973) and these Proceedings.
[7] B. R. Appleton, L. C. Feldman, Atomic Collisions in Solids (eds. D. W. Palmer, M. W. Thompson and P. D. Townsend, North Holland, Amsterdam, 1970) p. 417.
[8] S. T. Picraux, W. L. Brown and W. M. Gibson, Phys. Rev. B6, 1382 (1972).
[9] J. H. Barrett, Phys. Rev. B3, 1527 (1971).

AXIAL AND PLANAR CHANNELING IN TiO$_x$ SYSTEM*

G. DELLA MEA, A. V. DRIGO, S. LO RUSSO, P. MAZZOLDI
and S. YAMAGUCHI**
Istituto di Fisica dell'Università
Unita del Gruppo Nazionale di Struttura della Materia
Padova, Italy

and

G. BENTINI, A. DESALVO[†] and R. ROSA
Laboratorio per la Chimica e Tecnologia
dei Materiali e Componenti per l'Elettronica
Consiglio Nazionale delle Ricerche, Bologna, Italy

ABSTRACT

A channeling study of TiO$_x$ crystals with different oxygen concentration has been performed. In this system oxygen atoms distribute in the octahedral interstitial sites of the h.c.p. metal lattice. The effect of the titanium and oxygen sublattice were separated by observing both the deuteron spectra backscattered from the titanium sublattice and the proton yield from the $^{16}O(d,p)^{17}O^*$.

The results concerning the channeling half angle are in satisfactory agreement with the theoretical predictions of Barrett. In the {0001} and {10$\bar{1}$0} planes, which consist entirely of titanium or oxygen atoms, flux-peaking effect in the yield of the nuclear reaction was always observed and this effect decreases with increasing oxygen concentration.

The axial channeling results for ten atomic percent of oxygen display a weak flux-peaking effect, while at higher concentrations we observe the angular dip of the oxygen sublattice.

A Monte Carlo program to simulate the effect of the interstitial impurities has been tried for a silicon crystal with increasing boron concentration. The results concerning the influence of the intersitital concentration on flux-peaking or

channeling are in qualitative agreement with our
experimental results in TiO_x h.c.p. structure.

Introduction

In recent years several channeling experiments have been
carried out on diatomic crystals [1,2,3,4]. The mean value of
atomic number and of atomic spacing has been introduced in the
Lindhard [5] equations to discuss experimental results.

In diatomic crystals with large difference either in atomic
number Z or in concentration of components, the steering power is
mainly determined by the high atomic number array. Consequently
the flux distribution along the low Z rows will be enhanced (see
for example Ref. [6],[7]) and high values of the minimum yield
and of dechanneling rate with respect to the low Z sublattice could
be expected. A systematic study on diatomic crystals having different concentrations of the low Z element may give valuable information on channeling phenomena.

Because of the wide range of interstitial solubility of oxygen
in titanium, up to the composition of $TiO_{0.5}$, TiO_x alloy was selected for the present study.

Deuterons at 1.0 MeV energy were chosen as projectile in order
to permit simultaneous observation of the channeling behaviour both
with respect to the titanium atoms, by means of wide-angle Rutherford scattering, and with respect to the oxygen atoms, by means of
the $^{16}O(d,p)^{17}O^*$ reaction.

In the primary solid solution of the Ti-O system an ordered
phase is present below a certain temperature, dependent on the
oxygen concentration. The ordered structure consists of antiphase
domains of several hundred Ångstroms diameter [9]. In the disordered TiO_x structure, oxygen atoms occupy fractionally the octahedral interstitial sites in the h.c.p. titanium lattice as shown
in Fig. 1. Lattice constants and atomic densities as function of
oxygen concentration are listed in Table I [9].

In what follows, the Bravais-Miller 4-index symbols associated
to the hexagonal system will be used to denote crystallographic
directions and planes. Figure 2 shows the atomic arrangements
for three axial directions of TiO_x lattice. As far as planar
channeling is concerned, this structure shows interesting features.
For some planar configurations such as the $\{11\bar{2}0\}$, each plane
contains both elements, whereas the $\{0001\}$ and $\{10\bar{1}0\}$ planes consist of a single atomic species as shown in Fig. 3.

In the present work a channeling study has been performed on
TiO_x crystals with five oxygen concentrations, i.e., Ti, $TiO_{.11}$,
$TiO_{.20}$, $TiO_{.28}$ and $TiO_{.39}$ to obtain information about the concentration dependence of channeling parameters such as critical angle
$\psi_{1/2}$, minimum yield, χ_{min}, and dechanneling rate.

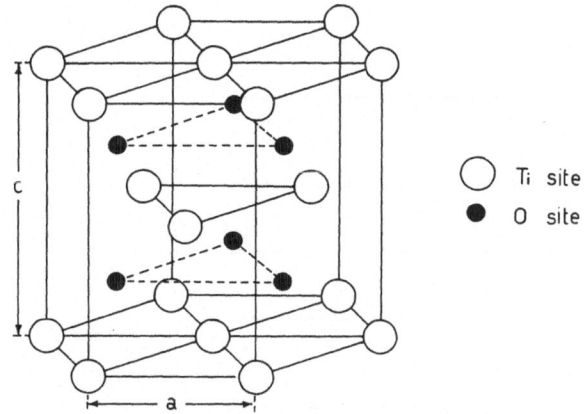

Fig. 1.   Structure of disordered TiO$_x$ alloy.   Open circles are
titanium atoms.   Solid circles indicate available sites for oxygen.
Oxygen atoms distribute statistically with the occupation probability: x≡O/Ti.

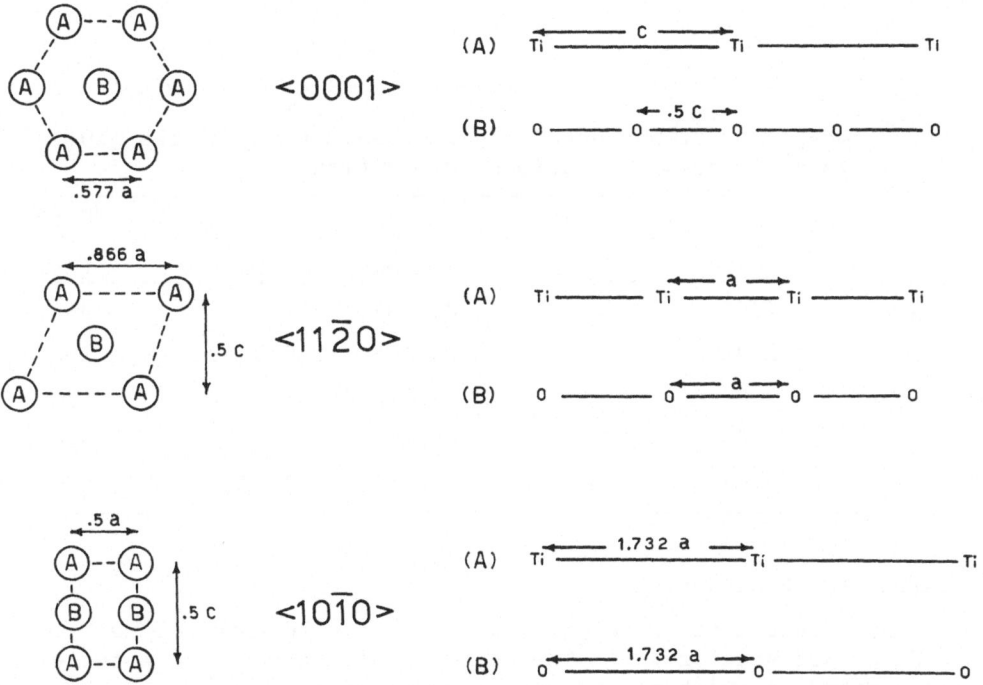

Fig. 2.   Atomic arrangements for the <0001>, <11$\bar{2}$0> and <10$\bar{1}$0>
axial directions in TiO$_x$.   The arrangements shown on the left are
"end views" of the channel and letters refer to the individual
strings shown on the right.

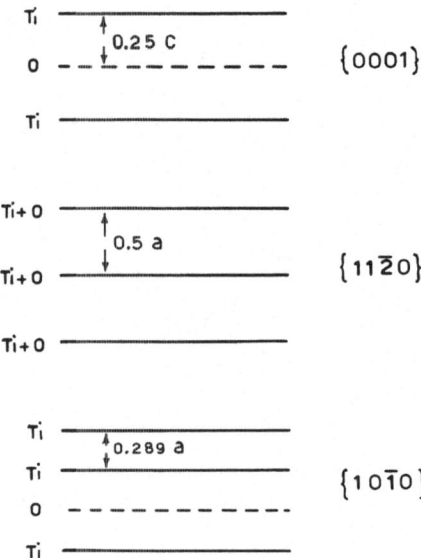

Fig. 3. Planar atomic configurations and interplanar distances in TiO_x.

Table I

Lattice constants, atomic densities and Debye temperatures [10] of TiO_x alloys are listed for various compositions.

	Lattice Constants(Å)		Atomic Densities(Atoms/Å^3)		Debye Temp. (°K)
	a	c	Ti Atoms	O Atoms	
Ti	2.95	4.68	0.0567	————	363
Ti-$O_{0.11}$	2.96	4.72	0.0558	0.0062	390
Ti-$O_{0.20}$	2.97	4.75	0.0553	0.0111	450
Ti-$O_{0.28}$	2.97	4.77	0.0549	0.0155	520
Ti-$O_{0.39}$	2.97	4.79	0.0546	0.0216	600

Experimental Procedure

The measurements were carried out on the Legnaro 5.5 MeV Van de Graaff accelerator. The crystals were mounted on a three axis goniometer which allowed orientation of the sample with an accuracy better than 0.05°. All meausrements were made at room temperature. A monoenergetic beam of 1 MeV deuterons was used; typical beam currents ranged from 10 to 20 nano amps, and the beam divergence was less than 0.03°. Backscattering deuterons and

protons from the $^{16}O(d,p)^{17}O^*$ nuclear reaction were detected by
means of two solid state surface barrier detectors. A nickel foil
thick enough (about 5 µm) to stop the backscattered deuterons was
inserted in front of the proton detector to avoid the deuteron high
flux. The acceptance angles of the detectors were chosen in such
a manner that the proton counting rate was sufficiently high without appreciable pulse pile-up in the deuteron detector. Crystalline
samples of TiO$_x$ were grown using the zone melting technique described
elsewhere [11,12]. Before installation in the goniometer chamber
the crystals were chemically polished. The samples have been annealed to obtain an ordered structure but as for the TiO$_{.11}$ specimen
there is the possibility to have a mixture of disordered and ordered
structure.

Data Analysis

    In order to obtain the information on the channeling behaviour
as a function of depth it is necessary to convert the energy scale
of spectra into an equivalent depth scale. In the present study
the stopping power for TiO$_x$ was obtained from the Bragg rule, which
states that the stopping power for the X$_m$Y$_n$ compound is given by

$$\left(\frac{dE}{dz}\right)_{X_m Y_n} = m\left(\frac{dE}{dz}\right)_X + n\left(\frac{dE}{dz}\right)_Y$$

where $\left(\frac{dE}{dz}\right)_X$ and $\left(\frac{dE}{dz}\right)_Y$ are the stopping powers for the X and Y elements respectively. Unfortunately no systematic data on energy
loss in titanium are yet available. Hence the values of $\left(\frac{dE}{dz}\right)_{Ti}$
was evaluated approximately from the relation

$$\left(\frac{dE}{dz}\right)_{Ti} = \frac{Z_{Ti}\, \rho_{Ti}}{W_{Ti}} \frac{W_V}{Z_V \rho_V} \left(\frac{dE}{dz}\right)_V$$

where Z, ρ and W correspond respectively to the atomic number, density and atomic weight of titanium and vanadium. The values of
$\left(\frac{dE}{dz}\right)_O$ and $\left(\frac{dE}{dz}\right)_V$ used in this calculation were taken from the stopping power tables by Williamson et al. [13].
    Recent measurements [14,15] report discrepancies with the data
of ref. [13] and a breakdown of the Bragg rule for various com-
pounds [16,17]. Nevertheless the deviations for deuterons in our
energy range should be less than 10%.
    In this work the deuteron backscattering spectra were analyzed
by assuming the channeling stopping power equal to the random one.
The depth analysis was limited to only 1 µm because the low energy
portion of these spectra contains the contribution from the oxygen

atoms. The surface χ_{min} values were obtained by extrapolating the dechanneling curves to zero depth.

The analysis of proton spectra was limited at the peak energy. The $^{16}O(d,p)^{17}O^*$ nuclear reaction cross-section presents a maximum near 970 keV energy [18,19]. Thus the proton peak energy corresponds to a depth ranging from 0.30 to 0.35μm, varying with oxygen concentration.

Experimental Results

Planar Channeling

The composition dependence of the yield of both the elastic scattering from the titanium sublattice and the $^{16}O(d,p)^{17}O^*$ reaction was studied for the $\{1\bar{2}10\}$, $\{0001\}$ and $\{10\bar{1}0\}$ planes whose atomic arrangements are shown in Fig. 3. Figure 4 presents an example of planar yield for both processes obtained from TiO$_{.28}$ crystals. In the $\{1\bar{2}10\}$ plane, where each plane contains titanium and oxygen atoms, the backscattering yield curve shows the same channeling behaviour as that observed for the $^{16}O(d,p)^{17}O^*$ reaction.

In the $\{0001\}$ and $\{10\bar{1}0\}$ planes in which each plane consists entirely of titanium or oxygen atoms, however, the two processes exhibit completely different orientation dependence. The scattering curves from the titanium sublattice show always the expected channeling effect, while the nuclear reaction yield from oxygen sublattice always gives a peak for planar alignment.

The steering power of the pure oxygen planes appears to be too weak to maintain a steered trajectory. The oxygen planes may

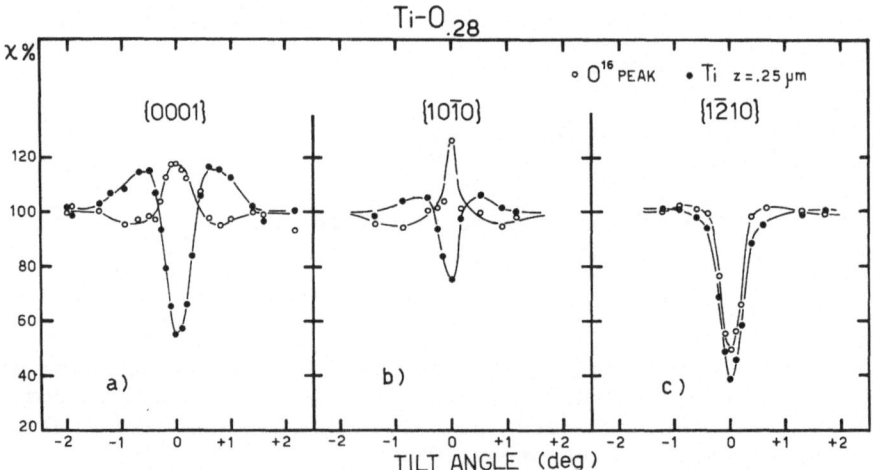

Fig. 4. Channeling normalized yields for: (a) $\{0001\}$, (b) $\{10\bar{1}0\}$, and (c) $\{1\bar{2}10\}$ planes of TiO$_{.28}$ crystals. o: proton yield; •: deuteron backscattering yield.

be considered simply as interstitial atoms and the flux-peaking effect is evident also even at the highest oxygen concentration investigated. The measured values of $\Psi_{1/2}$ and χ_{min} are listed in Table II and III. A comparison with the theoretical values is given in the following section.

Because our depth analysis is limited to only 1.0μm, we cannot state if the dechanneling is a linear or an exponential function of the depth. Assuming exponential dechanneling [20], we derived the half-thickness for dechanneling, $X_{1/2}$, which is listed in Table IV.

Table II

Measured and calculated critical angles for planar channeling. Calculated values (a) and (b) are given in parentheses below the corresponding measured ones. Experimental values for the oxygen sublattice correspond to a mean depth of 0.3 μm. F.P. means that flux peaking is observed.

Sublattice		Critical Angles $\Psi_{1/2}$ for Planar Channeling				
		Ti	Ti-O$_{0.11}$	Ti-O$_{0.20}$	Ti-O$_{0.28}$	Ti-O$_{0.39}$
{0001}	Ti (a)	0.32	0.30	–	0.22	0.26
		(0.26)	(0.25)	(0.26)	(0.26)	(0.26)
	(b)	(0.37)	(0.35)	(0.35)	(0.35)	(0.35)
	O	–	F.P.	–	F.P.	F.P.
			(NO CH.)	(0.01)	(0.03)	(0.05)
{11$\bar{2}$0}	Ti (a)	0.21	0.26	0.25	0.23	–
		(0.17)	(0.18)	(0.19)	(0.20)	(0.21)
	(b)	(0.29)	(0.28)	(0.28)	(0.28)	(0.28)
	O	–	0.26	0.22	0.21	–
			(0.14)	(0.16)	(0.17)	(0.18)
{10$\bar{1}$0}	Ti (a)	0.20	–	–	0.17	–
		(0.15)	(0.14)	(0.14)	(0.15)	(0.15)
	(b)	(0.27)	(0.27)	(0.27)	(0.27)	(0.26)
	O	–	–	–	F.P.	F.P.
			(NO CH.)	(0.01)	(0.03)	(0.05)

Table III

Measured and calculated minimum yield for planar channeling. Calculated values are given in parentheses below the corresponding measured ones. Experimental values for the oxygen sublattice correspond to a mean depth of 0.3 μm.

Sublattice		Minimum Yield χ_{min}% for Planar Channeling				
		Ti	Ti-O$_{0.11}$	Ti-O$_{0.20}$	Ti-O$_{0.28}$	Ti-O$_{0.39}$
{0001}	Ti	19	40	–	55	50
		(25.5)	–	–	–	–
	O	–	128	–	118	113
			(NO CH.)	(100)	(94)	(91)
{11$\bar{2}$0}	Ti	53	49	36	32	31
		(32.9)	(32.5)	(30.4)	(29.7)	(27.5)
	O	–	66	42	50	38
			(43.2)	(39.8)	(37.9)	(35.5)
{10$\bar{1}$0}	Ti	45	–	–	69	83
		(32.1)	–	–	–	–
	O	–	–	–	127	123
			(NO CH.)	(100)	(94)	(90)

Axial Channeling

A study of the composition dependence of the yield of both the backscattering from the titanium sublattice and of the nuclear reaction from the oxygen sublattice was performed for the <0001> and <11$\bar{2}$0> axes. Along both directions Ti and O atoms lie on separated sets of rows, as seen in Fig. 2

In the h.c.p. lattice there are so many low index planes that we found difficulties to tilt the crystal in a direction away from the intersecting planes. Hence the sample was step tilted across the axis within a lattice plane.

Normalized yields thus obtained are shown in Figs. 5, 6, and 7. The yields are normalized at the random value.

As it can be seen from these figures, the channeling dips from the oxygen sublattice are quite distinct from the titanium sublattice dips. The titanium yield exhibits always a normal channeling dip. On the other hand the yield curves from oxygen are very

Table IV

Dechanneling half-thickness $x_{1/2}$ for planar channeling relative to the titanium sublattice.

	Dechanneling Half-Thickness $x_{1/2}$ (µm)				
	Ti	Ti-O$_{0.11}$	Ti-O$_{0.20}$	Ti-O$_{0.28}$	Ti-O$_{0.39}$
{0001}	3.8	1.9	–	1.4	1.0
{11$\bar{2}$0}	1.7	1.3	1.7	1.5	–
{10$\bar{1}$0}	1.7	–	–	0.9	0.8

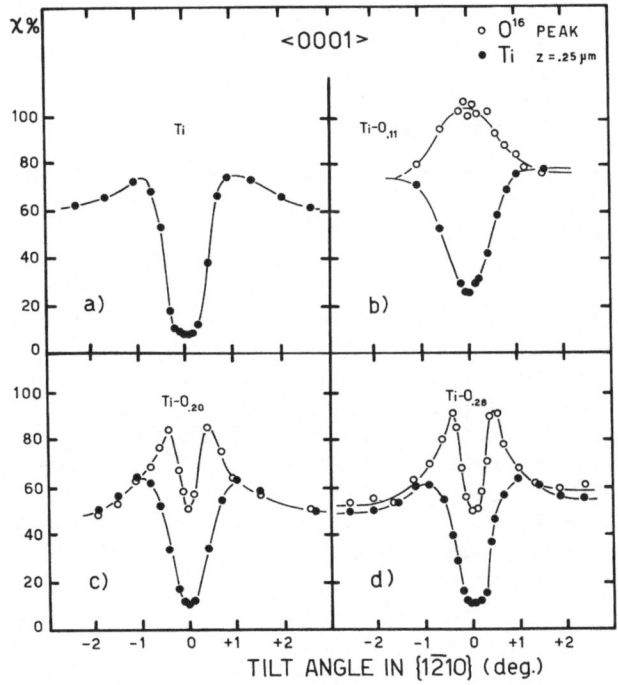

Fig. 5. Channeling normalized yields for the <0001> axis at various compositions: a) Ti; b) TiO$_{.11}$; c) TiO$_{.20}$; d) TiO$_{.28}$. Tilt angle is within the {1$\bar{2}$10} plane. o: proton yield; •: deuteron backscattering yield.

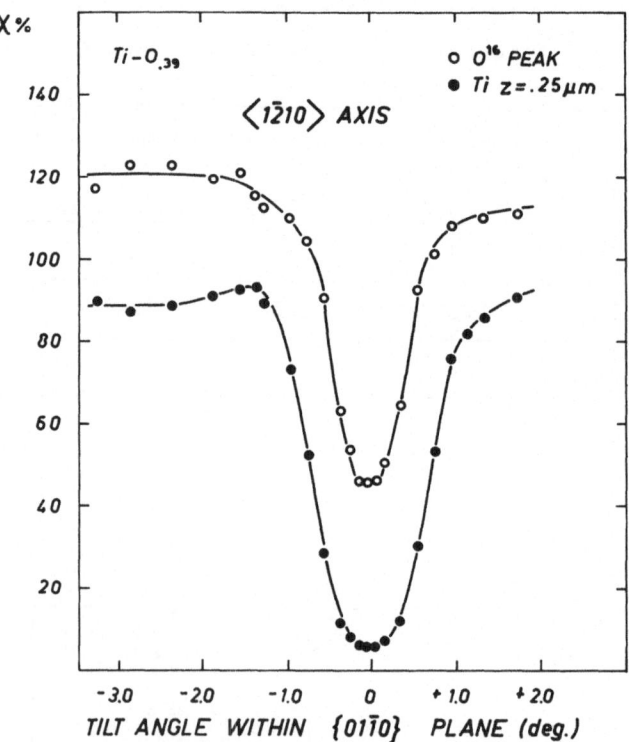

Fig. 6. Channeling normalized yields for the $\langle 1\bar{2}10\rangle$ axis of $TiO_{.39}$.
Tilt angle is within the $\{01\bar{1}0\}$ plane. o: proton yield;
●: deuteron backscattering yield.

sensitive to the oxygen concentration. For both axes at $TiO_{.11}$
composition the nuclear reaction shows no angular dip at all.
Moreover for the $\langle 1\bar{2}10\rangle$ axis we clearly observe a flux-peaking
effect, but lower than that for the $\{0001\}$ plane.

Angular dips for the oxygen sublattice are observed only on
specimens at higher oxygen compositions. The measured values for
$\psi_{1/2}$ and χ_{min} are listed in Table V and VI.

Care must be taken in comparing the experimental $\psi_{1/2}$ values
with the theoretical ones because of the particular angular tilt.

Discussion

    Planar channeling - critical angles

The theoretical values of $\psi_{1/2}$ listed in Table II were obtained
according to: (a) the Barrett [21] formula modified by Gemmell
and Mikkelson [22] for polyatomic crystals and (b) the Lindhard

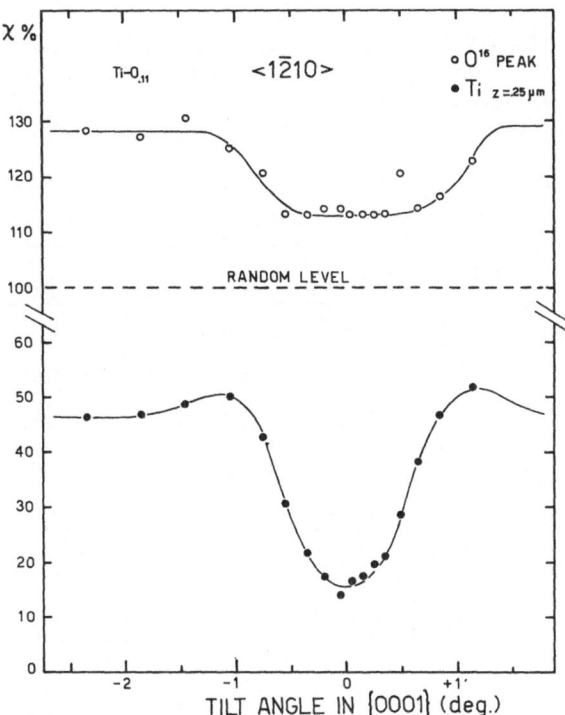

Fig. 7. Channeling normalized yields for the <1$\bar{2}$10> axis of TiO$_{.11}$.
Tilt angle is within the {0001} plane.   o: proton yield;
●: deuteron backscattering yield.

formula. Amplitudes for thermal vibrations in TiO$_x$ at room temperature were taken from the x-ray results by Schoening and Witt
[10].

        Titanium sublattice: as it can be seen in Table II, Lindhard's
formula gives values higher than the experimental ones, while
Barrett's formula gives lower values. Generally a better agreement
is found with Barrett's calculation.

        The general predicted trend is that $\psi_{1/2}$ is almost unchanged
by increasing the oxygen concentration because there are two com-
petitive processes due to the increase of the Debye temperature
and of the lattice constants. The experimental values show the same
constant trend within the estimated experimental errors (±0.05°).

        Oxygen sublattice: the experimental values of Table II refer
to a mean depth of .3μm, so care must be taken in the comparison
with the theoretical ones. Calculations were performed only using
Barrett's formula. For the "mixed" {1$\bar{2}$10} plane the calculated
critical angles are lower than those for titanium as consequence
of the greater thermal vibration of oxygen atoms. Such difference is

Table V

Measured and calculated critical angles for axial channeling. Calculated values (a) and (b) are given in parentheses below the corresponding measured ones. Experimental values for the oxygen sublattice correspond to a mean depth of 0.3μm. F.P. means that flux-peaking is observed.

	Sublattice		Critical Angles $\Psi_{1/2}$ in Axial Channeling				
			Ti	Ti-O$_{.11}$	Ti-O$_{.20}$	Ti-O$_{.28}$	Ti-O$_{.39}$
<0001>	Ti		0.58	0.69	0.69	0.70	–
		(a)	(0.54)	(0.55)	(0.57)	(0.59)	(0.61)
		(b)	(0.80)	(0.81)	(0.86)	(0.88)	(0.90)
	O			F.P.?	0.24	0.27	0.30
		(a)	–	(0.11)	(0.17)	(0.21)	(0.27)
		(b)		(0.25)	(0.31)	(0.41)	(0.50)
<11$\bar{2}$0>	Ti		0.72	–	–	0.70	0.77
		(a)	(0.68)	(0.69)	(0.73)	(0.76)	(0.79)
		(b)	(1.00)	(1.00)	(1.03)	(1.10)	(1.12)
	O			F.P.	–	0.35	0.42
		(a)	–	(0.10)	(0.15)	(0.21)	(0.26)
		(b)		(0.22)	(0.23)	(0.37)	(0.44)

experimentally observed as it can be seen from Fig. 4(c). For the {0001} and {10$\bar{1}$0} planes the calculated critical angles are very small. This fact together with the predicted minimum yield (see Table III) confirms the experimental observation that the steering power of the oxygen planes is too low to produce a channeled trajectory.

Planar channeling – minimum yield

As for the planar minimum yield a theoretical expression in good agreement with experimental data is not available. Thus for the evaluation of planar minimum yield we used Lindhard's estimate

$$\chi = 2\ r_{min}/d_p$$

Table VI

Measured and calculated minimum yield for axial channeling. Calculated values are given in parenthese below the corresponding
measured ones. Experimental values for the oxygen sublattice correspond to a mean depth of 0.3μm.

	Sublattice	Minimum Yield (χ_{min}) in Axial Channeling				
		Ti	Ti-O$_{0.11}$	Ti-O$_{0.20}$	Ti-O$_{0.23}$	Ti-O$_{0.39}$
<0001>	Ti	6.0 (3.6)	19 (3.2)	6.5 (3.0)	6.0 (2.9)	7.0 (2.8)
	O	–	104 (40)	52 (21)	48 (14)	30 (10)
<11$\bar{2}$0>	Ti	2.5 (2.3)	10 (2.1)	– (2.0)	6.0 (2.0)	4.5 (2.0)
	O	–	113 (52)	– (28)	66 (19)	46 (14)

where r_{min} was calculated from the relation $V(r_{min})=E\psi_p^2$. For the
{1$\bar{2}$10} plane the theoretical prevision indicate a decrease of minimum yield with increasing of oxygen concentration and a higher value
of the χ_{min} for oxygen than for titanium atoms. Such results are
clearly related to the previous $\psi_{1/2}$ calculations. The experimental
datum confirms such trend, but shows a larger decrease of χ_{min}.
Such discrepancy could be explained with two hypotheses:
  i):   a larger increase of the Debye temperature with increasing
        oxygen concentration;
 ii):   thermal vibration anisotropy due to the presence of oxygen
        atoms.
    For "monoatomic" planes as {0001} and {1010} the experimentally
observed increase of titanium χ_{min} with increase of oxygen concentration should be connected to the multiple scattering from oxygen
planes which appear to exert a significant dechanneling effect even
close to the surface. The importance of such a process is confirmed
by the significant increase in the half-thickness for dechanneling
in these planes.
    For the "mixed" {11$\bar{2}$0} plane the dechanneling rate seems to
be constant as seen from Table IV.
    The oxygen χ_{min} calculations for the {0001} and {10$\bar{1}$0} planes
suggest that channeling is not detectable experimentally and a flux-
peaking effect is always observed in the yield of the nuclear reaction. The flux concentration γ was deduced from the relation
$\gamma=(Y_O-Y_{Ti})/(1-Y_{Ti})$, where Y_O and Y_{Ti} are the channeling yields

normalized to the random value for oxygen and titanium respectively.

The titanium yield was taken as the mean value between the surface and the 0.3μm depth. The results for the {0001} plane at various compositions are: $TiO_{.11}$, $\gamma=1.5$; $TiO_{.28}$, $\gamma=1.4$; $TiO_{.39}$, $\gamma=1.3$. The flux concentration appears to decrease with increasing oxygen content and the value extrapolated to zero oxygen concentration turns out to be about 1.6. A calculation has been performed using the theoretical expression given by Kumakhov corrected for the beam divergence [23]. At a distance from the half-way plane equal to the Thomas-Fermi length we obtained $\gamma=1.54$, in good agreement with our result. The decrease of γ suggests that the flux-peaking effect is destroyed by a high interstitial concentration (O/Ti>1). On the other hand, at these high concentrations, the steering power of the oxygen planes could be sufficient to produce the channeled trajectories. This γ decrease may explain the difference between our results and those for the {100} plane of UO_2 [2] and for the {110} plane of $BaTiO_3$ [22].

Axial channeling - critical angle

The theoretical values of $\psi_{1/2}$ listed in Table V were obtained according to: (a) the Barrett formula modified by Gemmell and Mikkelson [22] for polyatomic crystals, and (b) the Lindhard formula corrected for the influence of thermal vibrations following Andersen's treatment [24). As for the titanium sublattice, Barrett's formula gives an overall better agreement with the observations. The unexpected increase of the experimental values of $\psi_{1/2Ti}$ along the <0001> axis in the alloyed samples can be attributed to the particular tilting direction [25]. Namely the χ_{min} of the {1$\bar{2}$10} plane decreases very rapidly and strongly influences the channeling dip. This discrepancy is not observed for the $\psi_{1/2,0}$ probably because of the large shoulders of the axial-to-planar transition.

As for the oxygen sublattice, the mean atomic distances along the oxygen rows, used for the $\psi_{1/2}$ calculation, are reported in Table VII. In this evaluation we take an average over a 0.3μm length. The values thus obtained exhibit no difference between ordered and disordered structure, because of the existence of anti-face domains. The experimental results show, at least qualitatively, the predicted increase with increasing oxygen concentration. However, a larger discrepancy from the Barrett values is observed for the <11$\bar{2}$0> axis. This difference could be explained by the ordered arrangement of oxygen atoms of our samples. Indeed the mean inter-atomic distances inside a single domain are one half of those listed in Table VII for the <11$\bar{2}$0> axis. The critical angles calculated using the shorter interatomic distance are: $\psi_{1/2}=.32°$ for $TiO_{.28}$ and $\psi_{1/2}=.39°$ for $TiO_{.39}$ in good agreement with the experimental observation. We notice that for the <0001> axis the differences between ordered and disordered oxygen interatomic distances are very small, at least at high compositions, to give any detectable difference. This result suggest the possibility to detect long range using the channeling technique.

Table VII

Mean axial interatomic distances for various Ti-O$_x$ alloys.

| Sublattice | | Mean Atomic Distances in Ti-O$_x$ (Å) | | | | |
		Ti	Ti-O$_{0.11}$	Ti-O$_{0.20}$	Ti-O$_{0.28}$	Ti-O$_{0.39}$
<0001>	Ti	4.68	4.72	4.75	4.77	4.79
	O	–	21.26	11.87	8.46	6.16
<11$\bar{2}$0>	Ti	2.95	2.96	2.97	2.97	2.97
	O	–	26.67	14.85	10.53	7.63
<10$\bar{1}$0>	Ti	5.11	5.14	5.15	5.15	5.15
	O	–	46.22	25.70	18.26	13.34

A study on both ordered and disordered alloys is in progress. For TiO$_{.11}$ specimens the possibility to have an ordered phase is yet an open question. Even for the disordered crystals, however, the theoretical critical angles indicate the possibility of channeling from the oxygen rows. On the other hand our results show no channeling at all. According to Lindhard [5] a qualitative condition for the validity of the continuum theory is that $r_{min} > \psi_1 d$.

This relation is not satisfied for the TiO$_{.11}$ samples if we put r_{min} equal to a_{TF} for oxygen, d equal to the mean interatomic distance of the oxygen row and ψ_1 equal to the titanium critical angle by considering that the distribution of the transverse energy near the surface is mainly due to the titanium string.

Axial channeling – minimum yield

Using the empirical formula proposed by Barrett minimum yields for titanium lattice were evaluated. Eriksson and Davies [2] derived a procedure to calculate χ_{min} for a polyatomic crystal considering the steering power of each atomic specie. Following their procedure, the general form of χ_{min} value for a diatomic crystal (A and B rows) are expressed as follows:

$$(\chi_{min})_A = \left| Nd\pi \ r_{min}^2 \right|_{A \ row} + K \left| Nd\pi \ r_{min}^2 \right|_{B \ row}$$

where A and B refer to the two types of atoms present. The factor K is a measure of the relative scattering power of the atomic species and is given by the ratio of $Z_B d_A / Z_A d_B$. Oxygen minimum yield was calculated following this procedure and using the previously calculated titanium minimum yield.

For the titanium sublattice, the predicted minimum yields do not change significantly with the oxygen concentration. This trend is confirmed experimentally but for the $TiO_{0.11}$ specimens. The anomalously high minimum yield observed in these samples may be connected to the flux-peaking effect. In fact at this concentration the oxygen atoms do not behave as a continuum string but act as individual scattering centers as discussed previously. Thus the dechanneling effects will be increased even close to the surface.

As for the yield from oxygen sublattice, the theoretical prevision indicate strong decrease of minimum yield with increasing of oxygen concentration. The observed minimum yields are much larger than the predicted values even if we take into account that the experimental values refer to a mean depth of .3µm. The reason of this discrepancy may be due to the multiple scattering from oxygen atoms as well to the flux enhancement inside the channel discussed previously. However, the ratio between the experimental minimum yield from oxygen and that from titanium turns out to be about $(1.3\pm.2)K$ in qualitatively agreement with the estimate of Eriksson and Davies.

We think that more measurements are necessary to clarify the surface dechanneling mechanism in polyatomic crystals. In particular our data suggest that flux-peaking strongly influence the surface dechanneling and that the domain boundary can exert a significant dechanneling effect. The principal features of the oxygen sublattice behaviour are presented also by the computer results.

Computer Results

The computer model has been described in detail elsewhere [25,26]. Since our model was devised for the <110> channel in silicon, we put a boron atom in an interstitial site, which gives approximately the same ratio of atomic numbers as between oxygen and titanium. The hexagonal <0001> axial channel of titanium was simulated by the <110> channel of silicon, with a boron atom in the center. Tilting along the $\{1\bar{2}10\}$ plane in titanium was simulated by tilting along the $\{110\}$ plane in silicon, which can easily be shown to be equivalent.

Tilting along the $\{0001\}$ plane of titanium was simulated as follows. The plane perpendicular to a <110> axis in the silicon cubic lattice is still of the $\{110\}$ type. For sake of clarity assume the original <110> axis, corresponding to the titanium hexagonal axis, to be the [110]. Viewing that latter axis through the [110] axis, which lies in the (110) plane (and corresponds in our model to the titanium basal plane), the interstitial lying at the center of the [110] axis lies now at one of the four major sides of the [110] channel. Therefore planar channeling along $\{0001\}$ plane in titanium was simulated through $\{110\}$ planar channeling but with the interstitial on the side of the axial channel.

The particle flux was measured as usual by dividing the channel area into grid squares and recording the particles passing

at a given depth [27]. The area of the grid is 3.7×10^{-3} in units
of channel area, which corresponds to dimensions of the order of
a_{TF}^2. The total number of ions sent in the crystal was 100: to
improve statistics we averaged over a thickness of 600 Å corresponding to $\sim$150 atomic spacings. Moreover, taking into account the
symmetry of the axis, we summed over the four equivalent grid
squares. For sake of simplicity the interstitals were assumed to
be equally spaced at an appropriate distance to give the required
concentration.

The results are shown in Table VIII. The flux is of course
normalized to the random value and is given both at the interstitial
and lattice sites; two computer values are shown, one for the grid
square of side of $\sim a_{TF}$ and the other one for time larger, i.e.,
of side $\sim 2a_{TF}$. The values are computed for a depth of 0.3μm, which,
taking into account the different stopping powers of silicon and
titanium, corresponds to the value of 0.25μm of the experiments.
In any case no important differences as a function of depth in the
mean values of the flux (i.e., averaged over the well known depth
oscillations) appear beyond 0.2μm.

In general the agreement is better with computer values in the
largest grid square, which can partly be due to better statistics
and partly to the inherent limitations of the model, which tends
to underestimate the "nuclear encounter probability" as discussed
in our previous paper [25]. The general trend is, however, depicted
also by the smaller grid, except for axial channeling at 10% concentration, where the values at the interstitial site are too small.

From Table VIII we see that computer data support the experimental results, in particular:

i) In the axial case we have a maximum flux at the interstitial
   site, of the order of random value, for 10% concentration,
   which turns, however, in a yield dip at higher concentrations
   (20%)

ii) On the contrary in the planar case the dip does not still
    appear up to the concentration 20% and the flux remains higher
    than random.

iii) The transition between axial and planar channeling gives rise
     to a much higher shoulder at the interstitial than at the
     lattice site, ranging up to about the random value.

In conclusion computations show the rather general value of
the experimental results with respect to the role of an interstitial
sublattice and indicate that this kind of experiments can be more
useful for problems such as atom location.

Conclusion

The present investigation in TiO$_x$ alloys presents a systematic
channeling study of a substance having different concentrations of
interstitial atoms. The observed channeling behaviour provides
proof that planar flux-peaking is progressively destroyed by increasing of interstitial concentration. For the axial case we

Table VIII

Computer flux at a depth of 3000 Å both at interstitial and lattice sites of 1 MeV deuterons channeled along the <110> axis in silicon. The tilting angles are all along the {110} plane containing the axis; atomic concentration of 10% corresponds to $AB_{0.11}$, of 20% to $AB_{0.25}$, and of 30% to $AB_{0.43}$, for the interstitial location, C and S mean center and side of the axial channel respectively; a_{TF} and $2a_{TF}$ mean the approximate side of the grid square.

Tilting Angle	Atomic Concentration	Interstitial Location	Normalized Flux					
			At Interstitial Site			At Lattice Site		
			a_{TF}	$\sqrt{2}a_{TF}$	Experimental Yield	a_{TF}	$\sqrt{2}a_{TF}$	Experimental Yield
0°	10%	C	0.35	0.85	1.00	0.03	0.07	0.25
2°	10%	C	0.25	0.40	0.75	0.40	0.45	0.75
0°	20%	C	0.15	0.40	0.50	0.05	0.10	0.10
0.4°	20%	C	1.10	0.90	0.90	0.20	0.50	0.40
1°	20%	C	0.45	0.50	0.70	0.55	0.60	0.65
2°	20%	C	0.20	0.40	0.55	0.20	0.40	0.55
2°	10%	S	1.50	1.65	1.30	0.35	0.40	0.45
2°	20%	S	1.40	1.45	1.20	0.50	0.55	0.60
2°	30%	S	1.05	1.35	1.15	0.65	0.70	0.60

observed the evolution of the oxygen behaviour from that of individual interstitial atoms to that of continuum strings with decreasing of the mean interatomic distance. This observation is confirmed by a computer simulation using a Monte Carlo program. Moreover, critical angle measurements for the oxygen sublattice suggest the possibility to detect the long-range order of the structure. However, more measurements are needed to test this possibility.

References

*Work performed at Laboratori Nazionali di Legnaro, Italy. Research sponsored in part by Gruppo Nazionale di Struttura della Materia del Consiglio Nazionale delle Ricerche.

**On leave from the Research Institute for Iron, Steel and Other Metals, Tohoku University, Sendai, Japan.

†Istituto Chimico Facoltà di Ingegneria - Università di Bologna.

[1] S. T. Picraux, J. A. Davies, L. Eriksson, N. G. E. Johansson and J. W. Mayer; Phys. Rev. 180, 873 (1969).

[2] L. Eriksson and J. A. Davies, Ark. Fys. 36, 439 (1969).

[3] H. Matzke, J. A. Davies and N. G. E. Johansson, Can. J. Phys. 49, 2215 (1971).

[4] F. W. Clinard, Jr. and W. M. Sanders, J. Appl. Phys. 43, 4937 (1972).

[5] J. Lindhard, Mat. Fys. Medd. 34, 14 (1965).

[6] B. Domeij, G. Fladda and N. G. E. Johansson, Radiation Effects 61, 155 (1970).

[7] J. U. Andersen, O. Andreasen, J. A. Davies and E. Uggerhøj, Radiation Effects 7, 25 (1971).

[8] S. Andersson, B. Collén, U. Kuylenstierna and A. Magnéli, Acta Chem. Scand. 11, 1641 (1957).

[9] S. Yamaguchi, J. Phys. Soc. Japan 27, 155 (1969).

[10] F. R. L. Schoening and F. Witt, Acta Cryst. 18, 609 (1965).

[11] M. Hirabayashi, M. Koiwa and S. Yamaguchi, "The Mechanism of Phase Transformation in Crystalline Solids," Institute of Metal Monograph and Report Series n. 33 (Institute of Metal, London).

[12] S. Yamaguchi, K. Hiraga and M. Hirabayashi, J. Phys. Soc. Japan 28, 1014 (1970).

[13] C. F. Williamson, J. P. Boujot and J. Picard, Report CEA R3042 Saclay (1966).

[14] P. D. Bourland, W. K. Chu and D. Powers, Phys, Rev. B 3, 3625 (1971).

[15] W. K. Chu and D. Powers, Phys. Rev. 187, 478 (1969).

[16] P. D. Bourland and D. Powers, Phys. Rev. B 3, 3635 (1971).

[17] D. A. Thomson and W. D. Mackintosh, J. Appl. Phys. 42, 3969 (1971).

[18] G. Amsel and D. Samuel, Anal. Chem. 39, 1689 (1967).

[19] G. Amsel, J. P. Nadai, E. d'Artemare, D. David, E. Girard and J. Moulin, Nucl. Instr. and Meth. 92, 481 (1971).

[20] D. S. Gemmell, Rev. Modern Phys., to be published.

[21] J. H. Barrett, Phys. Rev. 13, 1527 (1971).
[22] D. S. Gemmell and R. C. Mikkelson, Radiation Effects 12, 21 (1972).
[23] M. A. Kumakhov, Radiation Effects 15, 85 (1972).
[24] J. U. Andersen, Mat. Fys. Medd. 36, 7 (1967).
[25] G. Della Mea, A. V. Drigo, S. Lo Russo, P. Mazzoldi, G. Bentini, A. Desalvo and R. Rosa, Phys. Rev. B 7, 4029 (1973).
[26] A. Desalvo, R. Rosa and F. Zignani, Journ. Appl. Phys. 43, 3755 (1972).
[27] R. B. Alexander, G. Dearnaley, D. V. Morgan, J. M. Poate, D. Van Vliet, Proc. European Conf. Ion Implantation, Reading (1970), p. 181.

ENERGY DEPENDENCE OF THE SURFACE MINIMUM YIELD FOR AXIAL CHANNELING*

G. DELLA MEA, A. V. DRIGO, S. LO RUSSO, P. MAZZOLDI,
G. CORNARA, S. YAMAGUCHI**
Istituto di Fisica dell'Università
Unità del Gruppo Nazionale di Struttura della Materia
Consiglio Nazionale delle Ricerche, Padova, Italy

and

G. G. BENTINI, G. CEMBALI, F. ZIGNANI[†]
Laboratorio per la Chimica e Tecnologia
dei Materiali e Componenti per l'Elettronica
Consiglio Nazionale delle Ricerche, Bologna, Italy

ABSTRACT

The energy dependence of the axial channeling minimum yield has been investigated in diamond type structure using silicon <110> and germanium <111> crystals. Experimental results have been compared with the Monte Carlo results obtained by Barrett. The energy dependence of surface minimum yield was observed and found in satisfactory agreement with the Barrett's predictions. The effective number of surface layers has been evaluated.

Charged particles incident within a critical angle to a crystal axis or plane are steered and channeled by a series of gentle collisions. Therefore, when the ion beam is exactly aligned with an axial or planar direction, the normalized nuclear encounter probability has a minimum value, proportional to the so called minimum yield χ.

Lindhard [1], using the continuum model, predicted that the row minimum yield, just below the surface, would be

$$\chi = Nd\pi(\rho_{\perp}^2 + a^2_{TF})$$ (1)

where N is the atomic density of the crystal, d the row spacing, $\rho_\perp^2$ the mean square amplitude of the thermal vibrations normal to the string and a_{TF} Thomas-Fermi screening length.

Barrett [2], using a Monte Carlo computer program to follow the trajectories of energetic ions in a lattice, obtained an energy dependence for the row minimum yield. His results can be fitted by the expression

$$\chi = \chi_0 (1 + \frac{\psi_{1/2}^2 d^2}{\kappa^2 \rho_\perp^2 / 2})^{1/2} \tag{2}$$

and $$\chi_0 = (C(\Delta) N d \pi \rho_\perp^2 + C'(\Delta) N d \pi a_{TF}^2) \tag{3}$$

where $\psi_{1/2} = KR\psi_1$ (the product KR is a function of the thermal vibration and ψ_1 is the Lindhard's critical angle), $C(\Delta)$ and $C'(\Delta)$ are coefficients dependent on the beam divergence Δ, and κ is an adjustable parameter. Barrett found, for $\Delta < 0.1°$, $C(\Delta) = 3.0 \pm 0.2$, $C'(\Delta) = 0.2 \pm 0.1$ and $\kappa = 2.2$.

Komaki et al. [2] investigated the temperature dependence of row minimum yield in silicon and germanium crystals. The energies of the incident particles were higher than 1.0 MeV so that the energy dependence of χ should have been negligible.

The values they found for C and C' were respectively 3.4 ± 0.2 and 0.4 ± 0.07 for germanium <111> and 2.6 ± 0.2 and 0.00 ± 0.04 for silicon <110>. Isotropic thermal vibrations were assumed.

In this work we give row minimum yield measurements as a function of energy, in the range of E/Z between 30 and 2,000 keV, where Z is the atomic number of the incident particles.

Beams of H^+, $^4He^+$ and $^{10}B^+$ ions and single crystals of silicon and germanium as targets have been used. The study has been performed on the 400 keV ion accelerator of LAMEL-CNR in Bologna and on the Legnaro 5.5 MeV Van de Graaff accelerator. The incident beam was collimated by slits and annular collimators of various sizes, arranged at distances such that the maximum beam divergence was always kept better than $0.03°$. The samples used were dislocation-free p-type Si about $3000\Omega \cdot cm$ and n-type Ge about $5\Omega \cdot cm$. Just before the measurements the specimens were etched, washed and dried in a nitrogen stream. The energy distribution of the backscattered particles was measured by means of a silicon surface barrier detector, positioned at about $160°$ with respect to the incident beam.

To avoid radiation damage effects on the samples the analyzed region of the crystal has been changed at each measurement.

Indeed for germanium we observed an increase of the minimum yield of about 50% over the undamaged value, after an irradiation dose of $2500\mu C/cm^2$ produced by 2.0 MeV ^4He ions with random direction. The same effect has been observed and extensively studied by Campisano et al. [4]. The doses used for each experimental

point were always kept lower than 5μC/cm² at low energy and 50μC/cm²
at high energy. These values were chosen because radiation damage
was not detectable in these conditions.

The surface yield of the aligned spectra was obtained by extrapolating the dechanneling curves to zero depth.

Figures 1 and 2 show respectively the minimum yield along the
<111> direction of germanium and along the <110> direction of sili-
con, as a function of E/Z. In both cases the predicted increase of
the minimum yield with decreasing of the E/Z value is shown. The
error bars shown in the figures represent the statistical error, by

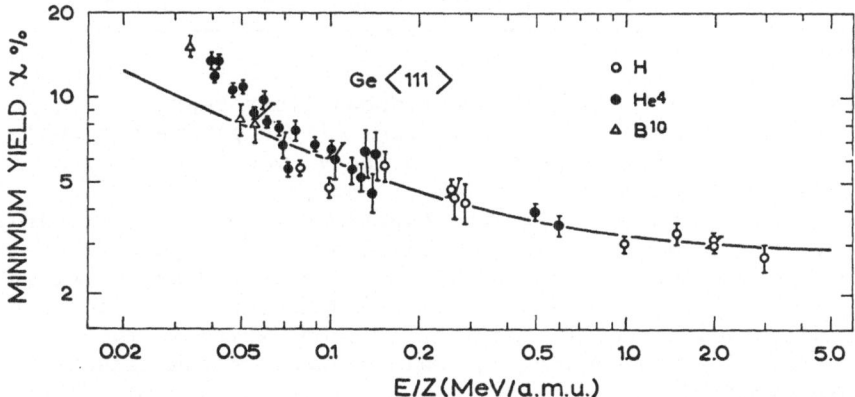

Fig. 1. Dependence of minimum yield χ on energy for germanium <111>.
The continuous line represents the values calculated from eq. (2)
(see also Table I).

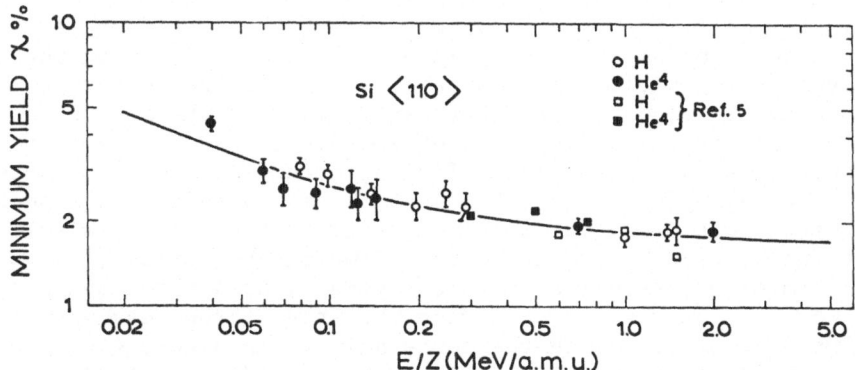

Fig. 2. Dependence of minimum yield χ on energy for silicon <110>.
The continuous line represents the value calculated from eq. (2),
(see also Table I). The experimental data of Fontell et al. [5] are
reported for comparison.

Table I

Values of the parameters used in eq. (2) to fit the experimental
data. The mean square amplitudes of the thermal vibrations are
obtained from x-ray experiments [6] and KR from Barrett's [2] calculation. (*)Data calculated from ref. [3] for comparison with
our data. (**)Data calculated from ref. [3] using the constants
given by Barrett's [2].

	$\chi_o\%$	κ	$\rho_{\perp}^2(cm^2)$	KR	$\chi_o^{(*)}\%$	$\chi_o^{(**)}\%$
Ge <111>	2.8	1.0	$1.47\ 10^{-18}$	.77	$3.9\pm.3$	$3.25\pm.35$
Si <110>	1.7	1.2	$1.14\ 10^{-18}$	.88	$1.8\pm.2$	$2.40\pm.35$

far the most important one. The continuous lines represent the χ
values obtained from eq. (2). The values of the parameters used
are reported in Table I, together with the χ_o values calculated
from eq. (3) using the data of ref. [2] and [3].

The k values that best fit the experimental points are 1.0 ± 0.1
and 1.2 ± 0.1 for germanium <111> and silicon <110> respectively.

The higher values of χ_o observed at high energy by Komaki et al.
[3] in germanium could be due to some radiation damage of the surface, as discussed above; in fact, in silicon, which is much less
sensitive to radiation damage, our values are in close agreement
with their results.

In Figure 1 we observe that χ increases steeper than the predictions, based on eq. (2), for energies lower than 70 keV; this trend
is hardly detectable in Figure 2. On the other hand, the value
assumed by the energy dependent term in eq. (2), calculated for
germanium at 70 keV, corresponds to the value at 50 keV calculated
for silicon. Therefore, to see in silicon <110> the same trend
we see in germanium, we need to investigate χ at lower energies.

The effective number of atom layers which contribute to the
surface scattering yield has been obtained by the area under the
surface peak. In Figure 3 we report an example of the method used
to obtain this peak: the random spectrum was scaled to the extrapolated surface value of the channeling spectrum and then subtracted
from the channeling spectrum. The surface peak obtained is reported
on the bottom left of Figure 3. The energy of the particles scattered at the surface is indicated by arrows on the scale.

The number of effective surface layers L, as defined by Barrett
[2], is obtained by the ratio between the area of the peak normalized to the random yield and the row spacing. In Figure 4 we report the data so obtained versus E/Z of the incident particles.

Fig. 3. An example of the method used to obtain the effective
number of surface layers: ● random spectrum; o aligned spectrum.
The dashed line represents the scaled random spectrum. The obtained surface peak is reported on bottom left of the figure.

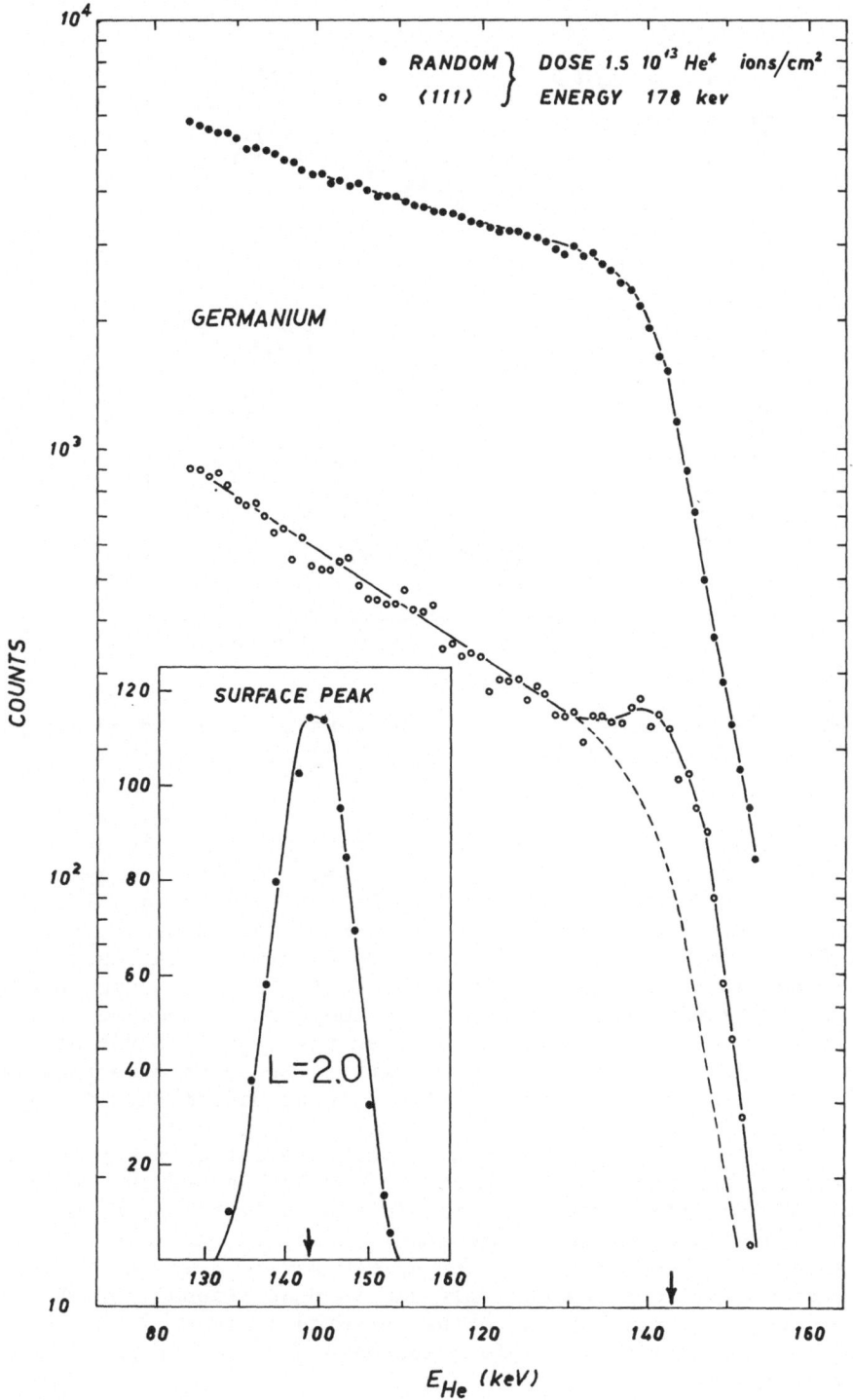

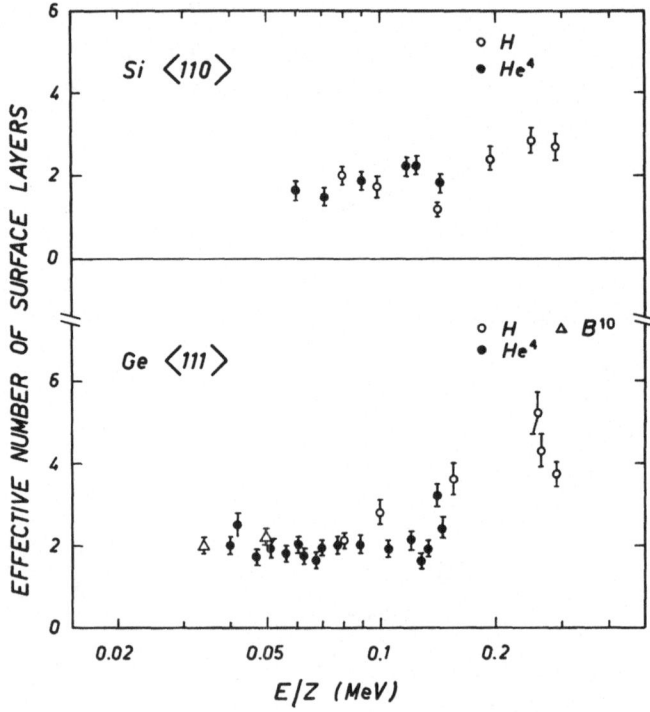

Fig. 4. Effective number of surface layers as function of energy
for Ge <111> and Si <110>.

Following Barrett [2] the screening by the first layer at low
energies is totally effective and hence L=1. At low energy we observe the constant trend predicted by Barrett, but at a value higher than unity. The mean values of L in this region are 1.8±0.3
and 2.0±0.3 for Si <110> and Ge <111> respectively. At energies
higher than about 150 keV L values increase with increasing energy.
However, we cannot state the functional dependence because we would
need better energy resolution at high energy. Thermal vibrations
could increase this limit value; in particular we observe that the
mean square thermal vibration is higher in germanium than in silicon.

Another possible explanation is that the difference between
experimental and predicted L values is due to the presence of about
one amorphous oxide layer. If this is the explanation, a contribution (χ_3) to the minimum yield should be taken into account.
The calculated χ_3 contribution for E/Z=40 keV is 3.2% and 0.7% for
germanium and silicon respectively and is proportional to 1/E.
However, the χ_3 contribution to the measured minimum yield is not
sufficient to justify the rapid χ increase observed in germanium
at low energies.

Very similar deviations were observed by Barrett [2] for tungsten. Recently Kubota and Ohtsuki [8), using Lindhard's theory, presented an analytical expression of the minimum yield for axial channeling which gives rise to an energy dependence of χ not far from that of eq. (2) for $\rho > d\psi_1$; for $\rho < d\psi_1$ they find an exponential behaviour that could explain our experimental results.

As for the value of the k parameter, we find nearly the same value for silicon <110> and germanium <111>. The small difference could be attributed to an incorrect value of ρ.

In Barrett's [2] calculation the value k=2.2 for a <111> tungsten crystal is reported. This value should not be dependent on the lattice structure [7].

At the present time we have no ideas as to why our value is so different from that reported in ref. [2]. Further investigations on different lattice structure are in progress to understand this point. Moreover, the behaviour of the minimum yield at low energy could be clarified by changing the condition $\rho < d\psi_1$, acting on the target temperature.

Acknowledgments

We are most grateful to J. H. Barrett for stimulating comments about this work. We would like to thank Y. M. Ohtsuki for sending us valuable unpublished materials. We also appreciate the assistance of R. Lotti during the measurements.

References

*Work performed at Laboratori Nazionali di Legnaro, Italy and Laboratorio per la Chimica e Tecnologia dei Materiali e Componenti per l'Elettronica, Bologna, Italy.

**On leave from the Research Institute for Iron, Steel and Other Metals, Tohoku University, Sendai, Japan.

†Istituto Chimico Facolta di Ingegneria, Universita di Bologna.

[1] J. Lindhard, Mat. Fys. Medd. Dan. Vid. Selsk. 34, 14 (1965).

[2] J. H. Barrett, Phys. Rev. B3, 1527 (1971).

[3] K. Komaki, F. Fujimoto, M. Ishii and H. Nakayama, Phys. Lett. 37A, 271 (1971).

[4] S. U. Campisano, G. Foti, F. Grasso and R. Rimini, Appl. Phys. Letters 21, 425 (1972).

[5] A. Fontell, A. Arminen and E. Leminen, Rad. Effects 12, 255 (1972).

[6] B. W. Batterman and D. R. Chipman, Phys. Rev. 127, 690 (1962).

[7] J. H. Barrett, Private communication.

[8] Y. Kubota and Y. M. Ohtsuki, Phys. Lett. 43A, 521 (1973).

EXPERIMENTAL STUDY AND STOCHASTIC INTERPRETATION OF OSCILLATORY EFFECTS IN BACKSCATTERING SPECTRA IN PLANAR CHANNELING[*]

F. ABEL, G. AMSEL, M. BRUNEAUX and C. COHEN
Groupe de Physique des Solides de l'Ecole Normale Supérieure
Tour 23, Université Paris VII, 2 Place Jussieu
75221 Paris Cedex 05, France

ABSTRACT

Backscattering experiments with 1.9 MeV He ions were performed to study the oscillating trajectories of particles in planar channels of iron crystals. Strong maxima in planar aligned spectra are observed. Spectra registered for particles entering the crystal with angles of incidence with respect to the plane greater than the critical angle also exhibit strong maxima. Calculations which do not take into account multiple scattering predict qualitatively the experimental results. In order to obtain a quantitative agreement a stochastic model which accounts for the effects of multiple scattering is proposed and yields very good fits to the experimental results.

Oscillatory features of backscattering spectra of He[+] ions which undergo planar channeling in iron were recently observed and shortly reported [1]. An interpretation of the data based on the theory of stochastic processes was proposed in a preliminary report [2]. Our work on these phenomena which has been greatly extended in between is briefly summarized in this paper; the full details of both experimental and theoretical results will be published in the near future. The experiments were performed on iron crystals using 1.9 MeV helium beams. The backscattered particles were analysed with surface barrier detectors at θ_{lab} = 120° or 165°. The energy resolution was 15 keV corresponding to a depth resolution of respectively 100 Å and 150 Å. The samples were (001) oriented iron crystals. The oxygen content of the surface oxide layer

remaining on the samples after electrochemical polishing was determined by nuclear microanalysis using the O^{16} (d,p)O^{17} reaction.
This content was about 10^{16} oxygen atoms per square centimeter,
i.e., a thickness of about 20 Å, assuming Fe_2O_3. The (100) and
(110) planes were studied at 5° off the [100] axis. In both cases,
the aligned spectra exhibit four well resolved peaks behind the
surface peak [1]. The mean energy separation ΔE between two consecutive maxima is 2 to 4 times greater than the energy resolution,
according to the experiments. Spectra were also registered at
various angles of incidence (ϕ_0) with respect to the (110) plane.
For this plane, the half width at half minimum $\Psi_{1/2}$ of an angular
scan across the plane was found to be 19'. This scan was obtained
by integrating the yield corresponding to the first two peaks of an
aligned spectrum. Spectra were obtained for ϕ_0 = 0, 4', 10', 16'
and 22'. They all exhibit strong maxima which progressively damp
at low energies, i.e., at great penetration depths. The energy
intervals between two consecutive maxima is fairly constant in the
aligned spectrum. For all the other spectra, the same energy
interval is also found, except between the first two peaks. The
first maximum is shifted towards lower energies as ϕ_0 is increased.
It should be emphasized that for ϕ_0 = 22' the yield on the first
maximum is higher than the random yield by $\sim$20%.

The experimental results suggest that, <u>for all the spectra
registered</u>, the oscillatory effects observed arise only from a
group of particles having a well defined mean transverse energy
$E_\perp^0$, independent of ϕ_0. This is demonstrated by the constancy of
the energy separation between two consecutive maxima which corresponds to a constant mean half-wavelength λ characterizing the motion of the particles contributing to the oscillatory effects.
For the spectra registered at $\phi_0 \neq 0$, the observed shifts of the
first maxima are due to the fact that the particles belonging to
the transverse energy group of interest enter the crystal with
high impact parameters with respect to the plane. Consequently
they come near to the plane and can be backscattered only after a
certain penetration depth.

The mean value, λ, may be deduced from the experimental data
if the mean stopping power (i.e., averaged over one oscillation)
is known. This mean stopping power has been estimated by comparing
two aligned spectra registered at two different detection angles;
it appears to be very close to the "random" stopping power. We
found λ = 320 Å for the (100) plane and λ = 380 Å for the (110)
plane. The ratio $\frac{\lambda(110)}{\lambda(100)}$ is, as expected very near to the ratio
$\sqrt{\frac{d(110)}{d(100)}}$, d being the interplanar distance.

Some features of the experimental results can be predicted by
neglecting multiple scattering (i.e., assuming the conservation of
$E_\perp$) and using classical mechanics. The motion of a particle entering a crystal with an impact parameter y_0 and an angle of incidence
ϕ_0 with respect to a plane is determined by integrating the equation:

$$E(\frac{dy}{dx})^2 + U(y) + U(d-y) = E\phi_0^2 + U(y_0) + U(d-y_0) \qquad (1)$$

where E is the energy of the particle (considered as constant along the penetration depth x of interest) and U(y) a planar potential. We have integrated numerically equation (1) taking for U(y), as a first approximation, the standard planar potential proposed by Lindhard [3] which does not take into account lattice vibrations. A more realistic treatment, assuming that the particles do see a static lattice but perturbed by thermal vibrations, the position of the neighboring atoms being correlated (phonons), is being attempted. For a given angle of incidence of the beam with respect to the plane, the probability density s(y ; x) for a particle to be at a distance y from a plane at a given depth x in the crystal can be calculated once the particles trajectories are known, by integrating the contributions of particles with respect to their uniformly distributed entrance impact parameters. The backscattering yield at a depth x is then proportional to:

$$p(x) = \int_0^d s(y ; x) [g(y) + g(d-y)] \, dy \qquad (2)$$

where g(y) is the gaussian probability distribution of the lattice atoms perpendicularly to the planes at room temperature. p(x) was calculated for various values of ϕ_0 with respect to the (110) plane of iron crystals in the case of 1.9 MeV He$^+$ ions and compared to the experimental results. The main features of the experimental results (periodicity of the maxima and evolution of the first maxima with ϕ_0) appear in the behaviour of the calculated p(x). However, the mean half-wavelength λ extracted from p(x) is near 300 Å which is markedly smaller than the value calculated from the experimental results. Moreover, the damping of the maxima of p(x) is obviously weaker than experimentally observed. The qualitative agreement between p(x) and the experimental results illustrates the interest of the calculations which neglect multiple scattering; however, the quantitative discrepancies show the limits of this approximation and the necessity of a more refined analysis.

In the absence of multiple scattering the damping of p(x) is only due to the distribution of wavelengths of the particles as a function of their entrance impact parameter. In fact, as the particles of interest in our experiments come close enough to the planes to have a high backscattering probability, they also undergo multiple scattering which induces random changes in $E_\perp$. The periodicity of the structure observed in the experimental spectra suggests that the motion of the particles of interest is governed by the planar potential in the main part of their trajectories; however, when they come near to the planes multiple scattering has a dominant effect on these trajectories. A stochastic model [2], which accounts for these features will be briefly described here.

Let us consider an incident beam of monoenergetic particles
aligned with a crystallographic plane of a single crystal. Let
X_n be the non negative random variables with densities h_n (x)
associated with the depth position of the n-th trajectory oscillation extremum. If we neglect the angular spread of the particles
impinging on the first atomic layer of the crystal (behind the
amourphous oxide layer), which is taken as the origin of the depth
scale, we have $X_0 = 0$.

We call Z_j the random variables associated to the spread of
the lengths of the jth half trajectory oscillations around the
mean half wavelength λ. Thus, for $n \geq 1$:

$$X_n = n \lambda + \sum_1^n Z_j .$$ (3)

We introduce Y_n as the non negative random variables of densities
f_n (y) associated to the depth where a backscattering event occurs
in the vicinity of the nth extremum. We shall admit that Y_n can
be written as:

$$Y_n = X_n + L$$ (4)

where L is a centered random variable, independent of X_n, of density q(l). Hence, f_n is given by the convolution: $f_n = h_n \times q$.
The sum $\sum_{n=0}^{\infty} f_n(y)$ completely describes the shape of the backscattering spectra, once a depth—energy conversion scale is chosen.
The main problem is to determine from eq. (3) the probability laws
h_n (x). The latter may be calculated once the laws of Z_j and their
kind of dependence (i.e., the addition rules which apply to them)
are known. The laws of Z_j were chosen phenomenologically so as to
take into account the fact that the particles contributing to the
observed maxima in the spectra belong to a rather well defined
transverse energy group; this requires the introduction in the
model of a centering effect around the mean transverse energy of
the group (i.e., in the depth scale, around λ). This was achieved
by considering random variables Z_j which form a <u>stationary gaussian</u>
<u>Markov chain</u>. This means that the conditional density
t (z_{j+1} | Z_j, Z_{j-1}, ...) depends only on Z_j and may hence be
described by the transition probability density w (z_{j+1}; z_j); on
the other hand, the process being gaussian is fully determined [4]
by the variances σ_{Z_j} of the a-priori laws of the Z_j and by the
correlation coefficients r_n between successive Z_j. In the stationary case the single parameters σ_Z and r, independent of n, fully
describe the process. The conditional density w (z_{j+1}; z_j) is then
centered at r z_j with a variance $\sigma_Z\sqrt{1-r^2}$. The centering effect
arises from the fact that $|r z_j| \leq |z_j|$ as $0 \leq |r| \leq 1$. In our

case only positive values of r are to be considered.

It may be noticed that, for r = 1, if one has: Z_j = z, then, Z_2 = Z_3 = ... = Z_j = z. In this case the variance of X_n is σ_n = n σ_Z. This particular case is the formalization of the calculations performed assuming $E_\perp$ to be conserved. Hence, the closer r is to 1, the smaller is the randomizing influence of multiple scattering.

For any value of r it can be shown that h (x) is gaussian, centered on n λ with a variance:

$$\sigma_n = \sigma_Z \sqrt{n \frac{1+r}{1-r} - \frac{2 r}{(1-r)^2} (1-r^n)} \qquad (5)$$

Using these functions very good fits to the experimental results were obtained [2] both for the (110) and the (100) plane, introducing only second order corrections to take into account the slow increase of the population of the transverse energy group of interest due to the dechanneling of particles with initially low $E_\perp$. The fits to the experimental results have been obtained by an appropriate choice of the parameters r, σ_Z. We had to take a greater value of r for the (110) plane (r $\simeq$ 0.5) than for the (100) plane (r $\simeq$ 0.2). This is probably due to the fact that the potential of the (110) plane is higher than the potential of the (100) plane. Consequently the role of multiple scattering is lower in the case of the (110) plane (a lower dechanneling is experimentally observed) and the group of particles of interest is hence more "conservative."

The explicit relationship between the phenomenological parameters σ_Z and r and multiple scattering is under study.

References

*Work supported by the Centre National de la Recherche Scientifique (R C P N° 157).
[1] F. Abel, G. Amsel, M. Bruneaux and C. Cohen, Phys. Lett. A, 42A, n° 2, 165 (1972).
[2] F. Abel, G. Amsel, M. Bruneaux and C. Cohen, C. R. Acad. Sc. Paris B 276, 267 (1973).
[3] J. Lindhard, Mat. Fys. Med. Dan. Vid. Selsk 34, 14 (1965).
[4] W. Feller, "An Introduction to Probability Theory and Its Applications," Vol. II, John Wiley and Sons Inc. (1966) pp. 93-98.

INTERSTITIAL ATOM LOCATION IN SILICON BY SINGLE AND DOUBLE ALIGNMENT BACKSCATTERING OF MeV HELIUM IONS

K. MORITA[*] and H.D. CARSTANJEN
*Sektion Physik, University of Munich,
Munich, Germany*

ABSTRACT

Single alignment and double alignment back-scattering of helium ions from silicon have been simulated in a computer model. The dependence of the backscattering yield profiles on the tilt plane is examined for single alignment along <110> and the (110), (111), and (100) tilt planes for 2.1 MeV helium ions with a beam divergence of 0.04°. For off axis interstitial sites maxima in the yield profiles are found at angles which are characteristic of the particular site. In double alignment the yield profiles of 2.8 MeV helium ions backscattered from a silicon crystal with 5.10^{-3} interstitial silicon atoms per lattice atom are calculated for the combinations: <110> incidence channeling, <1$\bar{1}$0> blocking direction with detector tilt in (110) and <$\bar{1}$00> blocking direction with detector tilt also in (011). The former combination is sensitive to tetrahedral, hexagonal, and Yb positions and the latter to hexagonal. The lowest interstitial concentration to be detected is estimated to be 10^{-3}. The efficiency of the double alignment in determining the configuration of lattice defects is compared with that of single alignment.

Introduction

It is well known that channeling and blocking of swift ions in single crystals are valuable tools for solid state investigations [1]. The orientation dependence of the yields from close encounters

between swift ions and the atoms of a single crystal has been extensively used to locate the lattice sites of impurity atoms [2,3,4].
A detailed knowledge about ion flux and energy distribution in the
crystal is essential for the interpretation of the experimental
results. As shown recently [3,5,6] this knowledge can be obtained
by simulating the penetration of swift ions through a crystal lattice with a computer. This provides us with the details about the
flux and energy distribution at any penetration depth in dependence
on incidence angle and energy.

      In single alignment, channeling and blocking reduce the yield
of close encounters with atoms on normal lattice sites in axial
directions to 1 - 5% of the random yield and in planar directions to
20 - 40%. Due to flux peaking the yield from atoms between the
atomic rows can rise in axial channeling to ∿3 times the random
yield and in planar channeling to ∿2 times.

      An increase in sensitivity has been realized by double align-
ment which uses channeling and blocking simultaneously [7,8,9]. A
combination of incident axial channeling with axial or planar
blocking of the scattered projectiles reduces the yield from the
regular lattice atoms to ∿0.1% and ∿1%, respectively, of the random
yield. It raises the yield from displaced atoms in a channel by
a factor between 2 and 6. Therefore, in double alignment experiments the scattering yield from 10^{-3} self-interstitials per lattice
atom would be comparable with that from normal lattice atoms, which
illustrates that double alignment is the appropriate method for
locating interstitial atoms of about this concentration.

      In the present paper the tracing of the location of silicon
interstitials in a silicon crystal by a double alignment backscattering experiment with 2.8 MeV helium ions is simulated with
computer calculations. First, the dependence of the scattering
yield on crystal orientation is calculated for various possible
interstitial sites and (110), (111), and (100) incidence planes
in single alignment. Then the scattering yield profiles in double
alignment are derived for the assumed interstitial positions and
the combinations: incidence channeling direction <110>, blocking
direction <1$\bar{1}$0> with detector tilt in (110) and blocking direction
<$\bar{1}$00> with detector tilt also in (011). The results of the double
alignment calculations are compared with the single alignment calculations.

Calculation of Backscattering Yield

      The flux and energy distributions of helium ions in a silicon
lattice are calculated for (i) 2.8 MeV helium ions injected parallel
to <110> (incident channeling beam); (ii) 2.1 MeV and 2.0 MeV helium
ions injected in an angular region of 0 to 1.2° around <110> and 0
to 0.9° around <100> (emergent beam). The planes of incidence are

(110) in <100> and (110), (111), and (100) in <110>. The computer
model includes thermal lattice vibrations, electronic scattering,
electronic energy loss and an initial beam divergence. It is essentially the model used by Carstanjen and Sizmann [10]. The dependence of the electronic scattering and the electronic energy loss of
the projectile on its energy and impact parameter towards a lattice
atom is taken into account as derived by Van Vliet [11]. The interaction potential used is the Molière approximation to the Thomas-
Fermi potential and the scattering angle of each binary interaction
is calculated in the momentum approximation. The helium ions scattered through angles larger than 10° in a single deflection are
assumed to be dechanneled. Thermal vibrations of the lattice atoms
are simulated by displacements from their equilibrium position in
the traverse plane, according to a Gaussian probability of the
standard deviation 0.065 Å. The electronic stopping cross section
includes the contribution of excitations of closed shell electrons,
valence electrons and plasmons. Only the close collisions are used
in the scattering of helium ions by electrons.

The flux and the energy distributions of the ions in the transverse plane are obtained by recording the transverse coordinates
and the energy of every ion at fixed depths in the crystal.

With the flux and energy distributions the relative yield for
Rutherford scattering by impurity atoms at normal lattice positions
and at selected interstitial positions is calculated in single and
double alignment. On the transverse plane a certain interstitial
position may have different projections ("sites") which in the following are labelled with integers. The projections of the assumed
interstitial positions on the (110) and the (100) planes are illustrated in Fig. 1, where the notation of the various interstitial
sites is given in the legend.

The yield for such an interstitial position is obtained by the
weighted mean of all its different projections of equivalent sites.

The relative yield of helium ions from an interstitial site
at incidence angle ψ_{in} in single alignment is given by:

$$R_s(E,\psi_{in}) = N \sum_i \iint W_c(r,z) \; \sigma(\bar{E}(r,z)) \; f(r-r_i)drdz \qquad (1)$$

where r is the coordinate in the transverse plane, $W_c(r,z)$ is the
flux distribution of the helium ions in the crystal at traveling
depth z, $\bar{E}(r,z)$ is the mean energy of the ions at the position (r,z),
$f(r-r_i)$ is the probability distribution of the atoms in the transverse plane around the interstitial site r_i and N is the normalizing
factor determined by the experimental conditions. The summation
extends over all equivalent interstitial sites i in the transverse
plane. In the present study we assume for any atom in the lattice
$f(r) = (r/\rho^2)\exp(-r^2/\rho^2)$, ρ = the amplitude of atom vibration in
the transverse plane. The integration over r extends over the transverse plane.

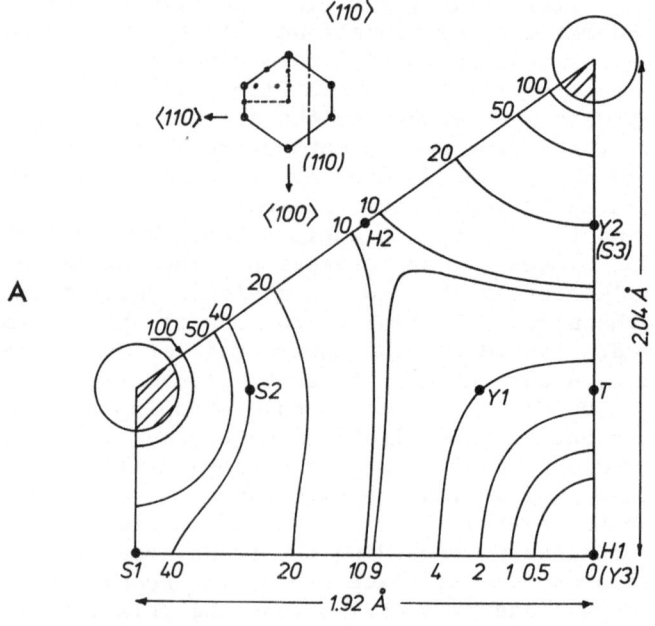

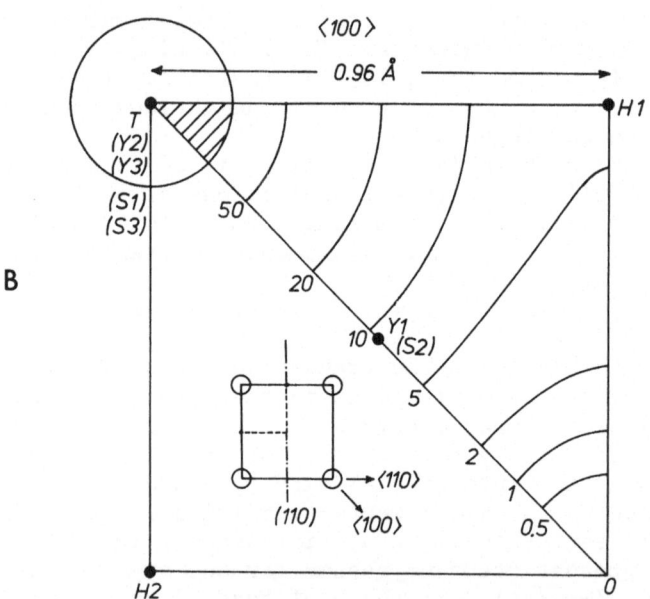

Fig. 1. Net continuum potential (eV) for helium ions in silicon
and various interstitial sites in the (110) transverse plane (a)
and in the (100) transverse plane (b). The large circles represent

In double alignment the relative scattering yield at detector angle ψ_{out} under the condition that the energy loss of the ions is small, is given by:

$$R_D(E, \psi_{out}) = N\sigma(\bar{E}) \sum_i \iint W_c(r,z)W_b(r',z')g(r-r_i,z-z_i)drdz \quad (2)$$

where r and r' are the coordinates in the planes transverse to the incidence channeling direction z and the observation blocking direction z', respectively. $W_b(r',z')$ is the probability that ions emerge from the crystal at detector angle ψ_{out} relative to the channel axis after having been scattered at point (r',z'). The function $g(r-r_i, z-z_i)$ is the three-dimensional probability distribution of an atom around the position (r_i, z_i), and $\sigma(\bar{E})$ is the scattering cross section for helium ions. In the present calculations $W_b(r',z')$ is replaced by the flux distribution of the helium ions channeled in the direction of observation, according to the reversibility rule between channeling and blocking phenomena [14]. The flux distributions W_c and W_b are averaged over depth intervals of 500 layers in order to reduce statistical fluctuations and to shorten the computer time. The maximum error which can be caused by this averaging procedure is less than 12%.

The relative scattering yield R_I, from a silicon crystal containing interstitial atoms is obtained by the superposition of the yield, R_{DI}, from the interstitial atoms and the yield, R_{DL}, from the normal lattice atoms.

$$R_I(E, \psi_{out}) = R_{DI}(E, \psi_{out}) + R_{DL}(E, \psi_{out})/No \quad (3)$$

Here N_o is the concentration of the interstitial silicon atoms.

Profiles of Backscattering Yields

Figure 2 shows the orientation dependence of the backscattering yield in single alignment (beam divergence 0.08°) for the lattice and various interstitial sites in <110> for 2.1 MeV and in <100> for 2.0 MeV helium ions. Each yield profile has a different but characteristic shape which can be used for an unambigious atom location. In <110>(Fig. 2a) the yield for the hexagonal H1 sites (at

---

the silicon lattice atoms (their radius is Thomas-Fermi screening radius). The small dark circles represent the projections of the interstitial sites onto the transverse plane: tetrahedral T (1/2, 0, 0), (1/2, 1/2, 1/2), (1/4, 1/4, 3/4); hexagonal H (H1, and H2): (3/8, 1/8, 7/8), (1/8, 3/8, 7/8), (3/8, 3/8, 5/8), (1/8, 1/8, 5/8), ytterbium position [2]Y (four Y1, Y2, and Y3) : (T) ± (1/8, 0, 0) and split <100> configuration [13]S(S1, four S2, and S3) (L) ± (1/8, 0, 0), (L) is the coordinate of the lattice position.

channel center) decreases drastically and rapidly with ψ_{in} increasing. For the off axis sites H2, Y1, Y2, S1, and S2 the yield
passes through a maximum with increasing ψ_{in}. In <100> (Fig. 2b)
the yield for H1 sites also passes a maximum. The yield profiles
for the tetrahedral T sites in <110> and for H2 and Y1 sites in
<100>, in spite of being off axis sites, do not show a maximum
which is assumed to be characteristic for off axis sites. In order
to examine the conditions where a maximum due to off axis sites
appears, the scattering yield from the first 500 layers was calculated in <110> for the tilt planes (110), (100), and (111), (beam
divergence 0.04°). These results are shown in Fig. 3. For a (110)

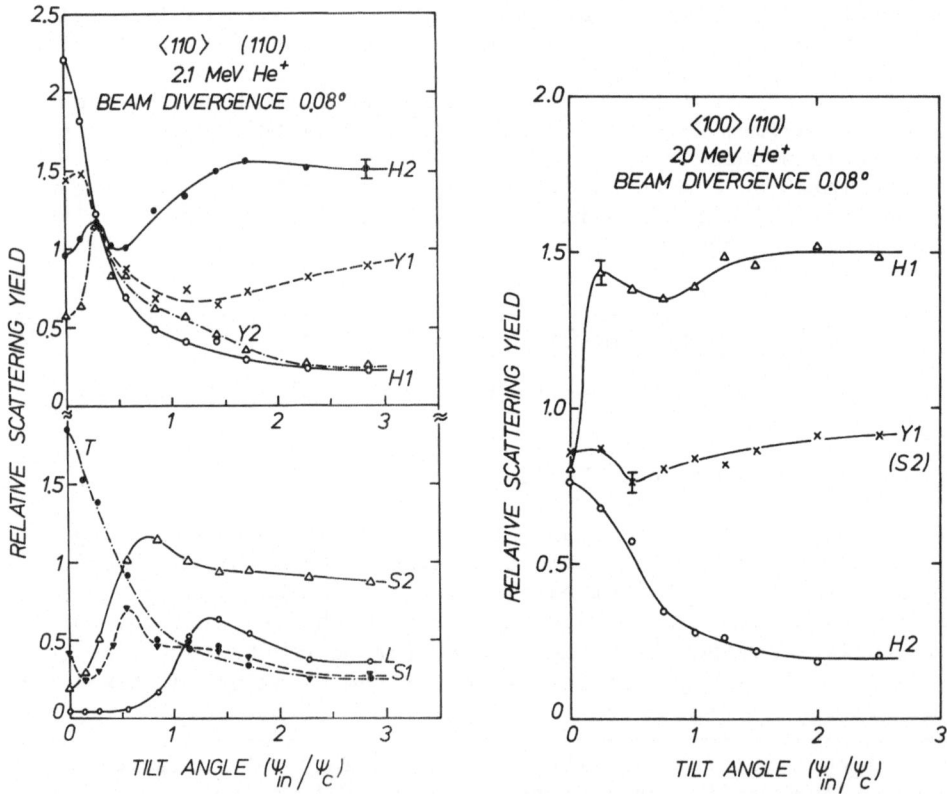

Fig. 2. Variation of the relative scattering yield from interstitial atoms (random = 1) with the tilt angle (ψ_{in}/ψ_c) against <110>
(a) and <100> (b) at beam divergence 0.08° in single alignment. ψ_c
is the experimental critical angle which is 0.52° in <110> and
0.44° in <100>. L represents the lattice positions. The symbols
refer to the interstitial sites in Fig. 1.

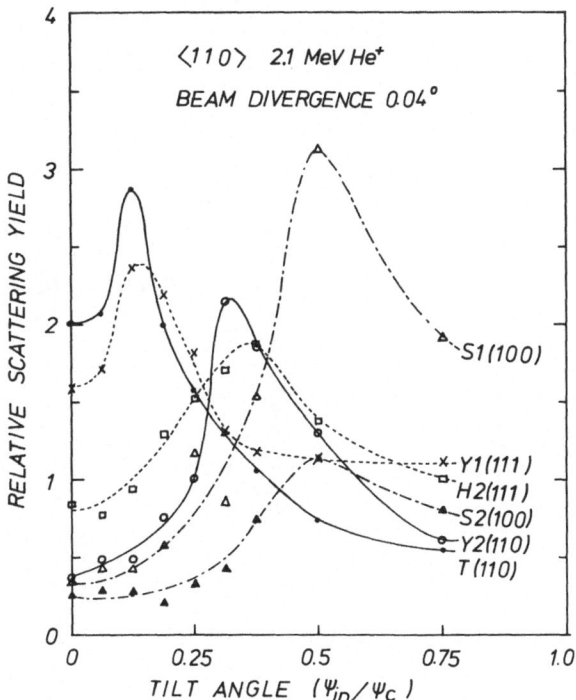

Fig. 3. Variation of the relative scattering yield from interstitial atoms (random = 1) with the tilt angle (ψ_{in}/ψ_c) against the <110> axis at beam divergence 0.04° in single alignment. ψ_c is the experimental critical angle. The symbols refer to the interstitial sites shown in Fig. 1 and the symbols in parentheses represent the tilt planes of (100), (111), and (110).

tilt plane the yield profile of T sites has a sharp maximum, however, no maximum for a beam divergence of 0.08° (cf. Fig. 2a). The site Y2 shows for 0.04° beam divergence a higher maximum than for 0.08° (Fig. 2a). However, for the Y1, H2, and S1 sites no maximum is found. For a (100) tilt plane both sites S1 and S2 show a maximum at about the same tilt angle of 0.26°. The sites H2 and Y1 have maxima at 0.19° and 0.075° for a (111) tilt plane. These results will be discussed in the next section.

The yield profiles for the different interstitial positions, obtained from the weighted mean of their projections on the transverse planes, are shown for <110> and <100> in Fig. 4. In <110> they do not show the characteristic features of the profiles of the single sites (cf. Fig. 2 and 3). However, in <100> the characteristic features remain. Therefore, the combination of both directions can be used to identify a given interstitial position provided

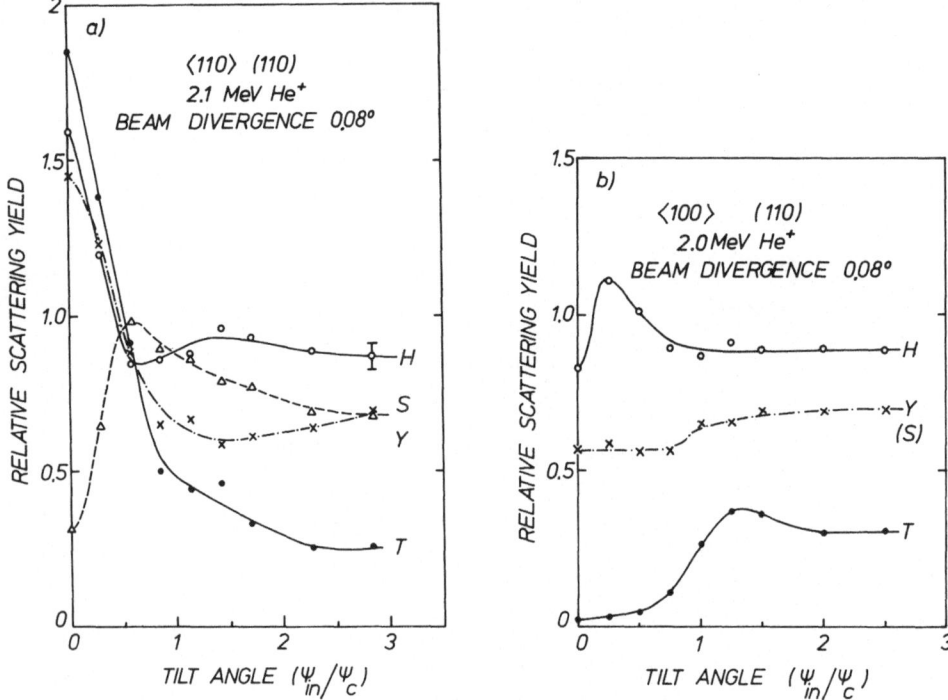

Fig. 4. Variation of the relative scattering yield from interstitial
atoms with the tilt angle against <110> (a) and <100> (b) in single
alignment. The scattering yield is the weighted mean of all equivalent sites. The symbols refer to different interstitial positions:
tetrahedral:T, hexagonal:H, Yb position:Y and split <100> configu-
rations:S.

that most of the impurity atoms are located on the same type of
position.
        The yield profiles for interstitial sites in double alignment
are similar to that in single alignment as can be seen from Eq. (2)
and are therefore not explicitely shown. Figure 5 gives the final
scattering yield profiles in double alignment for the combinations:

        (a) <110> – <1$\bar{1}$0> (110) and (b) <110> – <$\bar{1}$00> (011).

One finds (Fig. 5a) that in double alignment the yields for T, H,
and Y positions are enhanced at small angles. The yield profile
for H positions is more pronounced than in single alignment and can
be better distinguished from the other positions. The coincidence
of the profiles for Y and S positions in <100> – as observed in
single alignment – has disappeared (cf. Fig. 5b). The reduction
in the backscattering yield from the lattice atoms ("minimum yield")

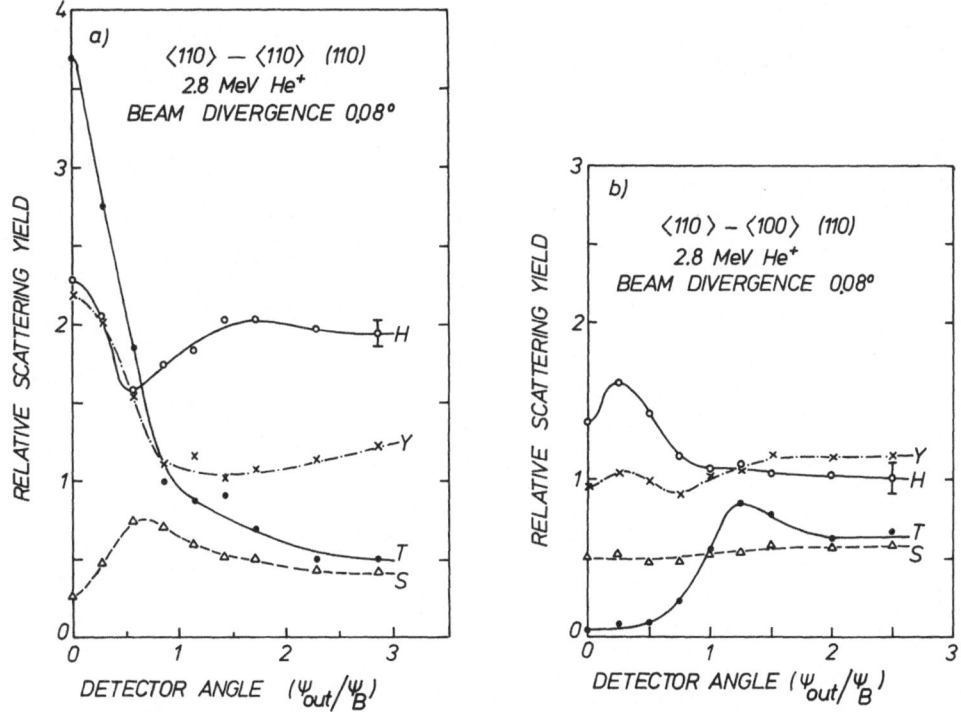

Fig. 5. Variation of the relative scattering yield from interstitial atoms with the detector angle against <110> (a) and <100> (b) in double alignment. In both figures the incidence channeling direction is <110>. The detector angle is normalized to the blocking angle ψ_B ($\psi_B = \psi_c$).

is $\sim 3 \times 10^{-4}$ for the axis and $\sim 3 \times 10^{-3}$ for the plane. This is low enough to make interstitial atom location possible.

Based on the above results of the double alignment, the yield profiles for a crystal with 5.10^{-3} interstitial atoms per lattice atom were computed according to Eq. (3). The profiles are shown in Fig. 6. The yields for perfect alignment are, as expected, of the same order of magnitude as those at large detector angles. Comparing the yields for the different positions along the same exit direction of a scattered projectile one does not find striking differences, except the strong peak for T positions and the pronounced minimum for S positions along <110> (Fig. 6a), or the maximum for H positions at an off axis angle and the pronounced minimum for T positions in <100> (Fig. 6b). Moreover, the profiles change in general with the concentration of interstitial atoms, in particular, at small angles. However, a comparison of the yield <u>profiles</u> for both combinations clearly allows to distinguish between the different positions mentioned.

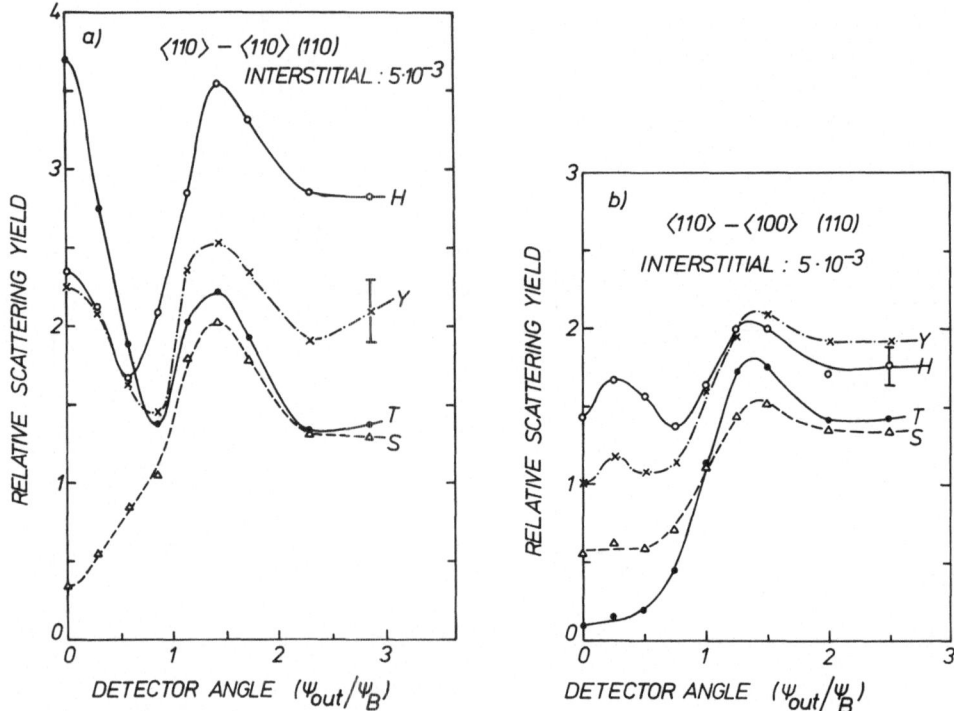

Fig. 6. Variation of the relative scattering yield from a silicon crystal with the detector angle against <110> (a) and <100> (b) in double alignment. In both figures the incidence channeling direction is <110>. The concentration of interstitial atoms is 5.10^{-3}.

Discussion

The maxima in the yield profiles for off axis sites (cf. Figs. 2 and 3) can be explained according to Morgan and Van Vliet [6].

For an angle of incidence ψ_{in} the points r within the potential energy contour $U(r) - E\psi_{in}^2$ which is set up by the target atoms in a transverse plane are accessible to all projectile ions and are equivalent, provided that they are uniformly distributed within this accessible area. A maximum in the yield profile then occurs at an incident angle ψ_M where $E\psi_M^2 = U(r_i)$, r_i being the position of the off axis site.

The angles ψ_M where the maxima in the profiles occur, the corresponding transverse energies $E\psi_M^2$, and the equi-potential contour $U(r_i)$ for various interstitial sites are tabulated in Table I. One finds that the computer results derived from projectile trajectory calculations, giving $E\psi_M^2$, for Y2 and T sites in a (110) tilt plane, for S1 and S2 sites in a (100) tilt plane and for Y1

Table I. Peak position ψ_M from the detailed computer calculations, transverse energy $E\psi_M^2$, and potential energy $U(r_i)$ from analytical energy contour calculations for various interstitial sites in the (110) transverse plane.

Interstitial site	Tilt plane	Peak position ψ_M(radian)	$E\psi_M^2$ (eV)	$U(r_i)$ (eV)
Y1	110	0.0015	3.6	4
	111	(0.0015)	(3.6)	
H2	110	0.0027	14	9.5
	111	(0.0033)	(23)	
Y2	110	0.0029	18	21
	110	(0.0030)	(19)	
S1	110	0.0054	58	42
	100	(0.0045)	(43)	
S2	110	0.0067	90	40
	100	(0.0045)	(43)	
T	110	(0.0011)	(2.4)	2.5

The values in parentheses are calculated for a beam divergence of 0.4° and a depth interval between 0 and 960 Å, without parentheses for a beam divergence of 0.08° and a depth interval between 0 and 3840 Å.

sites in a (111) plane are in good agreement with the equi-potential contours $U(r_i)$ derived analytically from a static lattice, cf. Fig. 1.

It is remarkable that a pronounced yield maximum is obtained for off axis sites, whenever the tilt plane through the considered off axis site passes areas with considerably lower potential than at the site. This is obvious in Fig. 2b for <100> and a (110) tilt plane where for instance the yield profile of H1 sites shows a maximum, whereas that of H2 sites does not.

The yield profiles for the first 1000 Å of penetration depend strongly on the scanning plane across a channel which indicates that statistical equilibrium of the ion flux in the transverse planes has not been reached. The continuum model must not be used in this region.

The values of $E\psi_M^2$ for S1, S2, and H2 sites in Table I without parentheses and for H2 sites and a (111) tilt plane are larger than the values of the corresponding potential contours. The yield profiles of S1 and S2 sites do not show a pronounced maximum. This is probably caused by the improved statistical equilibrium in the flux at the beam divergence of 0.08° and by the larger penetration depth of 2000 layers. However, the larger value of $E\psi_M^2$ for H2 sites and a (111) tilt plane (beam divergence 0.04°) is not entirely understood. The fact that one does not find a maximum for T sites (Fig. 2a) may be due to the large beam divergence ($\Delta\psi$) of 0.08°, since then $E(\Delta\psi)^2 = 4$ eV is larger than the corresponding potential value.

In general, as shown in the previous section, all the different projections of an interstitial position onto the transverse plane under consideration contribute to the total scattering. However, for a particular choice of the channeling and blocking directions in double alignment one obtains at small angles, where the contribution of the scattering yield from the lattice atoms is small, the yield profile characteristic for a particular projection of a given lattice position. This is the case for the combination of channeling direction <100> and blocking direction <110>, where the two directions include an angle of 90°. The yield of only Y1 sites (for Y positions) and of only S2 sites (for S positions) is enhanced in <110>, since the other sites Y2 and Y3, S1 and S3 are hidden in the atomic row along the <100> axis. If then the analyzing detector is tilted in an appropriate plane one obtains identical curves, as shown in Figs. 2 and 3 for single alignment, at least for small angles ($\psi_{out} < \psi_B$). Combinations of this kind render it easier to distinguish the different positions. In such cases the method can be extended to even lower concentrations of interstitial atoms. Additional information about the atom locations can be obtained by combining planar channeling in incidence and axial blocking in observation. Experimental results will be presented elsewhere.

Acknowledgments

The authors wish to express their gratitude to Prof. R. Sizmann for many stimulating discussions and the critical reading of the manuscript. One of the authors (K.M.) is also grateful to the members of Prof. Sizmann's laboratory for their valuable help in the course of this work. We wish to thank the members of the Leibniz-Rechenzentrum, Bayerische Akademie der Wissenschaften, München for valuable cooperation in the computer work. This investigation has been supported by the Bundesministerium für Forschung und Technologie, FRG.

References

*Permanent address: University of Nagoya, Faculty of Engineering, Department of Nuclear Engineering, Nagoya, Japan.

[1] J. W. Mayer, L. Eriksson and J. A. Davies, Ion Implantation in Semiconductors (Academic Press, New York, 1970).
[2] J. U. Andersen, O. Andreason, J. A. Davies and E. Uggerhøj, Rad. Effects 7, 25 (1971);
J. U. Andersen, E. Laegsgaard, and L. C. Feldman, Rad. Effects 12, 219 (1972).
[3] H. D. Carstanjen and R. Sizmann, Phys. Letters 40A, 93 (1972).
[4] K. Tachibana, K. Morita, and N. Itoh, Solid State Commun. 9, 1425 (1971).
[5] D. Van Vliet, Rad. Effects 10, 137 (1971).
[6] D. V. Morgan and D. Van Vliet, Rad. Effects 12, 203 (1972).
[7] V. S. Kalikanskas, M. M. Malov, and A. F. Tulinov, Soviet Physics JETP 26, 321 (1968).
[8] E. Bøgh, Can. J. Phys. 46, 653 (1968).
[9] B. R. Appleton and L. C. Feldman, Atomic Collision Phenomena in Solids, ed. by D. W. Palmer et al., 1970, p. 417.
[10] H. D. Carstanjen, and R. Sizmann, Rad. Effects 12, 225 (1972).
[11] D. Van Vliet, AERE Report No. R 6395, Harwell, 1970.
[12] J. Lindhard, Kgl. Danske Vid. Selskab. Mat. Fys. Medd., 34, (1965) No. 14.
[13] J. W. Corbett, Rad. Effects in Semi-conductors, ed. by F. Vook (Plenum, New York, 1968) p. 3.
[14] J. U. Andersen, and E. Uggerhøj, Can. J. Phys. 46, 517 (1968).

SECTION X
DECHANNELING

VALIDITY OF THE STATISTICAL EQUILIBRIUM HYPOTHESIS FOR CHANNELING*

JOHN H. BARRETT

Solid State Division, Oak Ridge National Laboratory
Oak Ridge, Tennessee 37830, U.S.A.

ABSTRACT

Previous Monte Carlo calculations of large numbers of trajectories in crystals have given a minimum yield for axial channeling about three times larger than that calculated analytically from the assumption of statistical equilibrium. Experimental measurements are in good agreement with the Monte Carlo results. The Monte Carlo calculations also showed a depth dependence of the minimum yield with two very distinct peaks below the surface peak, the deeper of the two being the more prominent. Some limited studies of the source of difference between the two methods of calculation showed that trajectories whose closest approach to the rows was about one or two times the thermal vibration amplitude were making a much larger contribution in the Monte Carlo calculations than in the analytical ones.

Further Monte Carlo calculations reported here give enough additional information to furnish an understanding of the difference between the two methods and to show that this difference is due to the invalidity of assuming statistical equilibrium. The yield in the surface and near surface region is entirely due to ions striking the crystal very near row ends exposed at the surface. Detailed studies of trajectories show that the surface peak comes from the originally struck row and that the first sub-surface peak comes from the ring of nearest neighbor rows, both just as expected. However, the second sub-surface peak is rather surprisingly due to the ring of fourth nearest neighbor rows to the

originally struck row. Further analysis of the
trajectories shows that both the unexpected order
of neighbor rows associated with this latter peak
and its high intensity are due to a focusing effect
of the intervening rows which is dependent on the
transverse energy, a process very similar to assisted
focusing in replacement sequences. The repeated
occurrence of this focusing is the source of enhancement of minimum yield in the Monte Carlo calculations
over that in the analytical ones.

The presence of this focusing effect constitutes
a breakdown of statistical equilibrium. Furthermore,
the breakdown will affect the yield even beyond the
near surface region since all particles will be subject
to focusing as they undergo dechanneling and will be
exposed thereby to more nuclear multiple scattering
than the statistical equilibrium hypothesis predicts.
It is also to be expected that this focusing process
will be of importance in determining $\psi_{1/2}$, and, indeed,
it seems likely to be the source of the fact that
lattice Monte Carlo calculations give values of $\psi_{1/2}$
about 80% as large as those given by analytical and
single row Monte Carlo calculations. It appears that
most analytical calculations of channeling effects
need to be modified to take into account deviations
from statistical equilibrium.

[*]Research sponsored by the U. S. Atomic Energy Commission under
contract with Union Carbide Corporation.

AXIAL DECHANNELING, I. A THEORETICAL STUDY

H. E. SCHIØTT, E. BONDERUP, J. U. ANDERSEN and H. ESBENSEN
Institute of Physics, University of Aarhus
DK 8000 Aarhus C, Denmark

ABSTRACT

The paper describes a continuation of previous
theoretical work on axial dechanneling. A diffusion equation, in which the diffusion function is
strongly dependent on transverse energy, accounts
for the change with depth of the transverse energy
distribution. In our previous work the calculated
minimum yields were typically a factor of 2-3 too
low. In this paper we have evaluated the diffusion
function more carefully and found it necessary to
introduce changes in both the electronic and the
thermal components. As shown in the accompanying
paper "Axial Dechanneling, II. An Experimental
Study" theoretical predictions agree quite well with
backscattering experiments on Si and W and with transmission experiments on Si. The merits and limitations
of the simple model using a steady increase in transverse energy are illustrated by a comparison of full
angular scans obtained from the simple model, and
from the diffusion equation. Finally, the application
of the diffusion equation to the analysis of blocking
dips, obtained for the determination of lifetimes of
compound nuclei, is described.

Introduction

This report describes an attempt to improve the theoretical
work on axial dechanneling, presented at the Gausdal conference [1],
and several notions from Ref. 1 will be applied without further
explanation.

Reaction Yields

The problem consists in calculating, at any depth z in a crystal, the distribution $g(E_\perp, z)$ in transverse energy $E_\perp$. If such a calculation can be performed, the directional dependence of the yield of various reactions can be obtained. By $\Pi_{react}(E_\perp)$ we denote the suitably normalized probability for a particle of transverse energy $E_\perp$ to take part in a reaction of the type under consideration. The reaction yield $\chi(z)$ is then given by

$$\chi(z) = \int dE_\perp \ g(E_\perp, z) \Pi_{react}(E_\perp) . \tag{1}$$

Expressions for $\Pi_{react}(E_\perp)$ can be obtained in several cases. Let us mention two examples:

1) For a process, such as backscattering, requiring a very close encounter between projectile and target nucleus, the reaction probability, normalized to incidence in a random direction, is fairly well approximated by a step function

$$\Pi_{react}(E_\perp) = \left\{ \begin{array}{ll} 0 , & \text{if } E_\perp < E_{\perp,c}(\rho) \\ 1 & \text{otherwise} \end{array} \right. . \tag{2}$$

The value of the critical transverse energy, $E_{\perp,c}(\rho)$ is a function of ρ, the root mean square vibrational amplitude perpendicular to strings. With $\Pi_{react}(E_\perp)$ given by (2), the corresponding function $\chi(z)$ only depends on the tail of the distribution $g(E_\perp, z)$, i.e. the number of particles with a transverse energy exceeding $E_{\perp,c}(\rho)$.

2) A second example is transmission through a crystal of thickness z. For a fixed angle of incidence with a string one observes the transmitted angular distribution $\chi(\Psi_t, z)$ behind the crystal. If $\Pi_{react}(\Psi_t, E_\perp)$ denotes the probability per unit solid angle that a particle with transverse energy $E_\perp$ just below the back surface of the crystal will be detected at the angle Ψ_t, the distribution $\chi(\Psi_t, z)$ can be obtained from (1). For the transverse motion we denote by $A(E_\perp)$ the area accessible to a particle of transverse energy $E_\perp$. The reaction probability is then determined by

$$\Pi_{react}(\Psi_t, E_\perp) = \frac{1}{\pi} E \ A(E_\perp)^{-1} \frac{d}{dE_\perp} \ A(E_\perp - E\Psi_t^2) . \tag{3}$$

In the following the transverse unit cell is replaced by a circle centered on the string, with area $\pi r_o^2 = (N d)^{-1}$, where N is the atomic density of the crystal and d the spacing of atoms in the string. The continuum potential $U(\vec{r})$ is assumed axially symmetric, and furthermore the convention $U(r_o) = 0$ is used.

At not too small values of Ψ_t the reaction probability has a pronounced peak at transverse energies close to the value $E\Psi_t^2$. This means that, as opposed to a backscattering experiment, a measurement of the distribution $\chi(\Psi_t, z)$ gives information about $g(E_\perp, z)$ at all

values of $E_\perp$. In the accompanying paper, entitled "Axial Dechanneling, II. An Experimental Study," [2] theory is checked against
results from both backscattering and transmission experiments.

Diffusion function

For the change with depth in the distribution $g(E_\perp,z)$ we
arrived in Ref. 1 at the diffusion equation

$$\frac{\partial g(E_\perp,z)}{\partial z} = \frac{\partial}{\partial E_\perp} A(E_\perp) \, D(E_\perp) \, \frac{\partial}{\partial E_\perp} \frac{g(E_\perp,z)}{A(E_\perp)} \quad , \qquad (4)$$

expressing the change in probability density in terms of the derivative (divergence) of a probability current, which is proportional
to the derivative (gradient) of the density in phase space g/A.
Given an initial distribution $g(E_\perp,z = o)$, which for both backscattering and transmission experiments is obtained from the transmission through the front surface of the crystal, equation (4)
determines the distribution at any depth z.

In the typical case of MeV protons, the diffusion function
$D(E_\perp)$ increases by several orders of magnitude over the transverse
energy interval of interest. For low values of $E_\perp$ the diffusion
is determined by scattering by electrons, whereas force fluctuations
due to thermal displacements are responsible for the changes in $E_\perp$
at high transverse energies. In Ref. 1 fairly large discrepancies
between calculated results and experimental evidence were encountered.
In fact, corresponding to backscattering experiments the calculated
minimum yields were typically a factor of 2-3 too low. These
differences were, however, not felt to be too discouraging, since
the expression applied for the diffusion function was known to be
rather poor in certain transverse energy regions. The following
section contains a description of a more careful evaluation of
$D(E_\perp)$. As will become apparent, it has been necessary to introduce
changes in both the electronic and the thermal contributions to
the diffusion.

Damping

In the derivation of Eq. (4) energy loss was neglected. We
shall now briefly comment on the magnitude of the errors introduced
into the description through this approximation. If the energy of
a particle is changed from E to $E+\delta E$ the transverse kinetic energy
$E_{\perp,kin} = E\phi^2$ and thus the total transverse energy $E_\perp$ is changed by
the amount [3]

$$\delta E_\perp = (\delta E)\phi^2 = \frac{\delta E}{E} E_{\perp,kin} \quad . \qquad (5)$$

The ratio $E_{\perp,kin}/E_{\perp}$, averaged over the area $A(E_{\perp})$, depends on $E_{\perp}$. By setting it equal to 1 we obtain an upper limit for the change in transverse energy due to energy loss

$$|(\frac{dE_{\perp}}{dz})_{loss}| < |\frac{dE}{dz}| \cdot \frac{E_{\perp}}{E} . \tag{6}$$

A stopping power appropriate for channeled particles should be applied on the right hand side of (6).

At a transverse energy $E_{\perp}$ stopping results in an extra (negative) current $g(E_{\perp},z) (\frac{dE_{\perp}}{dz})_{loss}$. Thus, on a path length δz the number of particles in the interval $(E_{\perp}, E_{\perp} + dE_{\perp})$ changes due to slowing down by $\delta g dE_{\perp}$, where

$$\delta g dE_{\perp} = g(E_{\perp},z) (\frac{dE_{\perp}}{dz})_{loss}(E_{\perp}) \delta z - g(E_{\perp} + dE_{\perp},z)$$

$$\cdot (\frac{dE_{\perp}}{dz})_{loss} (E_{\perp} + dE_{\perp}) \delta z . \tag{7}$$

This means that Eq. (4) is modified to read

$$\frac{\partial g(E_{\perp},z)}{\partial z} = \frac{\partial}{\partial E_{\perp}} A(E_{\perp}) D(E_{\perp}) \frac{\partial}{\partial E_{\perp}} \frac{g(E_{\perp},z)}{A(E_{\perp})}$$

$$+ \frac{d}{dE_{\perp}} g(E_{\perp}) |(\frac{dE_{\perp}}{dz})_{loss}| . \tag{8}$$

This equation has also been obtained by Beloshitsky et al. [4]. In the cases investigated in Ref. 2 the inclusion of the "damping" term in Eq. (8) leads to changes in the calculated results of at most a few per cent, the changes in the yield curves being in opposite directions for backscattering and emission (blocking). Differences of this order of magnitude are barely significant at the present accuracy, and we decided to display results based on Eq. (4) rather than on Eq. (8). Since the former equation, as opposed to the latter, gives exact reversibility between backscattering and blocking, the calculated curves apply to either situation.

At a fixed value of $E_{\perp}$ the diffusion function increases with decreasing total energy E. To the extent that energy loss can be assumed independent of transverse energy, this effect is included in the equations (4) and (8). In all of the experiments to be reported in the accompanying paper the relative energy loss was small, and consequently slowing down was neglected in the determination of $D(E_{\perp})$.

Determination of $D(E_\perp)$

As shown in Ref. 1 the diffusion function is connected to the change $d<E_\perp>$ in average transverse energy for a group of particles which, with the same initial value $E_\perp$, move a distance dz in the axial direction,

$$\frac{d}{dz} <E_\perp> = \frac{1}{A(E_\perp)} \frac{d}{dE_\perp} A(E_\perp) D(E_\perp) , \qquad (9)$$

and

$$D(0) = 0 . \qquad (10)$$

The relation (9) may be derived from Eq. (4), and also from Eq. (8) if the left hand side of (9) is interpreted as the change in average transverse energy with neglect of the (negative) term from energy loss.

The function $\frac{d<E_\perp>}{dz} (E_\perp)$ receives additive contributions from scattering by electrons and from thermal displacements. The evaluation of the two contributions will now be considered in turn.

Scattering by electrons

We have, as in Ref. 1, applied the results on multiple scattering in an electron gas of a constant density ρ. For a projectile of high velocity v and low charge $Z_1 e$ the dielectric description leads to the following expression for the change in average square angle with distance z [5]

$$\frac{d<\phi^2>}{dz} = \frac{\pi (Z_1 e^2)^2}{E^2} \rho [\log \frac{2mv^2}{\hbar\omega_o} - 1] , \qquad (11)$$

where $(-e)$ and m are the charge and mass of the electron, respectively. The plasma frequency ω_o is determined by the density through

$$\omega_o^2 = \frac{4\pi e^2 \rho}{m} . \qquad (12)$$

For the electronic term $(\frac{d<E_\perp>}{dz})_e$ in (9) we take the right hand side of (11), averaged over the accessible region for a particle of transverse energy $E_\perp$:

$$\left(\frac{d<E_\perp>}{dz}\right)_e(E_\perp) = \frac{\pi Z_1^2 e^4}{E} \frac{1}{d\, A(E_\perp)} \int_{-d/2}^{d/2} dz$$

$$\cdot \int_{A(E_\perp)} d^2\vec{r}\; \rho(\vec{R})[\log \frac{2mv^2}{\hbar\omega_o(\vec{R})} - 1], \qquad (13)$$

where $\vec{R} = (\vec{r},z)$, the center of the atom being at origo. In Ref. 1 we applied for $\rho(R)$ the expression (2.6') in Ref.3, i.e. the atomic electron density $\rho_{st}(R)$, which corresponds to Lindhard's standard potential [3]. At distances $r\sim r_0$ from strings ρ_{st} underestimates the electron density by a factor of, perhaps, ~ 4. This is connected with the fact that atomic electrons outside a cylinder of radius r_0 around a string have been neglected altogether. The correspond-ing underestimate of $\left(\frac{d<E_\perp>}{dz}\right)_e$ at low transverse energies may be expected to lead to sizeable errors in certain cases since scatter-ing by electrons gives by far the dominating contribution to the diffusion at low values of $E_\perp$.

The integrand in (13) is approximately proportional to $\rho(R)$. Thus, for fixed r, $r\ll r_0$, the integral over z receives its main contributions from a small z-interval around z = 0. From this it follows that for small values of r it is not important that the density $\rho_{st}(\vec{r},z)$ is too low at large z. Within a central cylinder of radius r_c around a string the atomic electron density is there-fore still taken to be $\rho_{st}(\vec{r},z)$. Outside the cylinder the density is set equal to a constant ρ_0, the value of which is chosen such that the total number of electrons per target nucleus come out cor-rectly. Thus $\rho_0 = \rho_0(r_c)$ depends on r_c. It so happens that the integral $d^{-1}\int_{-d/2}^{d/2} dz\rho_{st}(r_0/\sqrt{2},z)$ is closely equal to $\rho_0(r_0/\sqrt{2})$ and therefore we ensure approximate continuity of the two-dimensional electron density by choosing the value $r_0/\sqrt{2}$ for the radius r_c. The ratio between ρ_0 thus calculated and $\rho_{st}(r_0)$ is always equal to 4. In the cases Si<110> and W<100>, to be considered in Ref. 2, the values of ρ_0 correspond to the densities obtained when 3.4 and 6.6 electrons, respectively, are uniformly distributed over an atomic volume.

The electron density, averaged along the string direction, is connected to the continuum potential through the two-dimensional Poisson equation. The change in electron density is therefore accompanied by a change in the continuum potential. For the sake of consistency, this change has been included in the description.

Scattering due to thermal displacements

In the transverse plane we consider a projectile passing an atom which is displaced a distance $\vec{s}$ from its ideal position. If $\vec{F}(\vec{r})$ denotes the transverse force derived from the continuum potential, the projectile at $\vec{r}$ will, in addition to the force leading to conservation of transverse energy, be acted upon by $\delta\vec{F}$, where

$$\delta\vec{F} = \vec{F}(\vec{r}-\vec{s}) - \vec{F}(\vec{r}) . \tag{14}$$

Summing the squares of the corresponding momentum transfers along the path of a projectile and expanding to lowest contributing order in s/r, Lindhard arrived at an expression for the thermal contribution $\frac{d<E_\perp>_{th}}{dz}$ ($E_\perp$) to the change in average transverse energy [3]. For MeV protons at low transverse energies the thermal component turns out to be negligible compared to the electronic contribution. For large values of $E_\perp$ the situation is reversed. Let $\hat{r}(E_\perp)$ denote the closest distance of approach to a string, i.e. $U(\hat{r}) = E_\perp$. At a transverse energy corresponding to $\hat{r}$ of the order of two to three times ρ, the diffusion due to thermal displacements rather suddenly takes over and becomes completely dominating at still higher values of $E_\perp$. Although the condition for applying Lindhard's expression is not well fulfilled at these small values of $\hat{r}(E_\perp)$, we still first attempted to use the formula for all transverse energies at which it gives a lower value than that corresponding to a projectile incident in a random direction. The argument was that since the diffusion in the thermal region is, at any rate, very rapid as compared to the diffusion in the electronic region, moderate changes in $d<E_\perp>_{th}/dz$ are not expected to lead to significantly different calculated yields $\chi(z)$ for close encounter processes in cases where the projectiles enter the crystal at a low angle of incidence. The decisive point for such a calculation is the slow increase in $E_\perp$, which a particle experiences when traversing the electronic regime. However, as will appear from the following discussion, we now believe that in our previous work $d<E_\perp>/dz$ was underestimated by nearly an order of magnitude at very high transverse energies. Errors of this magnitude do turn out to influence $\chi(z)$ appreciably.

First we consider the question of estimating the transverse energy at which $d<E_\perp>/dz$ reaches the random value. In a target with randomly distributed atoms $d<\phi^2>/dz$ receives the major contribution from elastic scattering of the projectile in a screened Coulomb potential and is approximately proportional to the logarithm of the ratio between the effective upper and lower limits of an integral over impact parameters, $\log(a/b_n)$ [3,6]. Here, a denotes an appropriate screening distance and $b_n = Z_1 Z_2 e^2/E$ the collision diameter, Z_2 being the atomic number of the target atoms, $(Z_2 >> Z_1)$. Since b_n/a is typically of the order of 10^{-3}, a sufficient condition for obtaining normal (random) multiple scattering in a crystal is that collisions with impact parameters small compared to a occur with the same probabilities as for random incidence and that they are uncorrelated. The first condition requires that the projectile appears with equal probability everywhere in the transverse plane, i.e. the projectile must be able to overcome the transverse energy barrier $E_{\perp,B}$ presented by the continuum potential. As an estimate of $E_{\perp,B}$ we take the height of the potential, which one obtains by averaging the standard potential over thermal displacements of the atoms on the string. With Ca being the screening radius in the standard potential this gives for all relevant values of the ratio a/ρ

$$E_{\perp,B} \simeq \frac{Z_1 Z_2 e^2}{d} \log \left(1 + \frac{2(Ca)^2}{\rho^2}\right) . \tag{15}$$

Secondly, correlations can only be of importance for impact parameters larger than or of the order of ρ and with $b_n/\rho \sim 10^{-3}$, we can neglect the contribution to multiple scattering from deflections corresponding to such impact parameters. We may thus require that $d\langle E_\perp\rangle/dz$ be equal to the random value for transverse energies exceeding $E_{\perp,B}$ in (15). With our previous procedure of obtaining $d\langle E_\perp\rangle/dz$ at high transverse energies, we found, at room temperature, for $\frac{d\langle E_\perp\rangle}{dz}$ $(E_{\perp,B})$ a value which for 1.6 MeV $H^+\to Si<110>$ and 2.0 MeV $H^+\to W<100>$ was 0.14 and 0.16 times the random value, respectively.

Our problem is now to find for $d\langle E_\perp\rangle/dz$ an expression, which in a reasonable manner bridges the gap between the value at $E_\perp \geq E_{\perp,B}$ and the Lindhard expression, which applies for transverse energies with $\hat{r}(E_\perp)$ larger than a few times ρ.

Written as a function of $\hat{r}$ the ratio $\gamma^{(1)}$ of the first term $d\langle E_\perp\rangle_{th}^{(1)}/dz$ in Lindhard's expansion and the random value of $d\langle E_\perp\rangle/dz$ is given by

$$\gamma^{(1)}(\hat{r}) = \frac{1}{2L_n} \frac{\rho^2}{(Ca)^2} \left[1 + \left(\frac{Ca}{\hat{r}}\right)^2 + \frac{2}{3}\right] \cdot \left[\left(\frac{\hat{r}}{Ca}\right)^2 + 1\right]^{-3} , \tag{16}$$

where L_n has the form of a logarithm, $L_n \sim \log(a/b_n)$.

A simple expression which attains the random value at $E_{\perp,B}$, and which coincides with $d\langle E_\perp\rangle_{th}^{(1)}/dz$ in the asymptotic limit of very large $\hat{r}$ is obtained when for the ratio γ we take

$$\gamma(\hat{r}) = \frac{1}{2L_n} \frac{\rho^2}{(Ca)^2} \left[\left(1 + \left(\frac{Ca}{\hat{r}}\right)^2\right)^f + \frac{2}{3}\right] \left[\left(\frac{\hat{r}}{Ca}\right)^2 + 1\right]^{-3} , \tag{17}$$

the constant f being determined such as to give for $d\langle E_\perp\rangle/dz$ the random value at $E_\perp = E_{\perp,B}$.

Before applying (17) we have to ascertain that we get reason-
able values, not only in the asymptotic limit $\hat{r}/Ca \to \infty$ but also in
the physically important region, in which the thermal contribution
becomes comparable to or larger than the electronic component,
i.e. for $\hat{r}$ of the order of two to three times ρ. In the cases
investigated, formula (17) gives, as indicated in Fig. 1, at room
temperature, values, which are in fairly good agreement with the
sum of the first two terms in the expansion in s/r, as long as the
second term is only a moderate correction to the first one. At
higher temperatures (17) tends to underestimate γ calculated to
second order. At large transverse energies the values obtained from
our procedure are well above the sum of the first two terms in the
expansion. An indication that this behaviour is essentially correct
we obtained from calculations with the transverse forces $F(r) \propto r^{-1}$
and $F(r) \propto r^{-3}$. In these two cases the expansion of $d\langle E_\perp \rangle_{th}/dz$
can be checked against an exact analytical expression, provided
the positions of the target atoms are restricted to regions of
space, which are inaccessible to the projectile.

At very high transverse energies a useful lower limit to $\gamma(\hat{r})$
is set by the probability $\Pi(\hat{r})$ for a target atom to be in the area

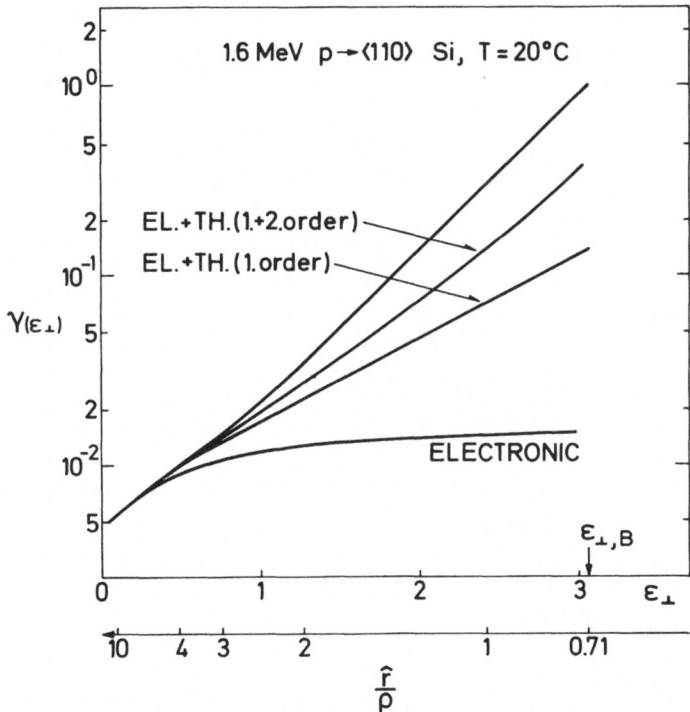

Fig. 1. $d\langle E_\perp \rangle/dz$ in units of the random value in various approximations. Thermal contribution to the uppermost curve is obtained
from (17).

$A(E_\perp)$. For Si<110> and W<100> at room temperature the right hand
side of (17) agrees quite well with $\Pi(\hat{r})$ in an interval $1 \lesssim \hat{r}/\rho \lesssim 2$,
when we apply the expression

$$\Pi(\hat{r}) = e^{-\hat{r}^2/\rho^2} . \tag{18}$$

In other cases, especially at elevated temperatures the ratio
$\Pi(\hat{r})/\gamma(\hat{r})$ can become considerably larger than unity for $\hat{r}/\rho$ of
order 1 - 2. We have therefore, at all transverse energies, applied
the larger of the two values $\gamma(\hat{r})$ and $\Pi(\hat{r})$ for the determination of
the diffusion function.

   As mentioned in the Introduction, a comparison with experimental results will be presented in Ref. 2. We therefore confine ourselves to showing, in Fig. 2, yields $\chi(z)$ of close encounter
processes as determined from numerical calculations in various
approximations. Curve I refers to an evaluation on the basis of
the original expression for $d<E_\perp>/dz$, curve II was obtained with
the electron density changed far away from strings, and curve
III corresponds to a situation in which we have modified both the
electronic and the thermal components of the diffusion in the

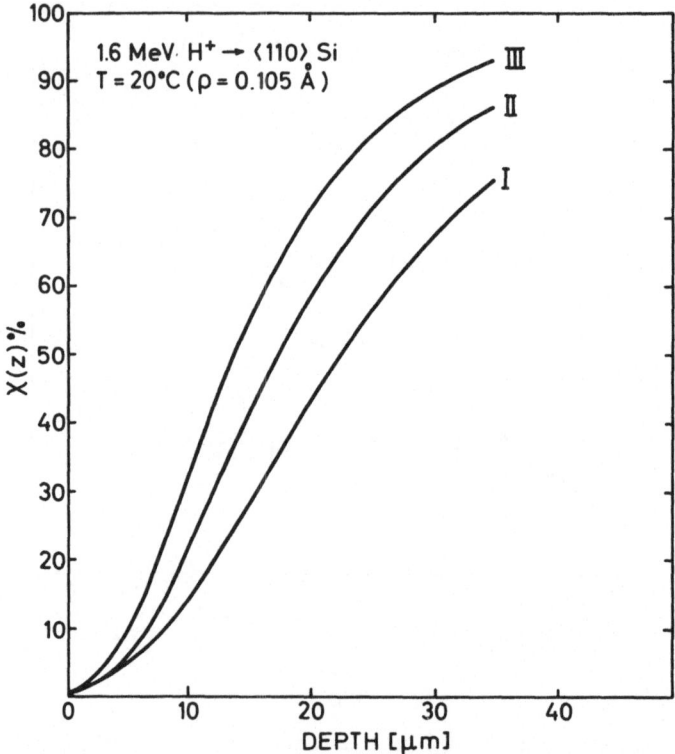

Fig. 2. Yields of close encounter processes calculated in various
approximations. Curve I with $d<E_\perp>/dz$ from Ref. 1. Curve II with
$d<E_\perp>_e/dz$ changed. Curve III with $d<E_\perp>_e/dz$ and $d<E_\perp>_{th}/dz$ changed.

manner described above. Both modifications are seen to be quite
important.

Steady Increase of $E_\perp$

    In Ref. 1, Fig. 4 we compared minimum yields $\chi(z)$ calculated
on the basis of the diffusion equation with the corresponding results obtained from the much simpler model with a steady increase
of transverse energy. Despite the complete failure of the intuitive
basis for this approximation [1] we found a rather good agreement
for all values of $\chi(z)$ up to ⌄25%. For larger minimum yields the
simple model gave a function $\chi(z)$, which increased much too rapidly
with depth. This situation is unchanged after the modification of
the diffusion function. Thus, as far as the calculation of minimum
yields for backscattering is concerned the two methods give a quite
similar response to changes in the diffusion function. As illustrated by the full angular scans in Fig. 3 the good agreement

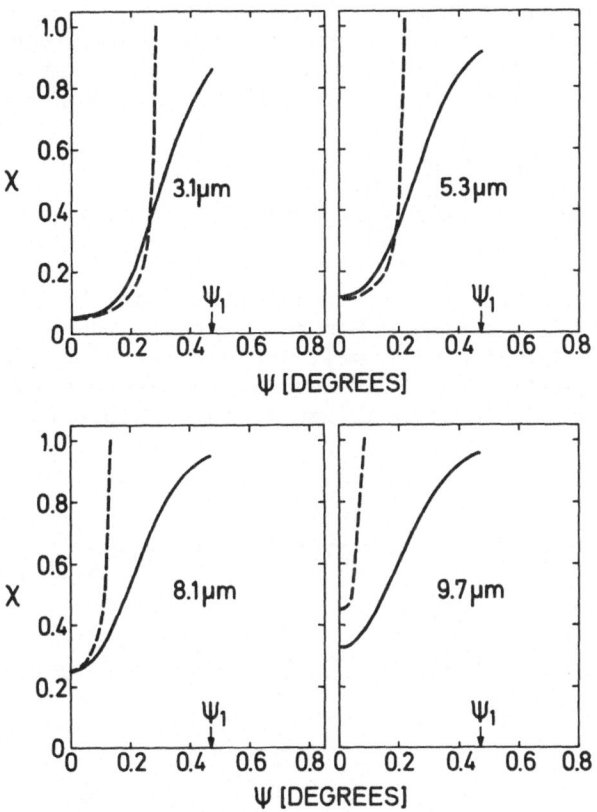

Fig. 3. Full angular scans: ——— from diffusion equation;
- - - - steady increase in transverse energy.

between backscattered yields as calculated from the diffusion
equation and from the simple model persists to angles of incidence
Ψ different from zero, provided we confine ourselves to yields
below $\sim$30%. The same figure clearly demonstrates the shortcomings
of the simple procedure for larger values of Ψ.

Lifetimes of Compound Nuclei

   A description of multiple scattering for channeled particles
can be important for many applications of channeling. In this
section we shall discuss, as an example, the application of blocking
to measure lifetimes of compound nuclei [7]. In such measurements
the effective crystal thickness is usually large enough for dechanneling to influence the dips appreciably, and it is therefore
important to investigate the resulting modifications of the procedure for lifetime extraction.
   According to the rule of reversibility, which applies when
"damping" is neglected, blocking dips are identical to angular scans
in backscattering experiments. To be specific, the blocking dip
$\chi(\Psi,z)$ belonging to emission at depth z can be obtained from (1),
when the distribution corresponding to bombardment at an angle of
incidence Ψ is inserted for $g(E_\perp,z)$ and the reaction probability is
given by

$$\Pi_{react}(E_\perp) = \frac{\pi r_o^2}{A(E_\perp)} \int_{\hat{r}}^{r_o} dP(r) \ , \quad E_\perp = U(\hat{r}) \ . \tag{19}$$

Here, $dP(r)$ is the differential probability for the nuclei to decay
at distance r from a string. For nuclei recoiling with transverse
velocity $v_\perp$ and decaying with well defined lifetime τ the distribution $dP(r)$ depends on two parameters, the average recoil distance
$v_\perp\tau$ and the thermal vibrational amplitude ρ. Introducing (19) into
(1) we obtain

$$\chi(z) = \int dE_\perp \ g(E_\perp,z) \frac{\pi r_o^2}{A(E_\perp)} \int dP(r)\Big|_{U(r) < E_\perp}$$

$$= \int_o^{r_o} dP(r) \int dE_\perp \frac{\pi r_o^2}{A(E_\perp)} g(E_\perp,z)\Big|_{E_\perp > U(r)} \ . \tag{20}$$

The expression

$$\chi(r,z) = \int dE_\perp \frac{\pi r_o^2}{A(E_\perp)} \cdot g(E_\perp,z)\Big|_{E_\perp > U(r)} \tag{21}$$

may be interpreted as the yield corresponding to well defined decay distance r. This is the function we need to know in order to extract the lifetime (or $v_{\perp}\tau$) from a measurement of $\chi(z)$.

At depth z = o the distribution $g(E_{\perp},0)$ is determined by the surface transmission, and for $\Psi = 0$ we obtain

$$g(E_{\perp},0) \; dE_{\perp} = d(\hat{r}^2)/r_o^2 \quad , \; E_{\perp} = U(\hat{r}) \quad . \tag{22}$$

Insertion of this distribution into (21) leads to

$$\chi(r,0) = \int_0^{r^2} d(\hat{r}^2)/r_o^2 \cdot \frac{\pi r_o^2}{\pi r_o^2 - \pi \hat{r}^2}$$

$$= \log(1 - r^2/r_o^2)^{-1}$$

$$\simeq r^2/r_o^2$$

$$= N d\pi r^2 \quad \text{for } r \ll r_o \; . \tag{23}$$

As a semi-empirical fit to results of computer simulations corresponding to particles at shallow depth, Barrett introduced a factor C $\simeq$ 2-3 on the right hand side of (23)[8]. This factor is believed to account for deviations from the picture of randomly distributed continuum strings at very small distances, $r \simeq \rho$. For somewhat larger distances, $\rho < r \ll r_o$, the factor should be close to unity [9,10].

With increasing depth $g(E_I,z)$ changes according to (4), and the corresponding change in $\chi(r,z)$ may be calculated from (21). It has sometimes been assumed that the increase in minimum yield due to a finite lifetime does not depend on crystal thickness, i.e. that dechanneling and lifetime effect give independent and additive contributions to the minimum yield. The extent to which such an assumption is valid is indicated in Fig. 4, which, for the case of 5 MeV protons emitted along a <110> direction in Ge, shows the minimum yield $\chi(r,z)$ as a function of r^2/r_o^2 at various depths z. The simple assumption is equivalent to having a set of parallel curves. Clearly, at small values of r the slope of the yield function χ is increasing with depth. At large depths an effective "Barrett factor" of C(z) $\simeq$ 2-3 is obtained. According to (20), (21) and (23) the minimum yield is given by

$$\chi(z) = C(z) \cdot \int dP(r) \cdot \frac{r^2}{r_o^2}$$

$$= C(z) \cdot N d\pi \; (\rho^2 + 2v_{\perp}^2\tau^2), \quad \text{for } v_{\perp}\tau \ll r_o \; . \tag{24}$$

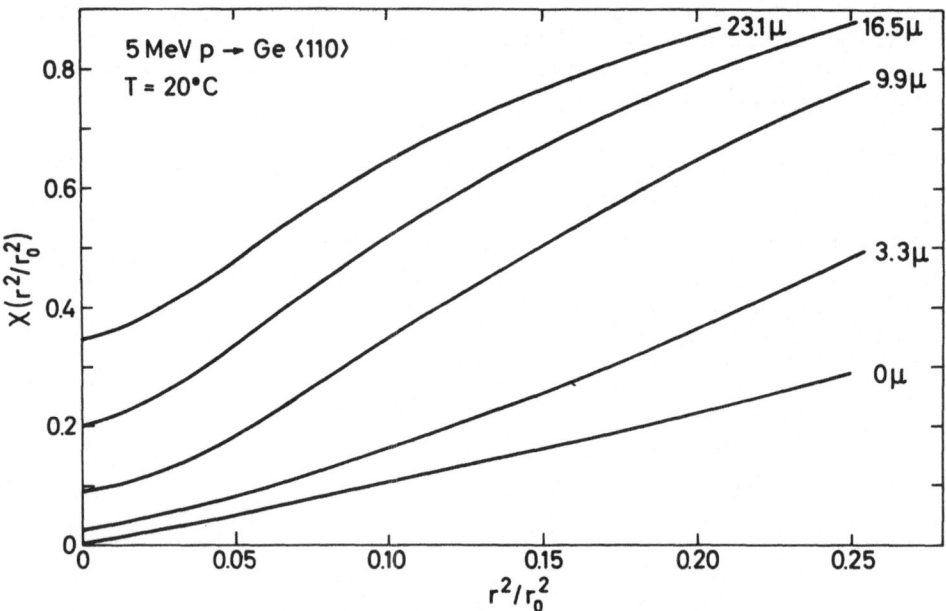

Fig. 4. Minimum yield as a function of decay distance at various depths.

Thus, the lifetime calculated from the increase in minimum yield due to a recoil displacement is proportional to $C(z)^{-1/2}$.

From equation (4) not only the minimum yield, but the full angular distribution $\chi(\Psi, z)$ may be calculated. Such a calculation is shown in Fig. 5 corresponding to the experiment by Clark et al. [7]. A simple exponential decay has been assumed, and the curves refer to various values of $v_\perp \tau/\rho$. The yield has been integrated from zero depth to the effective thickness of 10 μm. The experimental results agree fairly well with the calculated curve for $v_\perp \tau/\rho \sim 2-3$. In order to obtain an indication of the accuracy of the "additivity rule" mentioned above, the lifetime has been calculated from the measured change in minimum yield $\Delta\chi = 14.5\%$ for different maximum depths of emission. The results are tabulated in Fig. 5 and they show that knowledge of the crystal thickness may be of some importance.

Unfortunately, it is not clear whether the effective thickness of the crystal applied in the experiment was 10μm or 15μm [11]. For a thickness of 15μm the agreement between the experimental and theoretical full angular scans is somewhat better than shown in the figure. However, for the determination of the lifetime from the increase in minimum yield alone, such a change in thickness is not very important.

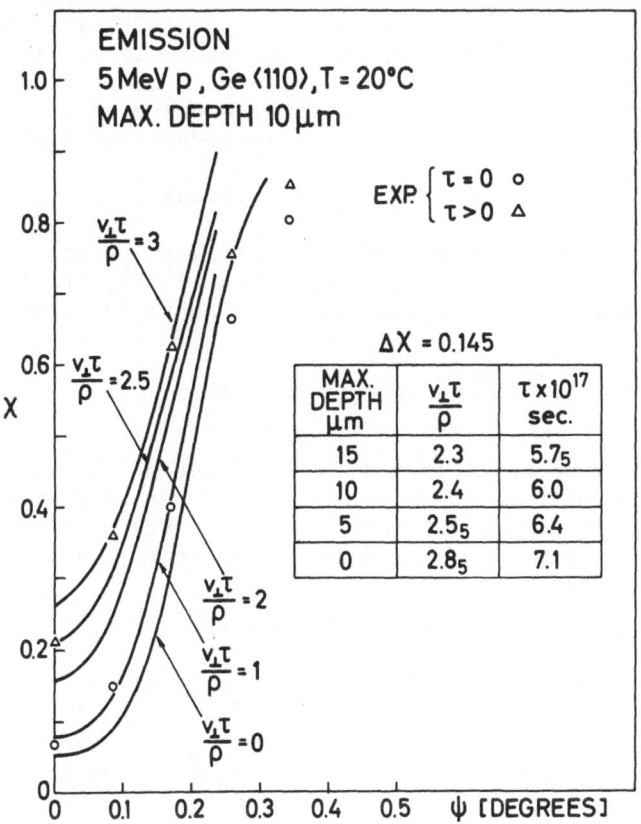

Fig. 5. Theoretical full angular scans for various values of $v_\perp\tau/\rho$ compared with experimental data from Ref. 7. Experimental points for $\tau = 0$ correspond to elastic scattering. The table displays, for various assumed crystal thicknesses, the lifetime evaluated from the change in minimum yield alone.

A lifetime of the order of 6×10^{-17} sec. is a factor of three larger than the one obtained by Clark et al. [7] through a different analysis.

Appendix

The thermal contribution to the change with depth of the average transverse energy is given in terms of the force fluctuations [3]:

$$\frac{d\langle E_\perp \rangle}{dz} = \frac{d}{4E} \cdot \langle \delta F^2 \rangle \; . \qquad\qquad (A1)$$

The geometry is illustrated in Fig. 6 which shows a projectile at $\vec{r}$ and a displaced atom at $\vec{s} = (x,y)$, the point 0 being the position of the string. The average on the right hand side of (A1) is over the distribution of string atoms, $P(\vec{s})$, and over positions $\vec{r}$ in the area accessible to particles of transverse energy $E_\perp$.

The magnitude of the fluctuation is found from Fig. 6 to be

$$\delta F^2 = F^2(1) + F^2(r) - 2F(1)F(r)\cos\Psi \ . \tag{A2}$$

Because of the axial symmetry of $P(\vec{s})$ only even powers of x/r and y/r need be retained in a series expansion of δF^2. Performing the expansion to lowest contributing order Lindhard [3] found the formula

$$<\delta F^2> = \frac{1}{2} <s^2> \ \cdot \ <F^2(r)/r^2 + F'^2(r)> \ . \tag{A3}$$

In this appendix we shall extend the calculation to obtain the next contribution to the force fluctuation. As indicated above, it is then necessary to include terms of the fourth order in the expansion of δF^2. First we expand in the quantity $1 - r$:

$$F(1) \simeq F(r) + (1-r)F'(r) + \frac{1}{2}(1-r)^2 F''(r)$$

$$+ \frac{1}{6}(1-r)^3 F''' + \frac{1}{24}(1-r)^4 F'''' + \ldots \tag{A4}$$

$$\cos\Psi = \frac{r-x}{1} \simeq (1-\frac{x}{r})(1-\frac{1-r}{r} + (\frac{1-r}{r})^2 - (\frac{1-r}{r})^3$$

$$+ (\frac{1-r}{r})^4 - \ldots) \ . \tag{A5}$$

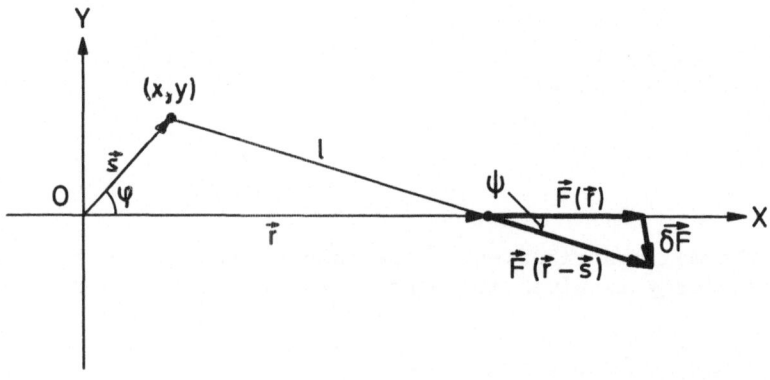

Fig. 6

Ultimately, we want to expand in the quantities x/r and y/r, and from Fig. 6 we find

$$1 = r\sqrt{(1-\frac{x}{r})^2+(\frac{y}{r})^2}$$

$$\simeq r-x + y^2/2r + xy^2/2r^2 + (4x^2y^2-y^4)/8r^3 + \ldots . \quad (A6)$$

In evaluating (A2) by means of (A4) and (A5), we drop all terms of fifth or higher order in x/r or y/r. This leads to

$$\delta F^2 = \frac{1-r+x}{r} \cdot 2F^2$$
$$+ (1-r)^2 \cdot F'^2 + \frac{1-r}{r} \cdot \frac{1-r+x}{r} (2rFF' - 2F^2)$$
$$+ (1-r)^3 \cdot F'F'' + (\frac{1-r}{r})^2 \cdot \frac{1-r+x}{r} (2F^2-2rFF'+r^2FF'')$$
$$+ (1-r)^4 \cdot (\frac{1}{3}F'F'''+\frac{1}{4}F''^2) . \quad (A7)$$

By introducing (A6) into (A7), we find terms of second and higher order in x/r and y/r. Since third order terms disappear when we average over ϕ they may be neglected. We therefore obtain the following formula, correct to terms of fourth order in x/r and y/r,

$$\delta F^2 = F^2 \cdot y^2/r^2 + F'^2 \cdot x^2$$
$$+ F^2(3x^2y^2/r^4 - \frac{3}{4} y^4/r^4) + FF'(y^4/2r^3 - 2x^2y^2/r^3)$$
$$+ F'^2(y^4/4r^2 - x^2y^2/r^2) + FF'' x^2y^2/2r^2$$
$$+ F'F'' \cdot 3x^2y^2/2r + F'F''' x^4/3 + F''^2 x^4/4 . \quad (A8)$$

Here, the force F and its derivatives are to be evaluated at the distance r. Averaging over the distribution $P(\vec{s})$ we obtain

$$<\delta F^2> = F^2 \cdot <s^2>/2r^2 + F'^2 \cdot <s^2>/2$$
$$+ F^2 \cdot 3<s^4>/32r^4 - FF' \cdot <s^4>/16r^3$$
$$- F'^2 \cdot <s^4>/32r^2 + FF'' <s^4>/16r^2$$
$$+ F'F'' \cdot 3<s^4>/16r + F'F'''<s^4>/8 + F''^2 \cdot 3<s^4>/32 .$$
$$(A9)$$

For a Gaussian distribution of atomic positions,

$$dP(s) = e^{-s^2/\rho^2} 2s\,ds/\rho^2 \,, \tag{A10}$$

we have $\langle s^2 \rangle = \rho^2$ and $\langle s^4 \rangle = 2\rho^4$ which can be introduced directly into (A9).

In order to obtain the ratio γ, cf. equation (16), we now average over the accessible area. The contribution $\gamma^{(1)}$, of first order in $\frac{\rho^2}{r^2}$, is given by equation (16), which in terms of $\varepsilon_\perp = 2E_\perp/E\Psi_1^2$ reads

$$\gamma^{(1)} = \frac{1}{2L_n} \frac{\rho^2}{(Ca)^2} (e^{\varepsilon_\perp} + \frac{2}{3}) (1-e^{-\varepsilon_\perp})^3 \,. \tag{A11}$$

The term $\gamma^{(2)}$ of second order in ρ^2/r^2, as obtained from (A9), has the form

$$\gamma^{(2)} = \frac{1}{2L_n} \frac{\rho^4}{(Ca)^4} \{e^{2\varepsilon_\perp} - 2e^{\varepsilon_\perp} + \frac{14}{5} - 4e^{-\varepsilon_\perp} - 7e^{-2\varepsilon_\perp}$$

$$+ 26e^{-3\varepsilon_\perp} - 24e^{-4\varepsilon_\perp} + \frac{36}{5} e^{-5\varepsilon_\perp} \} \,. \tag{A12}$$

Exact Calculation for Power Law Forces

As mentioned in the text an exact calculation of the force fluctuation in the case of transverse forces $F(r) \propto r^{-1}$ and $F(r) \propto r^{-3}$ can be performed.

Assume the force to be given by

$$F(r) = A/r^n \,. \tag{A13}$$

We utilize again equation (A2), and average at first over the angle ϕ. Clearly we then need to evaluate the integrals

$$\langle F^2(1) \rangle = A^2 \cdot \frac{1}{\pi} \int_0^\pi \frac{d\phi}{1^{2n}} = F^2(r) \cdot \frac{1}{\pi} \int_0^\pi \frac{d\phi}{(1+\frac{s^2}{r^2} - 2\frac{s}{r}\cos\Psi)^n}$$

$$\langle F(1)\cos\Psi \rangle = A \cdot \frac{1}{\pi} \int_0^\pi \frac{d\phi\cos\Psi}{1^n} = F(r) \cdot \frac{1}{\pi} \int_0^\pi \frac{d\phi(1-\frac{s}{r}\cos\phi)}{(1+\frac{s^2}{r^2} - 2\frac{s}{r}\cos\phi)^{\frac{n}{2}+\frac{1}{2}}} \cdot$$

$$\tag{A14}$$

Both integrals are simple if n = 1, or n = 3. In those cases we
find from (A2) and (A14) the results, valid for r > s,

$$\frac{<\delta F^2>}{F^2} = <1 - \frac{1-2(s/r)^2}{(1-(s/r)^2)^3} + \frac{6(s/r)^2}{(1-(s/r)^2)^5}> \text{ , for n = 3 (A15)}$$

$$\frac{<\delta F^2>}{F^2} = <\frac{(s/r)^2}{1-(s/r)^2}> \qquad\qquad \text{ , for n = 1.(A16)}$$

It is easily seen that an expansion of the right hand side of (A15)
and (A16) to fourth order in s/r yields results which agree with
(A9).

References

[1] E. Bonderup, H. Esbensen, J. U. Andersen, and H. E. Schiøtt,
     Rad. Eff. 12, 261 (1972).
[2] M. J. Pedersen, J. U. Andersen, D. J. Elliott, and E. Laegsgaard,
     Axial Dechanneling II. An Experimental Study, this conference.
[3] J. Lindhard, Mat. Fys. Medd. Dan. Vid. Selsk. 34, No. 14 (1965).
[4] V. V. Beloshitsky, M. A. Kumakhov, and V. A. Muralev, Rad.
     Eff. 13, 9 (1972).
[5] J. Lindhard, Mat. Fys. Medd. Dan. Vid. Selsk. 28, No. 8 (1954).
[6] N. Bohr, Mat. Fys. Medd. Dan. Vid. Selsk. 18, No. 8 (1948).
[7] G. J. Clark, J. M. Poate, E. Fuschini, C. Maroni, I. G. Massa,
     A. Uguzzoni, and E. Verondini, Nucl. Phys. A173, 73 (1971).
[8] J. H. Barrett, Phys. Rev. B3, 1527 (1971).
[9] Y. Hashimoto, J. H. Barrett, and W. M. Gibson, Phys. Rev.
     Letters 30, 995 (1973).
[10] R. P. Sharma, J. U. Andersen, and K. O. Nielsen, Nucl. Phys.
     A204, 371 (1973).
[11] E. Verondini, private communication.

Note Added in Proof

In the treatment of random multiple scattering as given by
Williams and Bohr [6] the angular distribution of an initially well
collimated beam is divided into a central, Gaussian part and tails
at large angles. The Gaussian results from small-angle multiple
scattering--a diffusion process--whereas the tails are due to large-
angle single-scattering events. In the text we applied for random
diffusion a value of D' corresponding to the expression in Ref. 3
for the mean-square angular deflection $<\phi^2>$ -- (Ref. 3, formulae
(4.1), (4.2) and (4.6)). This estimate includes the contribution
from single scattering and for thicknesses corresponding to the
measurements presented in Ref. 2, it turns out to be larger than

the value of $\langle\phi^2\rangle$ for the Gaussian part by as much as a factor of 2-3. Thus, in the calculations the random value of D' should be reduced by a factor 2-3. Single scattering may be included as an absorption term in the diffusion equation, but the effect on the results turns out to be negligible.

With the diffusion at $\hat{r}/\rho\gtrsim 1$ unchanged, the reduction of the random value of D' leads to a relative decrease in $\chi(z)$ of 10-25%, the larger change occuring at high temperatures. A further reduction would result if the exponent f in the function $\gamma(\hat{r})$ in (17) were determined with the reduced random value of D', i.e., with a reduced value of L_n. As noted in the text, however, for elevated temperatures $\gamma(\hat{r})$ underestimates the diffusion at large values of $\hat{r}/\rho$ even without this reduction.

Therefore, we now prefer to base the estimate of $\gamma(\hat{r})$ on the series expansion. We have tried the simple expression $\gamma = \gamma^{(1)}+\gamma^{(2)}(1+\alpha\gamma^{(2)}/\gamma^{(1)})$, where $\gamma^{(1)}$ and $\gamma^{(2)}$ denote the first two terms in Lindhard's expansion, cf. the appendix. A comparison with the exact calculations for forces $F(r) \propto r^{-1}$ and $F(r) \propto r^{-3}$ suggests for α a value $\simeq 1$. A variation of α by a factor of two leads to relative changes of $\lesssim 10\%$ in $\chi(z)$.

Reducing the random value of D' by a factor of 2-3 and setting $\alpha = 1$ we arrive at $\chi(z)$-curves which do not differ much from the ones presented in Ref. 2. At room temperature there is a reduction of 10-20%, whereas the values corresponding to the highest temperatures are virtually unchanged. For the transmission measurements these modifications in the diffusion function lead to an even better agreement between theory and experiment, because the reduction of the random diffusion results in higher intensity at angles $\phi \gtrsim \Psi_1$.

As these remarks indicate the uncertainty in $\chi(z)$ connected with the uncertainty in the diffusion function in the thermal and random regions may be $\sim 20\%$. The important point is, however, that theory and experiment now agree within this uncertainty in contrast to the large systematic discrepancies encountered previously [1].

AXIAL DECHANNELING, II. AN EXPERIMENTAL STUDY

M. J. PEDERSEN, J. U. ANDERSEN, D. J. ELLIOTT* and E. LAEGSGAARD
Institute of Physics, University of Aarhus
DK-8000 Aarhus C, Denmark

Introduction

Multiple scattering of channeled, positive particles may be described by a diffusion-like equation [1-3] with a diffusion function $D(E_\perp)$ which varies strongly with transverse energy. At low transverse energy, the main contribution comes from scattering by electrons, whereas at high transverse energy, $D(E_\perp)$ is dominated by force fluctuations due to thermal vibrations.

To test this description, we have investigated the dechanneling caused by multiple scattering by measuring as a function of penetration depth the minimum yield of Rutherford backscattering for MeV protons incident parallel to an axis on various crystalline targets. The results are compared to the calculations described in the accompanying paper [2a], "Axial Dechanneling, I. A Theoretical Study," referred to as I in the following.

As discussed by Campisano et al [4] at the Gausdal Conference in Norway, also the angular width of axial dips is strongly influenced by dechanneling, in particular close to the surface. Full angular scans have been measured for Si and W, and the depth dependence has been compared to the theoretical results from I.

Although dechanneling is the most important consequence of multiple scattering of channeled particles, measurements of the dechanneled fraction constitute a rather incomplete test of the theoretical treatment. A considerably more detailed check of the calculated distribution in transverse energy $g(E_\perp,z)$ has been obtained from measurements of the angular distribution of 1.6-MeV protons transmitted through $2 - 20$-μm Si crystals for incidence parallel to a <110> direction. The two types of measurements are illustrated in Fig. 1.

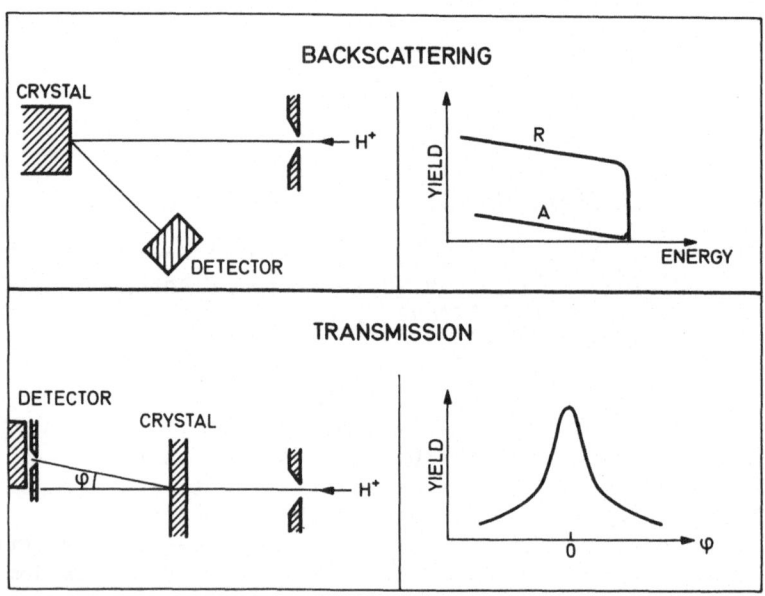

Fig. 1. Illustrations of the two experimental configurations. The energy spectra labelled R and A correspond to random and aligned incidence, respectively.

Backscattering

Procedure

In the conventional backscattering configuration, energy spectra of protons scattered through 135° were recorded for incidence parallel to an axis and for incidence in a random (i.e., non-channeling) direction. The incident beam was collimated to ±0.015°. The crystal was mounted in a goniometer and could be heated to ∿700°C. Heating causes two important problems. First, it becomes difficult to measure the beam current. Instead, the beam was monitored by detection of protons backscattered from a beam chopper. Second, the heat radiation from the target may introduce a high leakage current in the surface-barrier detector. This problem was solved by inserting a thin (0.1-μm) nickel foil in front of the detector.

The targets used for this investigation were all supposed to be good single crystals. Only in the case of Si, however, were the results reproduced on several different crystals (n and p type, ∿1kΩcm). Furthermore, the Si results are in good agreement with the carefully checked results in Ref. 5. In the case of W, the results were reproduced to within ∿5% in several runs after different surface preparations, and the crystal was known from previous measurements to be of very high quality. For the other cases (Fe, Nb, Mo, Ta), the crystal quality was less certain. The Ta crystal was chosen as the best of three samples, which gave rather different results.

The crystals were polished according to standard prescriptions. Immediately before use, the Si crystals were cleaned in HF acid. After each series of measurements at different temperatures, the thickness of the surface layer was measured by He backscattering. The results are given in Table I, expressed as the thickness of an equivalent amorphous layer of pure material. High-temperature measurements were discarded if they lead to a substantial increase in this thickness (e.g., 700°C Ta).

For Si, the influence of the surface oxide was estimated to be negligible from measurements of dechanneling in crystals with a much thicker oxide layer grown chemically on the surface. Although scattering in the surface layer may easily be included in the calculation, this has not been done for any of the comparisons made below.

The damage due to proton bombardment of the crystals was negligible in all cases investigated. In some cases, however, the surface analysis by He backscattering caused significant damage, and the beam spot had to be shifted after such an analysis (Nb and Mo).

Data treatment

In order to calculate the random fraction of the beam as a function of depth from the recorded energy spectra, the stopping power has to be known. In the energy region of interest, the energy dependence may be conveniently approximated by the expression

$$- \frac{dE}{dz} = (a + bE)^{-1} . \tag{1}$$

With the constants given in Table II, the measured stopping powers are reproduced to within 3%.

For channeled particles, the stopping is reduced by about a factor of two, but since particles in the random fraction of the beam, which determines the backscattering yield for aligned incidence,

Table I

Amorphous Surface Layers

Substance	Si	Fe	Nb	Mo	Ta	W
Equivalent Thickness	15±5 Å	15±3 Å	15±3 Å	12±3 Å	14±3 Å	5±1 Å

Table II

Stopping Power Constants

Material	a[μm/MeV]	b[μm/MeV^2]	Ref. for $\frac{dE}{dz}$
Si	9.68	14.73	6
Ta	4.81	3.95	7
W[†]	4.60	3.41	7

[†]Data for Au used

have had random stopping power for part of their path, it is not obvious which value should be used for the conversion of energy loss to depth.

At an informal meeting on dechanneling in Aarhus in 1970, it was decided to use random stopping power for the depth conversion in order to be able to compare data from different laboratories. For the present detailed comparison with calculations, however, a presumably more correct value for the aligned stopping power has been used. This choice is founded on the following simple argument which, without being explicitly stated, has been the basis for the energy–depth conversion in many previous publications.

Assume for simplicity a sharp transition between aligned and random beam, such that aligned particles have a well–defined stopping power, say, half the random value, and zero backscattering yield, whereas particles in the random beam have normal stopping and scattering yield. Further assume the backscattering cross section and the stopping power to be independent of energy. Now let z(E) denote the scattering depth for a particle which is backscattered immediately after dechanneling and which emerges from the crystal with energy E. All particles which are dechanneled before reaching this depth have the same probability of being backscattered with exit energy E, whereas particles dechanneled at depth z>z(E) must, if backscattered, emerge from the crystal with an energy smaller than E. Therefore the backscattering yield at energy E in the aligned spectrum is proportional to the random fraction of the beam at the depth z(E) [8].

According to this argument, an average channeled stopping power should be used in the energy–depth conversion for the path before backscattering. For the results to be presented in the following, a value close to half the random stopping has been assumed. Small corrections have also been applied for the energy dependence of the scattering cross section and the stopping power. The combined effect of all corrections is approximately equivalent to an increase of the depth scale by 20% from that derived from random stopping.

Many attempts have been made to measure the effective stopping power to be used for the conversion [8-10]. The results generally indicate an effective stopping power close to half the random value, except for small depths, where the main part of the dechanneling is due to poorly channeled particles having entered the crystal close to a string. The important distinction between this effective stopping power for energy-depth conversion and the average stopping for backscattered particles has, however, not always been realized [10,11].

The dechanneled fraction is derived from a comparison with the random energy spectrum. The shape of this spectrum may be calculated from the energy dependence of the stopping power and the scattering cross section. For large energy losses, the measured backscattering yield was significantly higher than calculated, probably due to plural scattering. Therefore only the upper part (E > 0.6 MeV for Si, E > 1 MeV in other cases) of the energy spectra has been used to calculate the dechanneling. In this region, the shape of the random spectrum agrees with calculations to within ∿5%.

Results

The experimental results for 1.6 MeV protons incident along a <110> direction in Si are in Fig. 2 compared with the results of the calculations described in I. A similar comparison is shown in

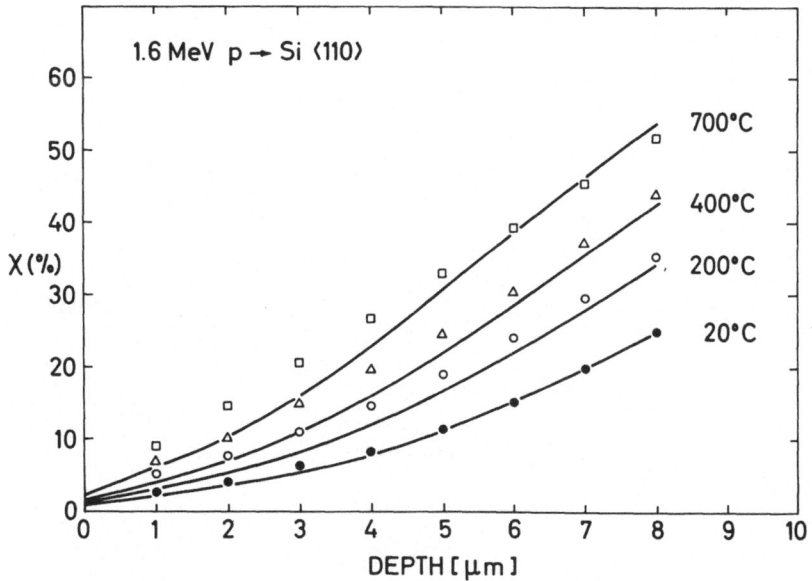

Fig. 2. Measured (points) and calculated dechanneling in silicon at various temperatures. The vibrational amplitudes were calculated from a Debye temperature of T_D = 543°K.

Fig. 3 for 2-MeV protons incident on W <100>. At room temperature, the agreement is perfect in both cases. For W also the temperature dependence is well described by the calculations, whereas for Si, the measured temperature dependence of the yield at small depths is somewhat stronger than calculated.

The dependence on the atomic spacing d in strings is shown in Fig. 4 for three axes in tungsten. The agreement between theory and experiment is excellent. For the highest-order axis, the calculated curve lies slightly above the measured points. This tendency becomes more pronounced for axes of still higher order (<410> and <510>) and may be due to the increasing importance of planar channeling.

For the other materials (Fe, Nb, Mo, Ta) investigated, there are larger discrepancies between measured and calculated yields, the latter being lower by a factor of 1.5 - 2. This might indicate large defect concentrations. The worst example is shown in Fig. 5.

The good agreement between the measured and the calculated dechanneling for Si and W suggests that the improved evaluation in I of the diffusion function is fairly realistic. Both the electronic and the thermal components are very important for the dechanneling rate, and it is desirable to test the theoretical estimates of these two contributions separately. Since the thermal scattering increases strongly with temperature, the measurements of the temperature dependence of dechanneling are mainly a test of

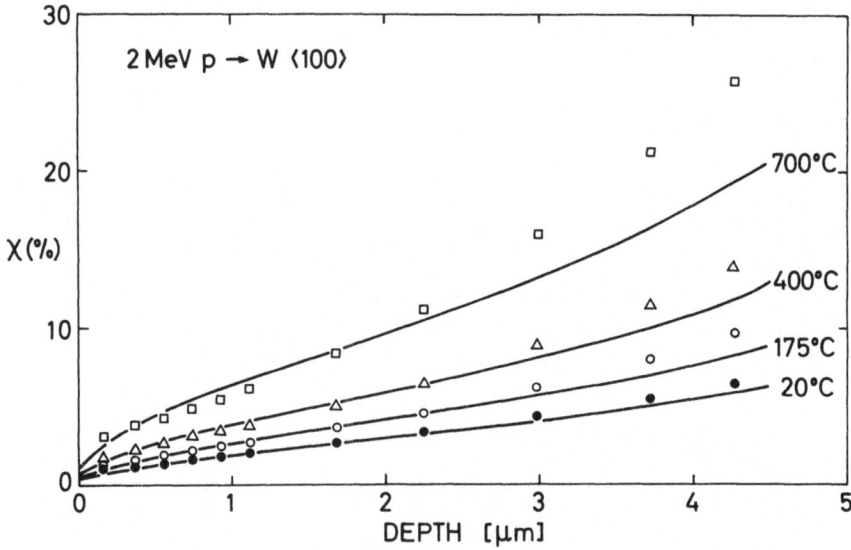

Fig. 3. Measured (points) and calculated dechanneling in tungsten at various temperatures. The vibrational amplitudes were calculated from a Debye temperature of T_D = 310°K.

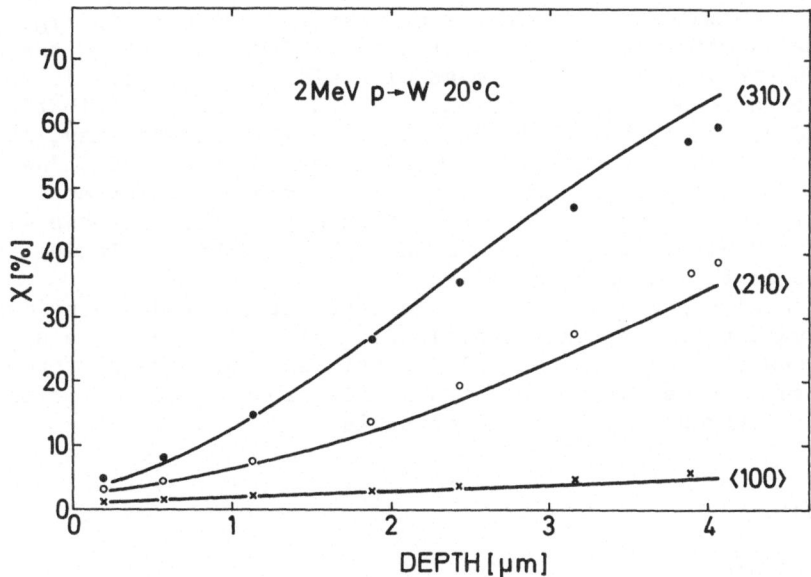

Fig. 4. Measured dechanneling along different axial directions in
tungsten--indicated by points--compared to calculations.

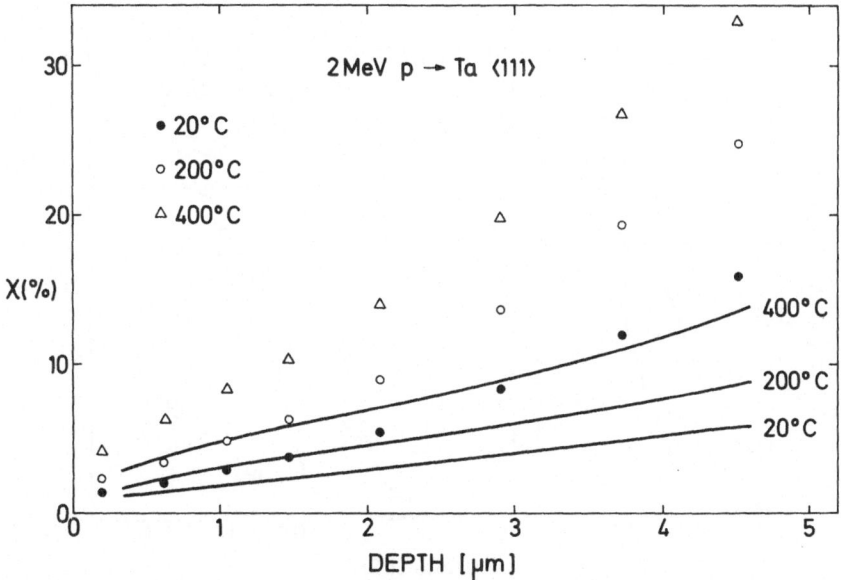

Fig. 5. Measured (points) and calculated dechanneling in tantalum
at various temperatures. The vibrational amplitudes were calculated
from a Debye temperature of T_D = 245°K.

the thermal component. Another such test is presented in Fig. 6. As thermal scattering dominates at high transverse energy, the calculated depth dependence of the angular width of the dips is very sensitive to changes in the estimate of the thermal contribution. It should be noted that there is a considerable uncertainty, at least ~10% at shallow depth, in measurements of the width, due to the influence of planar channeling. Furthermore, the effective stopping power for depth conversion should probably increase with increasing angle of incidence. Even with these uncertainties, the agreement in Fig. 6 is remarkable.

Since electronic scattering dominates at low transverse energy, the evaluation in I of this contribution can be tested by a study of the behaviour of well-channeled particles. We have performed such a study by measuring the angular distribution of particles transmitted through thin single crystals.

Transmission

Procedure

The thin silicon crystals were prepared according to the etching procedure described in Ref. 12. The thickness was determined both from spectra of backscattered protons and from the energy loss of particles transmitted in a random direction. From these

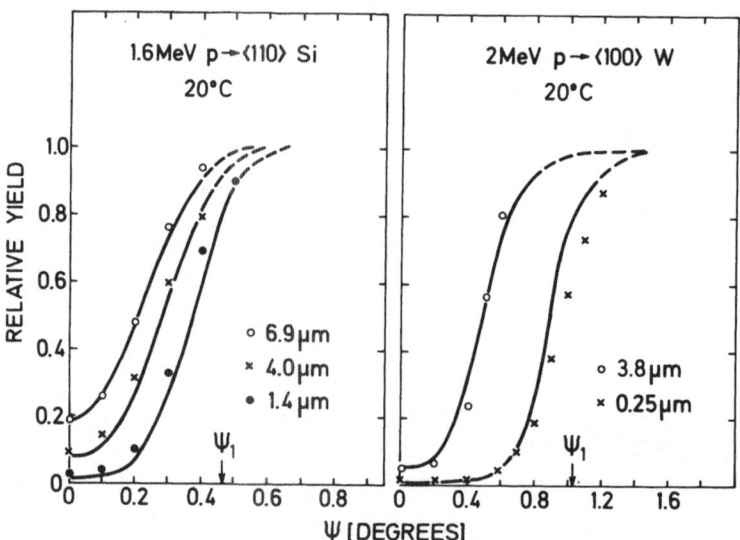

Fig. 6. Comparison of measured angular scans (points) with calculated dips (solid lines). At all angles the aligned stopping power was assumed to be half the random value for the energy-depth conversion.

measurements, the thickness is estimated to be uniform within
∿0.2 μm. The beam of 1.6-MeV protons was collimated to within
±0.1°, and after transmission parallel to a <110> direction, the
angular distribution of the particles was measured with a movable
counter. A detector placed at a fixed angle served as a beam
monitor.

Results

According to photographic measurements of angular distributions
of protons transmitted parallel to an axis (see, e.g., Ref. 13),
the intensity depends not only on the exit angle to the axis but
also on the azimuthal angle relative to major planes containing
the axis. Probably the contrast in such pictures is somewhat
misleading since in the present investigation it was found that
the aximuthal dependence is weak for not too large angles to the
axis. This is illustrated in Fig. 7. A scan with the movable
detector through the <110> axis in a "random" azimuthal direction
(not parallel to a major plane) was slightly asymmetric, and the

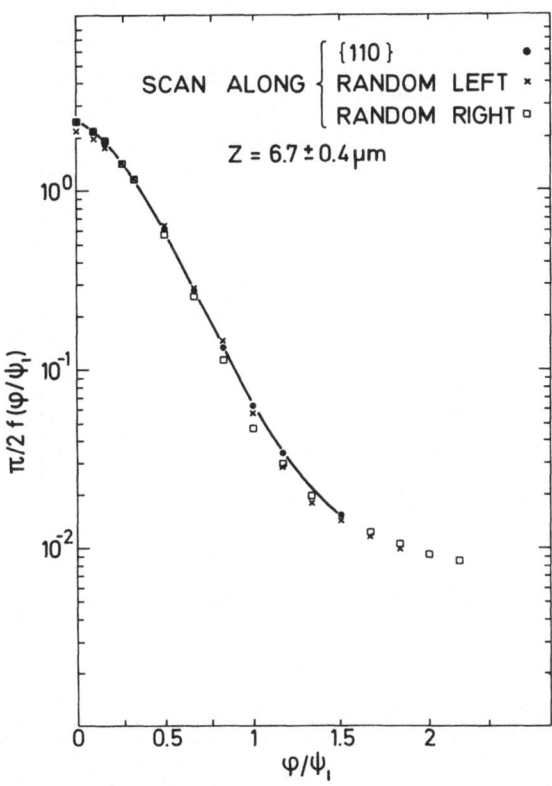

Fig. 7. Measured forward angular distributions for 1.6-MeV protons
incident along a <110> direction on a 6.7-μm thick silicon crystal.

"right" and "left" sides are shown. One side of a (symmetric) scan parallel to a <110> plane is shown for comparison. The differences between the three scans are only ∿10%.

In this and the following figures, the quantity $\frac{\pi}{2}f(\phi/\Psi_1)$ is plotted since it may conveniently be compared to the calculated distribution in transverse energy. The angle ϕ to the axis is measured in units of the characteristic angle Ψ_1 for axial channeling, given by [1]

$$\Psi_1 = (\frac{2Z_1Z_2e^2}{Ed})^{1/2} \tag{2}$$

with usual notation. The distributions have been independently normalized such that

$$\int_0^\infty f(x)2\pi xdx = 1 . \tag{3}$$

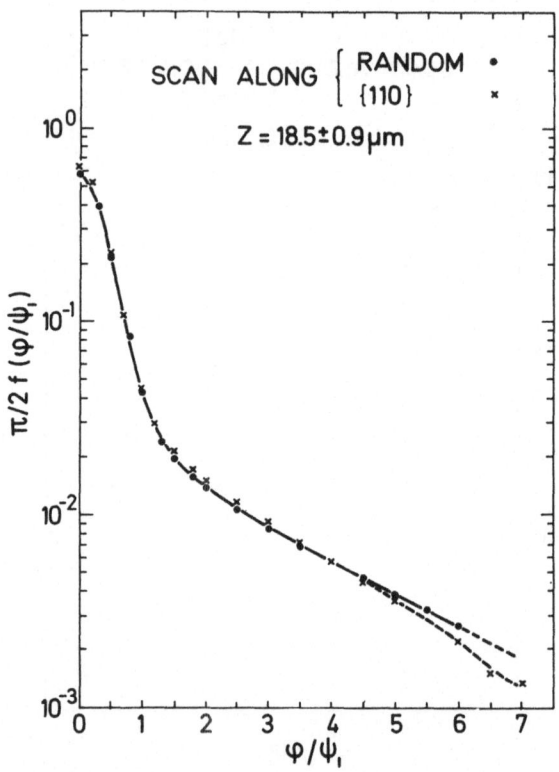

Fig. 8. Measured forward angular distribution for 1.6-MeV protons incident along a <110> direction on an 18.5-μm thick silicon crystal.

Similar scans for a thicker crystal, extending to much larger angles, are shown in Fig. 8. Again, there are only small differences between measurements in different azimuthal directions ($\lesssim$10%) except at very large angles. The dramatic change in slope at $\phi \simeq \Psi_1$ is a clear illustration of the strong reduction in multiple scattering for channeled particles.

The approximate azimuthal symmetry implied by these results indicates that the assumption in I of isolated and randomly distributed strings (i.e., neglect of planar effects) is probably quite good. Furthermore, we may conclude that the two-dimensional angular distribution is well represented by a one-dimensional scan through the <110> axis.

In Fig. 9 such scans are compared to the calculated distributions for three different crystal thicknesses. The agreement is good as small angles, but the calculated distributions tend to be low at angles $\phi \simeq \Psi_1$.

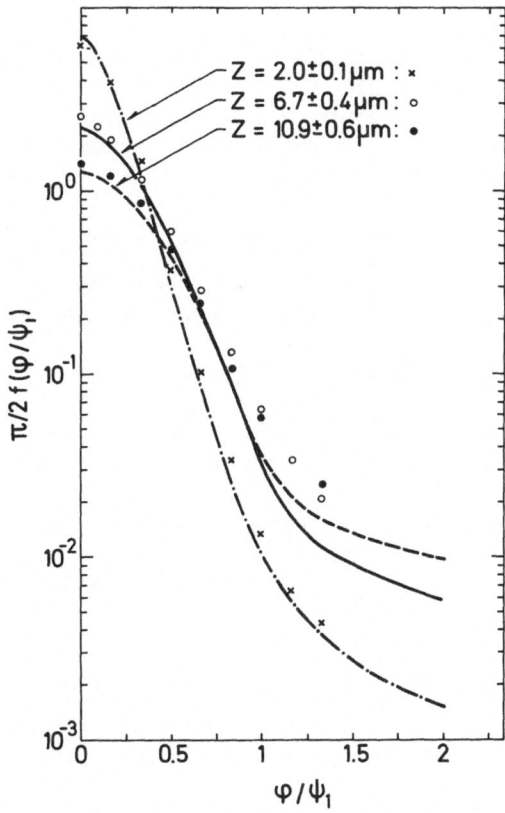

Fig. 9. Forward angular distributions for 1.6-MeV protons incident parallel to a <110> direction on silicon crystals of various thicknesses. The experimental results, indicated by points, are compared to the calculations described in I.

Conclusions

The good agreement with calculations for Si and W indicates
that the essential features of the dechanneling process are included
in the theoretical description, and that the estimates of thermal
and electronic scattering have now become realistic. It is important to establish whether the poor agreement for other materials
is in fact due to large defect concentrations. A somewhat surprising result of the transmission experiment is the weak influence of
planar effects. Also this point need to be elucidated by further
measurements.

References

*Present address: Mathematics and Physics Bldg., University
    of Sussex, Falmer Brighton BN1 9HQ, England.
[1] J. Lindhard, Kgl. Danske Videnskab. Selskab, Mat.-Fys. Medd.
    34, No. 14 (1965).
[2] E. Bonderup, H. Esbensen, J. U. Andersen, and H. E. Schiøtt,
    Rad. Eff. 12, 261 (1972).
[2a]H. E. Schiøtt, E. Bonderup, J. U. Andersen, and H. Esbensen,
    "Axial Dechanneling I. A Theoretical Study," this conference.
[3] V. V. Beloshitsky, M. A. Kumakhov, and V. A. Muralev, Rad. Eff.
    13, 9 (1972).
[4] S. U. Campisano, G. Foti, F. Grasso, I. F. Quercia, and E.
    Rimini, Rad. Eff. 13, 23 (1972).
[5] A. Fontell, E. Arminen, E. Leminen, Rad. Eff. 12, 255 (1972).
[6] C. Williamson and J. P. Boujot, CEA-2189 (1962)(ORTEC Manual
    for Surface Barrier Detectors).
[7] L. C. Northcliffe and R. F. Schilling, Nuclear Data Tables A7,
    233 (1970).
[8] A similar argument with the same result was given by D. Blanchin,
    J. C. Poizat, J. Remillieux, and A. Sarazin, Nucl. Instr.
    Methods 70, 98 (1969).
[9] K. Morita and N. Itoh, J. Phys. Soc. Japan 30, 1430 (1971).
[10] R. Hellborg, Physica Scripta 4, 75 (1971).
[11] F. Abel, G. Amsel, M. Bruneaux, and E. d'Artemare, J. Phys.
    Chem. Solids 30, 687 (1969).
[12] T. C. Madden, W. M. Gibson, IEEE Trans. Nucl. Sci. 11, 254
    (1964).
[13] G. Dearnaley, B. W. Farmery, I. V. Mitchell, R. S. Nelson, and
    M. W. Thompson, Proc. Int. Conf. on Solid State Physics Research
    with Accelerators, Brookhaven, 1967. BNL Report, 50083(C-52).

Note Added in Proof

A modification of the diffusion function in the thermal and random regions has been introduced (see note added in proof to Axial Dechanneling I). For the theoretical $\chi(z)$ curves, the modification leads to a relative reduction of 10–20% at room temperature, whereas the values corresponding to the highest temperatures are virtually unchanged.

For the transmission measurements, the theoretical distribution function is unchanged at low angles of emission, whereas the agreement between theory and experiment is improved at angles $\phi \gtrsim \Psi_1$.

DECHANNELING AND RECHANNELING CALCULATIONS

R. D. EDGE and Q. C. MURPHREE
University of South Carolina
Columbia, S. C. 29208, U.S.A.

Introduction

The mechanism whereby ions, channeled along a low index direction in a single crystal, are scattered out of that channel, and how they can feed back again, is a problem of considerable theoretical interest [1,2,3] as more experimental data becomes available [4,5,6].

We have completed a computer calculation giving the angular distribution of protons scattered off a string of gold atoms in the <110> orientation in the vicinity of the critical channeling angle under the assumption of the impulse approximation. As will be seen, the distributions present several interesting features and indicate how ions leave and enter the channel. The calculations can then be combined with previous calculations on the motion of ions near the center of the channel, and in the random medium, and using a diffusion calculation, an approximate general solution for the motion of protons both inside and outside the channel is possible.

Similar calculations can be also carried out using Monte Carlo techniques [7,8,9]. However, the computer time required to perform such calculations increases as the exponent of the depth of crystal penetrated by the ion, so that, unless radical simplifications are made, it is not possible to carry these calculations through to great depths. The semi-analytical approach we are using can easily be extended deep into the crystal and may be checked against Monte Carlo calculations for shallow depths.

Basis of Calculations

In our model, Molière's approximation to the Thomas-Fermi potential [10] was employed for the interaction of the incident proton with each atom of the string. The scattering angle was calculated using the impulse approximation [11,12,13] at the transverse plane through each atom, and used to find the trajectory to

the next atom. This produces an error of less than 1 percent for
scattering angles of less than 20° [14,15]. The (infinitely mas-
sive) atoms were considered located at the exact lattice positions.
For the calculations involving thermal vibrations, the Thomas-Fermi
potential was replaced by the Lindhard standard potential [16].

This, for a uniform string, is $U(r) = \dfrac{Z_1 Z_2 e^2}{d} \log \left| \left(\dfrac{Ca}{r}\right)^2 + 1 \right|$

where r is the distance from the string, d is the separation of
atoms of the string, a is the Thomas-Fermi radius of the string
atom, and $Z_1 Z_2$ are the atomic numbers of the incident ion and the
string atom. C is a constant $\sim\sqrt{3}$.

The trajectory of the ion was followed, using this technique,
in three dimensions, to obtain the angles of emission. All calculations were made at a proton energy of 334 keV.

Blocking, and Coplanar Trajectories

The blocking angle was found by starting the proton trajectory
at the site of a gold atom. The minimum angle with which such a
proton left the string at 334 keV was found to be 3.64° using the
Thomas-Fermi potential and 3.82° using the Lindhard potential.

Let us examine the trajectory of the proton incident in a
plane incorporating the string of gold atoms. For very small angles
of incidence, the collective effect of the string atoms ensures
that the emitted and incident angle are the same. As the angle of
incident increases, the atoms of the string, which the proton approaches closely, act individually to cause the angle of emergence
to oscillate [13,17].

Let us consider a proton approaching a string at an angle θ_{in},
released at a distance X_0 from the string large compared with the
range of atomic forces. The angle of scattering, θ_{out}, is plotted
as a function of X_0 in Figure 1 for 3° angle of incidence. Each
repetition corresponds to one complete period of the string. Hence,
one such period defines the scattering characteristics of the string
in the plane, for this angle, provided we start and finish far

enough away from the string. The differential of this curve $\dfrac{dX_0}{d\theta_{out}}$

as a function of θ_{out} provides the angular distribution of outgoing
protons originally incident at 3°. It will be seen from Figure 1
that it peaks at very large and very small angles of emergence.
As θ_{in} grows larger, the separation between the minimum and maximum
values of θ_{out} increases, until ultimately the proton breaks through
the string, and channeling can no longer be considered to occur.

To understand the phenomenon in three dimensions, Y_0 was varied as well as X_0. The incident trajectory of the proton was
shifted parallel to itself a distance Y normal to the plane including the string and the original trajectory, discussed above.
The geometry of this arrangement is shown in Figure 2.

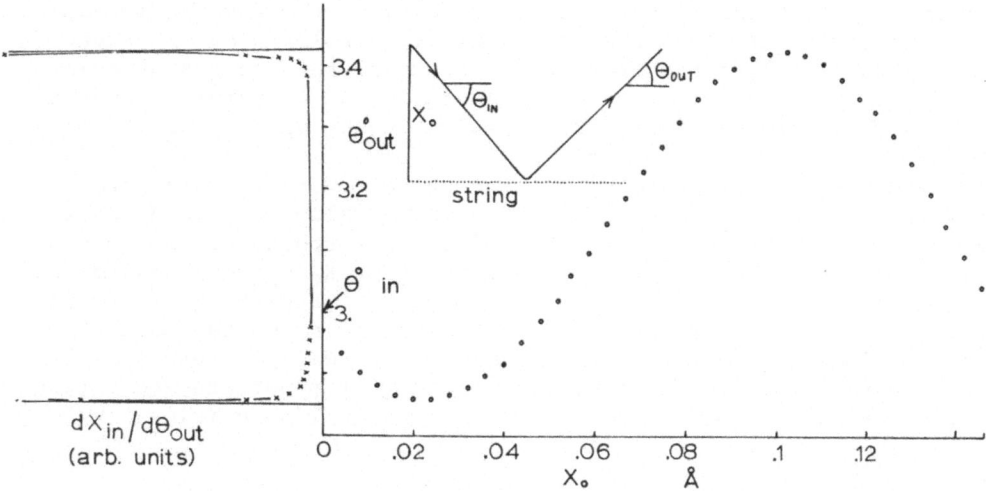

Fig. 1. The angle of emittance, θ_{out}, for a 3° angle of incidence, θ_{in}, as a function of X_o, the initial distance from the string at Z=0. Sting and trajectory are coplanar. A diagram of the geometry of the trajectory, and the angular distribution of the emitted particles are also shown.

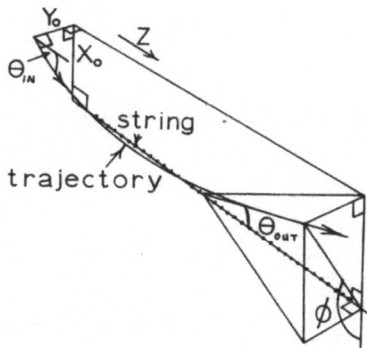

Fig. 2. Geometrical relationship between the initial conditions, X_o, Y_o and θ_{in}, and the final angles θ_{out} and ϕ.

Figure 3 shows what occurs when Y is increased in steps of
.01 Å for an incident angle of 2°. Because the particles are well
channeled at this angle, varying X has no effect. We have plotted
the angle of emergence to the channel θ_{out}, as the r variable, and
the azimuthal angle ϕ as the θ variable in a polar plot. For very
large values of Y, the proton is undisturbed, and emerges at point
B. In the plane of the string, it is reflected to point A, and in
between these two, it falls on a cone making an angle of 2° to the
string.

The curves for 3.5° angle of incidence are shown in Figure 4.
It will be seen that each point for a given value of Y found at 2°
has now spread out to a loop. Each loop corresponds to one complete period of the string, and each value of X_0 corresponds to one
point on the loop, until the cycle repeats. Furthermore, a specific
nuclear energy loss is associated with each point on the loop, being a maximum for large ϕ, and a minimum for small ϕ where ϕ is
the azimuthal angle. The protons still do not penetrate the string,
since they are incident somewhat below the blocking critical angle.

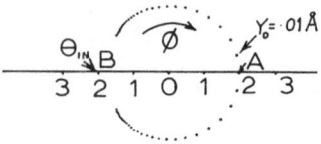

Fig. 3. Angular output pattern for particles incident at 2° to the
string, but at successively larger values of Y_0 such that ΔY_0 =
0.01 Å.

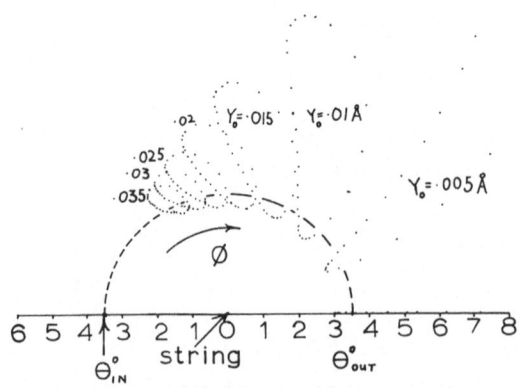

Fig. 4. Angular output patterns for particles incident at 3.5° to
the string through lattice points on the initial $X_0 Y_0$ plane such
that ΔY_0 = 0.005 Å, ΔX_0 = 0.01 Å. This diagram, and Figures 5 and
6 are repeated by reflection in the X axis.

Figure 5 shows the pattern at 4°. At this angle some particles
go directly through the channel but are emitted at large angles.
The loops are now much larger, overlap, and are not symmetrical.
The concavity of the loop away from the string enhances the scattering to larger rather than smaller angles. Note how the loops
converge on θ_{in} as Y increases. For each value of Y, θ_{out} again
has a maximum and minimum, with the greatest density of particles
at these extremes, as for Y=0.

At 6° (Fig. 6) many loops now lie completely inside one another,
and the majority of the particles penetrate the string for Y=0. We
are now well outside the region of channeling, but the collective
effects of the string are still obvious. Again, the unsymmetrical
nature of the curves encourages scattering to larger angles.

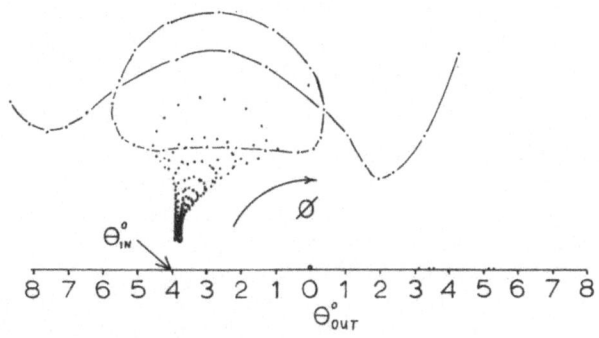

Fig. 5. Angular output pattern at 4° with $\Delta X_o = 0.01$ Å, $\Delta Y_o = 0.01$ Å.

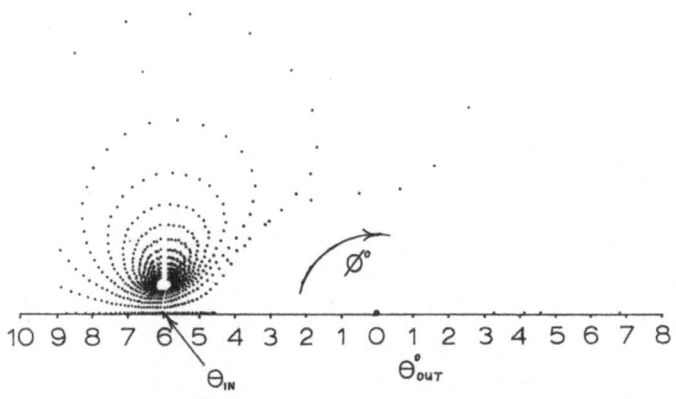

Fig. 6. Angular output pattern at 6° with $\Delta X_o = 0.005$ Å, $\Delta Y_o = 0.01$ Å.

Mean Angle of Emission, and Root Mean Square Scattering Angle

Since the strings are cylindrically symmetrical, we can, at least to a first approximation, ignore the azimuthal variation, and consider the average angle of emittance, $<\theta_{out}>$, and the root mean square deviation of θ_{out}. To find this, for a given incident angle, particles were introduced at points on a grid determined by X_O and Y_O. Values of X_O were selected to cover the range of one period of the lattice in equally spaced steps, (generally 75). Y_O was increased in steps of .001 Å to .1 Å. It was necessary to place limits on Y, because the potential has an infinite range. The cut off is somewhat arbitrary, but as will be seen later, agreement with experiment at larger angles indicated it was sufficient. Fig. 7 shows that the angular distribution for protons incident at 5° is a smooth curve, apart from a kink near $\theta_{out} = \theta_{in}$, which arises from the arbitrary cut off.

$<\theta_{out}>$ and $<\theta_{out}^2>$ were found by summing the calculated values of θ_{out} or θ_{out}^2 for the simulated particles, and dividing by the number of particles. Since the particles were initially regularly spaced on a lattice, this method provided accurate values for these quantities.

Fig. 8 shows $\dfrac{(<\theta_{out}> - \theta_{in})}{\lambda}$, which is approximately $\dfrac{\partial <\theta>}{\partial z}$ for small deviations, plotted against θ_{in}. λ is the mean free path between collisions in the z direction, along the channel.

We have assumed that the protons travel in straight lines between collisions, although a correction should be made for the focussing effects discussed by Barrett [18]. The mean free path on a plane normal to the row is $\dfrac{A}{2d}$, where A is the area of the plane per row, and d is the radius of the row, which would be 0.1 Å with our cut off. The mean free path in the z direction is then given by $\lambda = \dfrac{A}{2d\theta}$.

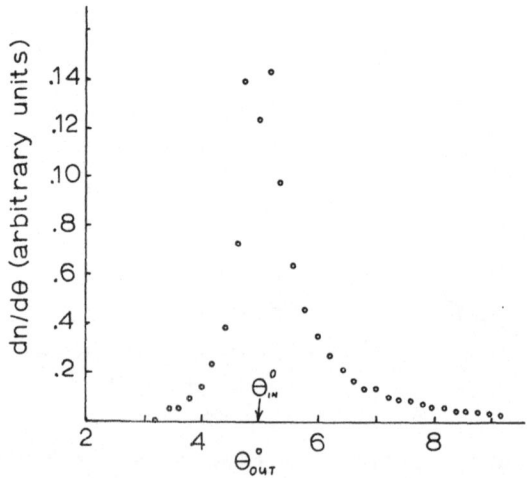

Fig. 7. Angular distribution in θ_{out} for protons incident at 5° to a string. Cut off at $Y_O = 0.1$ Å from the string.

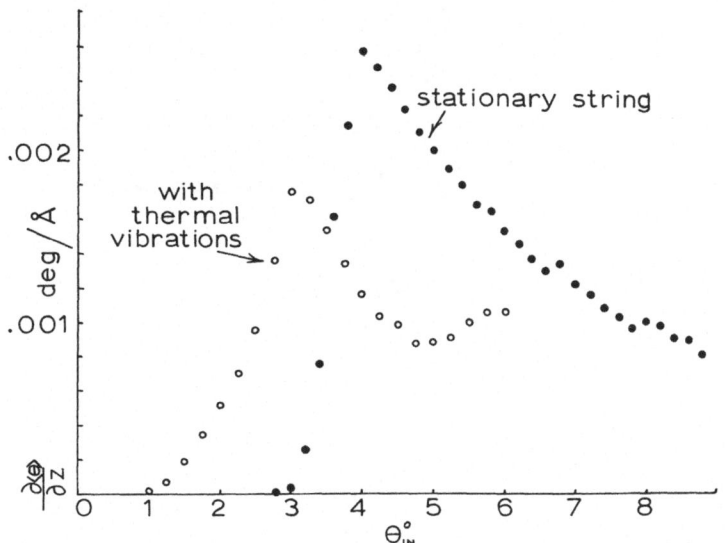

Fig. 8. $\dfrac{<\theta_{out}> - \theta_{in}}{\lambda}$ ($\sim \dfrac{\partial <\theta>}{\partial z}$) for a fixed string, and one for which the atom of closest approach was allowed a root mean square vibration in the X direction of 1.13 a_{TF}.

in the z direction is then given by $\lambda = \dfrac{A}{2d\theta}$.

Below about 2.5°, channeling ensures that $\theta_{out}=\theta_{in}$. Above this, the difference between $<\theta_{out}>$ and θ_{in} increases rapidly to a maximum of 0.9° at 4° decreasing surprisingly slowly as the angle increases. This corresponds to a constant flow of particles out of the channel, (such as might be produced by an attractive outside potential), and must be inserted in the diffusion equation.

Figure 9 shows $\dfrac{(<\theta_{out}^2>-<\theta_{out}>^2)}{\lambda} = \dfrac{<\Delta\theta_{out}^2>}{\lambda}$, which is approximately $\dfrac{\partial <\Delta\theta_{out}^2>}{\partial z}$. It will be seen that, below 3° $<\Delta\theta_{out}^2>$ is very small, passing through a maximum of .014 deg^2/Å at 4°. It then drops to an approximately constant value of .009 deg^2/Å.

At such large angles, the scattering should be approximately the same as in a random medium, where it can be compared with the calculations of Meyer [19] and the experimental results of Andersen et al. [20]. At these energies, the calculations of Molière [21] are not sufficiently accurate. The experimental results at 500 keV give a value of $\dfrac{\partial <\Delta\theta^2>}{\partial z}$ of 7×10^{-3} deg^2/Å at 1000 Å deep, whereas Meyer's value changes from 8×10^{-3} for 1000 Å, to 2.2×10^{-2} for 9000 Å.

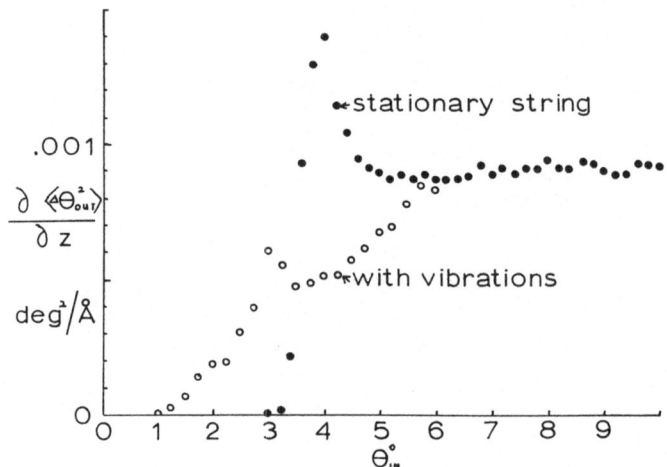

Fig. 9. $\dfrac{<\theta_{out}^{\ 2}> - <\theta_{out}>^2}{\lambda}$ $(\sim \dfrac{\partial <\Delta\theta_{out}^{\ 2}>}{\partial z})$ for a fixed string, and a

string vibrating as above.

Our results should be multiplied by a factor of approximately
two to compare with Meyer, because of the conversion from rectangular to polar coordinates. Our results then agree with those of
Meyer for thick targets, which might be anticipated from the nature
of the diffusion approximation, where $\dfrac{\partial <\Delta\theta^2>}{\partial z}$ is independent of
thickness.

Effect of Thermal Motion of the String Atoms

In order to study the effect of thermal motion on $<\theta>$ and
$<\Delta\theta_{out}^{\ 2}>$, distributions similar to those discussed previously were
obtained with the atom of closest approach displaced. The displacements were weighted according to a Gaussian distribution having a
mean square atomic displacement corresponding to the thermal motion.
Fig. 9 shows $\dfrac{\partial <\Delta\theta_{out}^{\ 2}>}{\partial z}$ using the Lindhard potential and a root
mean square vibration in the X direction only of $1.13a_{TF}$ [11].
This is a crude estimate, but it indicates the spreading of the
diffusion coefficient. Fig. 8 shows $<\theta>$ is spread also.
Near the center of the channel, the estimate provided by
Lindhard [16], based on the assumption of a large number of oscillating atoms which could be treated on a statistical basis, is
probably a reasonable assumption. However, this diverges rapidly
long before the edge of the channel is reached. It should be
possible to join our solution with that of Lindhard at some

intermediate position. At the very center of the channel, diffusion
occurs primarily through inelastic electron collision [8,16]. It
may be shown [8] that approximately $\dfrac{\partial <\Delta\theta_{out}^2>}{\partial z} \sim \dfrac{M_e}{M_p}\dfrac{1}{E}(-\dfrac{\partial E}{\partial z})$ (inelastic).

E here is the proton energy, M_p and M_e the proton and electron
masses and $\dfrac{\partial E}{\partial z}$ the rate of loss of energy by inelastic scattering of
electrons, which we assume [16] is one half of the rate of energy
loss in a random medium. The value in the random medium is
$5.5\times10^{-5}\text{deg}^2/\text{Å}$, which is 0.5% of the random nuclear scattering.
However, at the center of the channel, $\dfrac{\partial <\Delta\theta_{out}^2>}{\partial z}$, arising from

electron scattering, though small, is larger than that provided by
thermal motion. Based on Lindhard's estimate of electron density
and the effect of thermal motion, the electron scattering can pre-
cominate here, as shown in Figure 10. The equation used for the
thermal vibration scattering was [8]

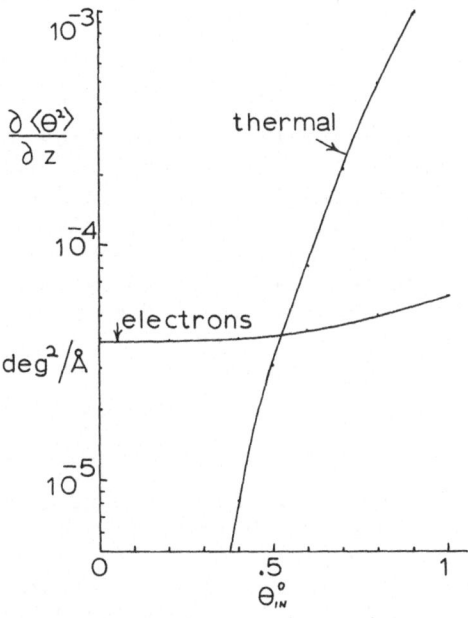

Fig. 10. $\dfrac{\partial <\theta^2>}{\partial z}$ in the vicinity of the channel center, estimated
using the electron density from the Lindhard standard potential for
the electron scattering, and the estimate for the thermal vibrations
given by Lindhard [16].

$$\frac{\partial <\theta^2>}{\partial z} = \frac{2\pi Z_1^2 Z_2^2 e^4}{E^2} \; N \; \frac{\rho^2_{rms}}{c^2 a^2} \; e^{\left\{\frac{2\theta^2}{\psi_1^2}\right\}} \left[1 - e^{\frac{-2\theta^2}{\psi_1^2}}\right]^3$$

where z_1, z_2, E and e have their usual significance. a is the the Thomas Fermi radius for gold, c is a constant occuring in the Lindhard potential, here taken to be $\sqrt{3}$, N is the number of atoms per unit volume, ρrms is the root mean square, vibrational ampli-

tude, and ψ_1 is the critical angle given by $\psi_1 = \left(\frac{2z_1 z_2 e^2}{Ed}\right)^{1/2}$,

d is the atomic separation in the string. We have now obtained

approximations for $\frac{\partial <\Delta\theta^2_{out}>}{\partial z}$ for all values of θ from the center of

channel to the random medium. No account has been taken of changes

in $\frac{\partial E}{\partial z}$ or E with depth.

Solutions to the Diffusion Equation

   In our case, the diffusion equation in θ, as particles are scattered back and forth in the channel, takes the form

$$\frac{\partial\rho}{\partial z} = \nabla \cdot (D\nabla\rho - F\rho)$$

where ρ is the density of particles as a function of depth z and D is the diffusion coefficient. This differs from the usual equation [1] by the addition of the term in F, which in a sense is equiva-lent to some type of force causing the particles to flow outward, the flux being a function of the density. Since we have already assumed the process of string scattering is cylindrically symmetri-cal, we may rewrite the equation without F as

$$\frac{\partial\rho}{\partial z} = \frac{\partial D}{\partial\theta}\frac{\partial\rho}{\partial\theta} + D\left(\frac{\partial^2\rho}{\partial\theta^2} + \frac{1}{\theta}\frac{\partial\rho}{\partial\theta}\right).$$

Note that, if we assume $D = \frac{k}{\theta}$ this reduces to $\frac{\partial\rho}{\partial z} = D\frac{\partial^2\rho}{\partial\theta^2}$, i.e.,
we may replace the polar equation with the more convenient one dimensional rectangular equation by suitably choosing D. We now
wish to relate D to $\frac{\partial <\Delta\theta^2_{out}>}{\partial z}$. The polar solution, for diffusion
of particles with the initial condition $\theta=0$, $z=0$ can be shown to be

$$\rho(\theta, \theta + d\theta)d\theta = \frac{2}{z\partial <\theta^2>/\partial z} \; \exp \frac{-\theta^2}{z\partial <\theta^2>/\partial z} \; \theta \; d\theta \; .$$

The corresponding linear one dimensional solution takes the form

$$\rho(\theta, \theta+d\theta)d\theta = \frac{1}{\sqrt{2\pi} \ z \ \partial\langle\theta^2\rangle/\partial z} \ \exp \frac{-\theta^2}{2z \ \partial\langle\theta^2\rangle/\partial z} \ d\theta$$

The solution to the diffusion equation, in both cases, has a dependence $e^{\frac{-\theta^2}{4Dz}}$. We shall assume we are always sufficiently far from the origin in determining $\langle\Delta\theta_{out}^2\rangle$, that the solution is approximately that of the linear case, so that

$$2D = \frac{\partial\langle\Delta\theta_{out}^2\rangle}{\partial z} \quad .$$

The simplest solution is the steady state solution, which would would occur very deep in the crystal. For this solution,

$$\frac{\partial\rho}{\partial z} = 0 \ \text{ and } \ D \ \frac{\partial\rho}{\partial\theta} = F\rho \quad .$$

Now

$$F = \frac{\partial\langle\theta\rangle}{\partial z} \quad ,$$

so

$$\rho = k \ \exp\left(\frac{2\partial\langle\theta\rangle/\partial z}{\partial\langle\theta_{out}^2\rangle/\partial z} \ \theta \right)$$

where k is a normalizing constant.

This result for ρ (θ) per unit solid angle is plotted in Figure 11 for a stationary string, and for the atom of closest approach vibrating in the X direction with 1.13 a_{TF} Å root mean square amplitude [11]. It is interesting that the flux commences to decrease from its random value well beyond the channeling region, and that, if thermal vibrations are included, there is an appreciable density of particles diffusing into the channel.

Conclusion

Several interesting results emerge from these calculations. Firstly, it may easily be seen from the angular distribution patterns that the collective effect of the string extends well beyond the usual region of channeling. Secondly, we may employ the angular distributions as a function of the angle of incidence on the

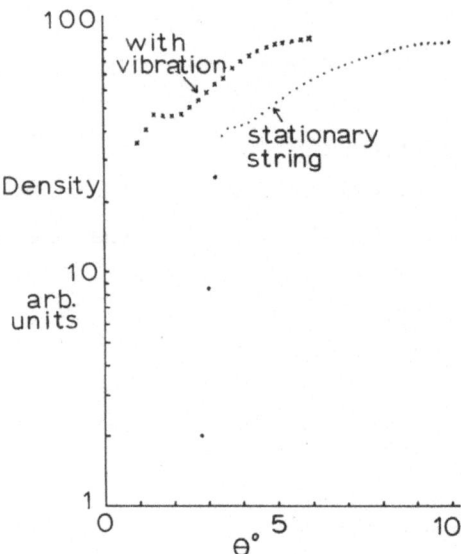

Fig. 11.   Steady state solution.   Density of particles across the
channel $(\frac{\partial n}{\partial \Omega})$ as a function of θ, for a stationary string and one
for which the atom of closest approach vibrated as before.

string to determine a diffusion coefficient D, and flux coefficient
F.   These may then be used to solve a diffusion equation for the
flux of particles into and out of the channel for a variety of
initial conditions.

We present the steady state solution to show that there is a
slight decrease in density just outside the usual channeling region,
and that there is an appreciable density within the channel only
if thermal vibrations are included.

The advantages of this type of approach over the Monte Carlo
calculations are, firstly, one can deal with great depths of the
crystal without exceptionally long computer times, and secondly,
it is conceptually easy to picture physically particles diffusing
into and out of the channel.

Acknowledgments

I wish to thank Drs. R. Sizmann and C. Varelas for many helpful discussions.

References

[1] E. Bonderup, H. Esbensen, J. U. Andersen and H. E. Schiøtt, Rad. Effects 12, 261 (1972).
[2] M. A. Kumakhov, Rad. Effects 15, 85 (1972).
[3] V. Beloshitsky, M. A. Kumakhov and V. A. Muralev, Rad. Effects 13, 9 (1972).
[4] S. U. Campisano, G. Foti, F. Grasso, M. LoSavio and E. Rimini, Rad. Effects 13, 157 (1972).
[5] K. Morita, Rad. Effects 14, 195 (1971).
[6] J. Mory and Y. Quéré, Rad. Effects 13, 57 (1972).
[7] M. T. Robinson and S. Oen, Phys. Rev. 132, 2385 (1963).
[8] Dirck Van Vliet, Rad. Effects 10, 137 (1971).
[9] John H. Barrett, Phys. Rev. B 3, 1527 (1971).
[10] G. Molière, Z. Naturforsch 2a, 133 (1947).
[11] C. Varelas and J. Biersack, Nuc. Inst. and Meth. 79, 213 (1970).
[12] J. U. Andersen and L. C. Feldman, Phys. Rev. B 1, 2063 (1970).
[13] R. D. Edge, C. Varelas and R. Sizmann, Rad. Effects 16, 95 (1972).
[14] Chr. Lehmann and G. Leibfried, J. Appl. Phys. 36, 2841 (1963).
[15] Chr. Lehmann and G. Leibfried, Z. Physik 172, 465 (1963).
[16] Jens Lindhard, Mat. Fys. Medd. Dan. Vid. Selsk 34, No. 14 (1965).
[17] C. Varelas and R. Sizmann, Rad. Effects 16, 211 (1972).
[18] J. H. Barrett, private communication and these proceedings.
[19] L. Meyer, Phys. Stat. Sol. (b) 44, 253 (1971).
[20] H. H. Andersen, J. Bøttiger and H. Knudsen, Rad. Effects 13, 203 (1972).
[21] G. Molière, Z. Naturf. 3a, 78 (1948).

MODIFIED DECHANNELING THEORY AND DIFFUSION COEFFICIENTS

M. KITAGAWA and Y. H. OHTSUKI

Department of Physics, Waseda University
Nishi-Ohkubo 4, 160 Tokyo, Japan

Introduction

When energetic ions are incident along a low index crystallographic direction, their motions are well confined inside a channel.
Due to a set of small angle collisions by electrons, lattice atoms,
defects in the crystal and so on, the transverse ion momentum increases, and this leads to dechanneling. Since the Lindhard theory
[1], the theoretical treatment of dechanneling has been developed
by many authors [2-5] according to the steady-increase model or
the diffusion model. Recently Bonderup et al. [3] and Björkqvist
et al. [4] have shown that by the steady-increase model it is
not possible to explain the experiments at higher dechanneled
fractions.

We note, however, that the above theories have been introduced
by some physical intuitions, which give rise to too simplified an
equation. In this paper, first we start from the generalized
Fokker-Planck equation in the sense of the redistribution of channeled ions, and derive a modified diffusion equation. In the modified equation, damping effects and two directional diffusion coefficients are included.

Modified Dechanneling Equation

Let $\phi(r_\perp, p; \Delta p)$ denote the transition probability that the ion
with momentum p suffers an increment Δp during the time interval
Δt at the point $r_\perp=(x,y)$ in the plane perpendicular to the channeling direction z. Since scattering processes in our system are
regarded as small angle scatterings, the behaviour of the channeled
ion in phase space is described in terms of the generalized Fokker-
Planck equation:

$$\frac{\partial f}{\partial t} + \underset{\sim}{v} \cdot \text{grad}_{\underset{\sim}{r}} f - \text{grad}_{\underset{\sim}{r}} U(r_\perp) \cdot \text{grad}_{\underset{\sim}{p}} f$$

$$= -f \, \text{div}_{\underset{\sim}{p}} (\langle \frac{\Delta p}{\Delta t} \rangle) - \langle \frac{\Delta p}{\Delta t} \rangle \text{grad}_{\underset{\sim}{p}} f + \frac{1}{2} \sum_{i=x,y,z} \frac{\partial^2}{\partial p_i^2} (\langle \frac{\Delta p_i^2}{\Delta t} \rangle f)$$

$$+ \sum_{i<j} \frac{\partial^2}{\partial p_i \partial p_j} (\langle \frac{\Delta p_i \Delta p_j}{\Delta t} \rangle f) + 0(\Delta p_i \Delta p_j \Delta p_k) \quad , \qquad (1)$$

where f is the one-particle distribution function of the channeled ion and <...> means the expectation value over the transition probability ϕ, that is

$$\langle \frac{\Delta p}{\Delta t} \rangle = \int \frac{\Delta p}{\Delta t} \phi(\underset{\sim}{r}_\perp, \underset{\sim}{p}; \Delta \underset{\sim}{p}) d(\Delta \underset{\sim}{p}) \quad . \qquad (2)$$

In the above, $U(r_\perp)$ is the continuum potential and $\sigma_{ij} = \frac{1}{2} \langle \Delta p_i \Delta p_j / \Delta t \rangle$ is the (i-j)th element of the diffusion coefficient.

In eq. (1), while $-\text{grad}_{\underset{\sim}{r}} U(r_\perp)$ means the binding force acting on the ion, $\langle \Delta p / \Delta t \rangle$ denotes the force due to inelastic processes, which are characterized by energy loss processes. If the damping force caused by energy losses acts against the ion motion, we easily obtain

$$\langle \frac{\Delta p}{\Delta t} \rangle = - S(r_\perp, E_z) \frac{\underset{\sim}{p}}{p_z} \quad , \qquad (3)$$

where $S(r_\perp, E_z)$ is the local stopping power [6-8] and $E \approx E_z$ the ion energy.

It is clear that eq. (1) describes the motion of ion inside the channel if we neglect the collision term, in which case f becomes the integral of motion.

In eq. (1), σ_{zz} is not the usual diffusion coefficient in our case. In fact, in high energy region, σ_{zz} is the higher power of the inverse of the ion velocity than that of $\langle \frac{\Delta p_z}{\Delta t} \rangle$, or σ_{xx} and σ_{yy}. In other words, for the z component of scattering processes energy losses are dominant rather than the diffusion. Therefore we can neglect σ_{zz}.

We consider the axial case. Although it would be possible to determine physical quantities for dechanneling in principle from eq. (1), we meet considerable mathematical difficulties to solve it. Therefore, we restrict ourselves to the steady-state process near the Lindhard equilibrium state [1], where it can be assumed that f does not contain x, y and t explicitly since channeled ions are well mixed in the transversal region [1,9]. Considering that the continuum potential $U(x,y)$ is an even function both for x and y, and that $z = \int_{E_z}^{E_0} \hat{S}^{-1}(E_z)dE_z$, and averaging over the x-y plane in a unit cell, we obtain the following equation in (p_x, p_y, E_z) space

$$f = \frac{\hat{S}(E_0)}{\hat{S}(E_z)} \cdot \frac{E_0}{E_z} u \, , \tag{4a}$$

$$-\frac{\partial u}{\partial E_z} = \frac{1}{2E_z} (\underset{\sim}{p}_\perp \cdot \text{grad}_{\underset{\sim}{p}_\perp} u) + \frac{1}{\hat{S}(E_z)v_z} \sum_{i=x,y} \frac{\partial^2}{\partial p_i^2} (\hat{\sigma}_{ii}u)$$

$$+ 2 \sum_{i<j} \frac{\partial^2}{\partial p_i \partial p_j} (\hat{\sigma}_{ij}u) \, , \tag{4b}$$

where the caret means averaged value over the x-y plane. E_0 is the incident ion energy.

Alternatively, we can write eq. (4b) in the form of the cylindrical coordinate $(p_\perp, \theta, E_z)$. Assuming that u is axially symmetric in $(P_\perp, \theta)$ space, we can neglect off-diagonal terms of diffusion coefficients. From eq. (4b) we obtain

$$-\frac{\partial u}{\partial E_z} = \frac{p_\perp}{2E_z} \frac{\partial u}{\partial p_\perp} + \frac{\partial^2}{\partial p_\perp^2} (\hat{\mu}_{\perp\perp} u)$$

$$+ \frac{1}{p_\perp} \frac{\partial}{\partial p_\perp} ((2\hat{\mu}_{\perp\perp} - \hat{\mu}_{\theta\theta})u) \, , \tag{5}$$

where $\hat{\mu}_{\perp\perp}$ and $\hat{\mu}_{\theta\theta}$ are mean-squared momentum spreads per unit energy loss, that is

$$\hat{\mu}_{\perp\perp} = \frac{1}{2} \langle \frac{\Delta p_\perp^2}{\Delta E_z} \rangle$$

$$\hat{\mu}_{\theta\theta} = \frac{1}{2} \langle \frac{\Delta p_\theta^2}{\Delta E_z} \rangle \quad (\Delta p_\theta = p_\perp \Delta\theta) \, , \tag{6}$$

Strictly speaking, $\hat{\mu}_{\theta\theta}$ does not vanish in our case, especially for plasmon excitation. In random system, $\hat{\mu}_{\perp\perp} = \hat{\mu}_{\theta\theta}$. Noting that the first term in the right hand side of eq. (5) is much smaller than the third term in our channeling condition where $\psi \lesssim \psi_c$ ($\psi = p_\perp/p_z, \psi_c$; the critical angle), we have the rearranged equation

$$-\frac{\partial u}{\partial E_z} = \frac{1}{p_\perp} \frac{\partial}{\partial p_\perp} \{ (p_\perp(\frac{\partial \hat{\mu}_{\perp\perp}}{\partial p_\perp}) + \hat{\mu}_{\perp\perp} - \hat{\mu}_{\theta\theta})u + p_\perp \hat{\mu}_{\perp\perp} \frac{\partial u}{\partial p_\perp} \} , \quad (7)$$

or as is usually used in statistical physics, eq. (7) may reduce for a steady-state process near the Lindhard equilibrium state as follows:

$$-\frac{\partial u}{\partial E_z} = \frac{1}{p_\perp} \frac{\partial}{\partial p_\perp} \{ p_\perp \hat{\mu}_{\perp\perp} [- \frac{d}{dp_\perp} (\ln u_{eq})] u \}$$

$$+ \frac{1}{p_\perp} \frac{\partial}{\partial p_\perp} (p_\perp \hat{\mu}_{\perp\perp} \frac{\partial u}{\partial p_\perp}) , \quad (8)$$

where u_{eq} is the equilibrium distribution in $p_\perp$ space corresponding to the Lindhard one.

In the above, for the steady-state process, $\hat{\mu}_{\theta\theta}$ does not appear explicitly in the differential equation. We, however, note the invariance of diagonal terms of $\hat{\mu}$ under the transformation in $p_\perp$ space, that is, $\hat{\mu}_{\perp\perp} + \hat{\mu}_{\theta\theta} = \hat{\mu}_{xx} + \hat{\mu}_{yy}$. Therefore we have to take into account $\hat{\mu}_{\theta\theta}$ to determine the magnitude of $\hat{\mu}_{\perp\perp}$ in view of each physical circumstance for phonon, single-electron and plasmon excitations.

We note that the first term in the right hand side of eq. (8) is not neglected in general in comparison with the second term which gives the usual diffusion equation widely used in dechanneling theories. In our case, we have from the standard potential [1],

$$\frac{d}{dp_\perp}(\ln u_{eq}) = - \frac{p_\perp}{M_1} \frac{2}{E_z \psi_c^2} \frac{A \exp(\frac{2E_\perp}{E_z \psi_c^2}) + 1}{A \exp(\frac{2E_\perp}{E_z \psi_c^2}) - 1} ,$$

$$A = \frac{C^2 a^2}{r_0^2} + 1 , \quad (9)$$

where M_1 is the ion mass, a is the Thomas-Fermi screening length, πr_0^2 is the area of the unit mesh in the x-y plane, $C=\sqrt{3}$, $E_\perp = p_\perp/2M_1$.

Recently, taking into account the damping effect, Björkqvist et al. [4] calculated dechanneled fractions according to the steady-increase model. In our theory, however, the damping effects are introduced in $\langle \Delta p/\Delta t \rangle$, which does not appear in the usual diffusion equation.

Dechanneling rate α per unit energy loss is defined as

$$\alpha = \frac{1}{F} \frac{\partial F}{\partial E_z} , \tag{10}$$

where $F = \int_0^{P_{\perp c}} f 2\pi p_\perp dp_\perp$, $(P_{\perp c} = p_z \psi_c)$. From eqs. (4a) and (10), the damping part α_{da} of dechanneling rate is obtained by

$$\alpha_{da} = -\frac{1}{E_z} - \frac{1}{\hat{S}(E_z)} \frac{d\hat{S}(E_z)}{dE_z} . \tag{11}$$

Concluding Remarks

A new type of dechanneling theory was developed by constructing the Fokker-Planck equation in the sense of the redistribution of channeled ions. A damping term and a new term in addition to the usual diffusion equation appear in the Fokker-Planck equation, which are not neglected in general. The Fokker-Planck equation includes two directional diffusion coefficients. It is noted that although $\hat{\mu}_{\theta\theta}$ does not appear explicitly in eq. (8), we have to take into account the process corresponding to $\hat{\mu}_{\theta\theta}$ to determine the magnitude of $\hat{\mu}_{\perp\perp}$, which contributes 50% at maximum.

Equation (11) shows that the contribution of the damping effect is of the same order with the experimental value of dechanneling rate, about $10^{-6} - 10^{-7}$ eV^{-1} for 1.5-MeV ^{1}H$^+$, ^{2}D$^+$ and ^{4}He$^+$ in the Ge <100> channel. For example, making comparison with the experiment mentioned above, damping effects contribute about 7% and 15% for respective case of ^{1}H$^+$ and ^{2}D$^+$, and more remarkable for the case of He ion. In Table I, α_M means Morita's quasiempirical result [2], in which the damping effect is not taken into account

TABLE I

Dechanneling rate for 1.5 - MeV ^{1}H$^+$, ^{2}D$^+$ and ^{4}He$^+$ ions in the Ge <100> axis at room temperature in units of 10^{-6} eV^{-1}. α_{exp} and α_M mean the experimental value of the dechanneling rate and Morita's quasiempirical result, respectively (Ref. 2). α_{da} means the damping part of dechanneling rate.

	α_{exp}	α_M	$\alpha_M + \alpha_{da}$
^{1}H$^+$	5.4±0.2	5.90	5.60
^{2}D$^+$	2.9±0.1	3.22	2.82
^{4}He$^+$	0.84±0.04	1.97 (He$^+$)	1.23 (He$^+$)
		0.99 (He^{2+})	0.30 (He^{2+})

and the temperature dependence of dechanneling rate, in other words, the nuclear part of the diffusion coefficient is determined experimentally. For the case of He ion, the effective charge number should be considered, and it is 1.58 as our result. Thus, as was suggested by Lindhard [1], we come to the conclusion that the damping effect may not be neglected for dechanneling, especially for the study of ion mass effect.

We did not discuss the detailed calculation of the diffusion coefficients $\hat{\mu}_{\perp\perp}$, which are considered by Ohtsuki separately at this conference.

References

[1] J. Lindhard, K. Dan. Vidensk. Selsk. Mat.-Fys. Medd. 34 (14) (1965).
[2] K. Morita, Radiat. Eff. 14, 195 (1971).
[3] E. Bonderup, H. Esbensen, J. U. Andersen, and H. E. Schiøtt, Radiat. Eff. 12, 261 (1972).
[4] K. Björkqvist, B. Cartling, and B. Domeij, Radiat. Eff. 12, 267 (1972).
[5] V. V. Beloshitsky, M. A. Kumakhov, and V. A. Muraler, Radiat. Eff. 13, 9 (1972).
[6] M. T. Robinson, Phys. Rev. 179, 327 (1969); Phys. Rev. B4, 1461 (1971).
[7] M. Kitagawa and Y. H. Ohtsuki, Phys. Rev. B5, 3418 (1972).
[8] Y. H. Ohtsuki and M. Kitagawa, Phys. Lett. A40, 313 (1972).
[9] Yu. V. Martynenko, Fiz. Tverd. Tela 13, 2580 (1971) [Sov. Phys.-Solid State 13, 2166 (1972)].

INELASTIC SCATTERING IN CHANNELING

YOSHI-HIKO OHTSUKI

Department of Physics, Waseda University
Nishi-Ohkubo 4, 160 Tokyo, Japan

Introduction

It is evident that the inelastic scattering plays an important role for channeling and dechanneling. For example, blocking pattern, star pattern and Kikuchi pattern are formed by such inelastic scatterings of channeling ions. The stopping power and the diffusion coefficient in dechanneling theory are also calculated in terms of inelastic scatterings of channeling.

In usual quantum theory, we treat such an inelastic scatterings by the golden rule of transition probability between the incident plane wave and the scattered plane wave (or spherical wave). However, in our case of channeling, we have to consider a local transition probability from the localized incident wave and average it according to the channeling path.

Local Transition Probability

When we consider inelastic scatterings, the crystal potential becomes non-local. The Schrödinger equation is written as

$$\nabla^2 \phi_o + \{K^2 - \frac{2m}{\hbar^2} V(\underset{\sim}{r})\}\phi_o$$

$$+ \frac{2m}{\hbar^2} \int A(\underset{\sim}{r},\underset{\sim}{r}')\phi_o(\underset{\sim}{r}') \, d\underset{\sim}{r}' = 0 \ , \tag{1}$$

where $V(r)$ means the crystal potential for elastic scatterings, and $\frac{\hbar^2 K^2}{2m}$ the energy of the incident ions. The non-local inelastic-scattering potential $A(\underset{\sim}{r},\underset{\sim}{r}')$ is given by

$$A(\underset{\sim}{r},\underset{\sim}{r}') = - \frac{m}{2\pi\hbar^2} \sum_{n\neq o} H'_{on}(\underset{\sim}{r}) \; H'_{no}(\underset{\sim}{r}') \; \frac{\exp(ik_n|\underset{\sim}{r}-\underset{\sim}{r}'|)}{|\underset{\sim}{r}-\underset{\sim}{r}'|} \; , \quad (2)$$

where $H'_{on}(\underset{\sim}{r})$ is the matrix element of the interaction Hamiltonian between the crystal and ion beams (its mass being M) with respect to o-th and n-th excited states of the crystal with the excitation energy $\frac{\hbar^2 k_n^2}{2m}$.

In previous paper [1], we have derived the localized absorption coefficient $\mu(y)$,

$$\mu(y) = \frac{1}{v_o} \int P(y;\underset{\sim}{K}'-\underset{\sim}{K}) \; d\underset{\sim}{K}' \quad\quad\quad\quad (3)$$

where

$$P(y;\underset{\sim}{K}'-\underset{\sim}{K}) = \frac{1}{4\pi^2\hbar^2 v_o} \sum_{n\neq o} e^{-iK'_y y} H'_{on}(K'_x,K'_y,K'_z-K)$$

$$\cdot \; H'_{no}(-K'_x,y,K-K'_z) \; \delta(E_n-E_o) \; . \quad (4)$$

In above, v_o and E_n are the velocity of incident ions and the energy of n-th excited states of the crystal and ion beams. $H'_{no}(Q_x,y,Q_z)$ means the Fourier component of $H'_{no}(x,y,z)$ with respect to x and z. We have selected z to be parallel to channeling axis.

It is worthy to note that the expression $P(y;\underset{\sim}{K}'-\underset{\sim}{K})$ is the differential transition probability of inelastic scatterings at the distance y from the atomic wall.

Back Scattering Yield

It is easy to calculate the back scattering intensity as a function of the depth z $I_B(z;\underset{\sim}{K}'-\underset{\sim}{K})$ by considering the position y to be a channeling path $y=f(z)$ in the above local transition probability $P(y;\underset{\sim}{K}'-\underset{\sim}{K})$

$$I_B(z;\underset{\sim}{K}'-\underset{\sim}{K}) = P(f(z);\underset{\sim}{K}'-\underset{\sim}{K}) \; . \quad\quad\quad (5)$$

For simplicity, we consider the nuclear interaction with Einstein model, and the small harmonic oscillation path,

$$\xi = \frac{d}{2} - y = f(z) = \xi_o \; \sin(\frac{2\pi}{\lambda} z) \quad\quad\quad (6)$$

where d and λ are the atomic wall separation and oscillation "wave length". In this case, we obtain [2]

$$P(y;\underset{\sim}{q}) = \frac{1}{4\pi^2\hbar^2 v_o^2} \sum_{g_\perp} S(\underset{\sim}{q},\underset{\sim}{q}+2\pi g_\perp) \; e^{2\pi i g_\perp y} \tag{7}$$

$$S(\underset{\sim}{q},\underset{\sim}{q}+2\pi g_\perp) = v_q \; v_{q+2\pi g_\perp} \left\{ e^{-M(2\pi g_\perp)} -e^{-M(q)-M(q+2\pi g_\perp)} \right\}$$

$$\cdot \; \delta(|K+q| - K + \frac{\Delta E}{\hbar v_o}) \; , \tag{8}$$

where $g_\perp$ means the reciprocal lattice vector and v_q the usual scattering factor respectively. $e^{-M(q)}$ is the Debye-Waller factor and ΔE the energy loss value.

When we consider two atomic planes with separation d, $I_B(z;\underset{\sim}{q})$ is given approximately as

$$I_B(z;\underset{\sim}{q}) = \frac{1}{2\pi^2\hbar^2 v_o^2} |v_q|^2 (\frac{d}{2\pi}) \sqrt{\frac{\pi}{\kappa(T)}}$$

$$\cdot \; e^{-\frac{d^2}{16\kappa(T)}} \left\{ 1 - \frac{1}{32}(\frac{\xi_o d}{\kappa(T)})^2 \; \sin^2(\frac{2\pi}{\lambda} z) \right\} \tag{9}$$

for $\kappa(T) >> \xi_o d/4$, $\kappa(T)$ being the usual Debye-Waller coefficient, which is proportional to the squared mean amplitude of the lattice vibration.

The oscillation of such a back scattering yield have been found experimentally by several authors [3]. It is worthy to note that we have no phenomenological parameter in our expression.

Local Diffusion Coefficient

One uses usually the diffusion-type equation in dechanneling theory [4]. A new type dechanneling theory [2] is presented using the Fokker-Planck equation by us at this congress. In such a theory, we have to give the local diffusion coefficient $D_i(y)$, which is derived in terms of the local transition probability;

$$D_i(y) = \frac{1}{2} \int q_i^2 \; P(y;\underset{\sim}{q}) \; d\underset{\sim}{q} \; , \tag{10}$$

$$(i = x \; or \; y)$$

where $\hbar q_i$ is the i-components of the momentum transfered.

Detailed calculations for $D_i(y)$ due to the nuclear part, the single electron part, and the plasmon excitation part have been calculated by Kitagawa and Ohtsuki [2], and Waho and Ohtsuki [5]. In Fig. 1, we show the nuclear diffusion coefficient $<\dfrac{dE_\perp}{dz}> = \dfrac{D_y(y)}{m}$ for Si {110} planar case of 10 MeV He$^+$ ions. The curve a and b are our calculations for different approximations. Curve c means the result of Lindhard [4]. In Fig. 2, we plot the electronic diffusion coefficient of 10 MeV He$^+$ ion in Si {110}. Curve a means our result and b the Lindhard's one. We compare various contributions in Fig. 3 with the plasmon excitations [2].

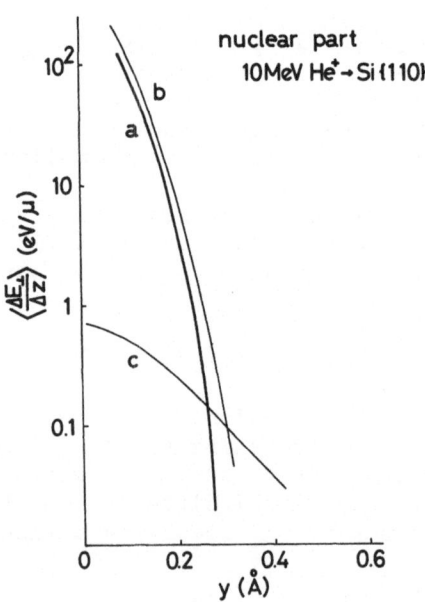

Fig. 1.   Nuclear diffusion coefficient for 10 MeV He ions in Si{110}.

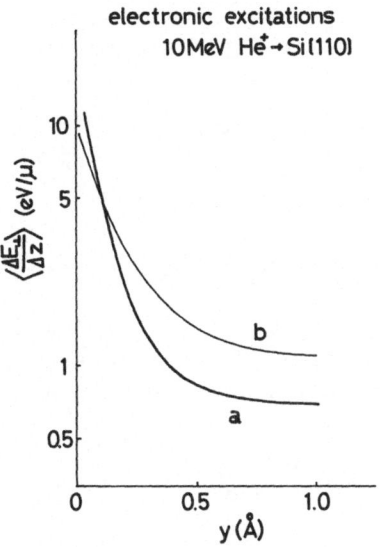

Fig. 2. Electronic diffusion coefficient for 10 MeV He ions in Si{110}.

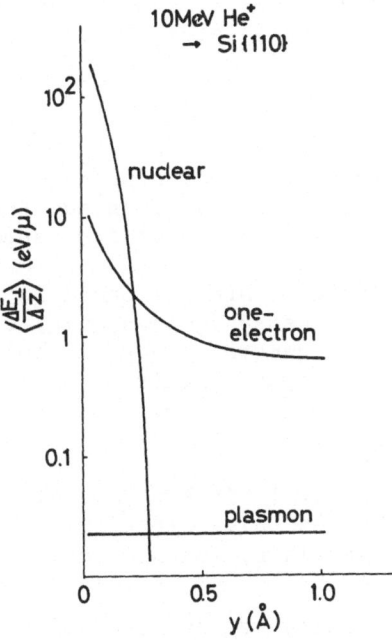

Fig. 3. Comparison with contribution from plasmon excitations.

Local Stopping Power

The local stopping power W(y) is also calculated from our
local transition probability, i.e.

$$W(y) = \frac{1}{4\pi^2 \hbar^2 v_o^2} \int dK' \, e^{-iK_y' y} \sum_{n=o} (E_n - E_o)$$

$$\cdot \, H'_{on}(K_x', K_y', K_z'-k) \, H'_{no}(-K_x', y, K-K_z') \, \delta(K'-k_n) \qquad (11)$$

In the case of the small angle scattering, or the straight-line
approximation, we may neglect K_x' and K_y' dependence in $\delta(\ldots)$.
And we can separate the integration over K_x' and K_y', we obtain

$$W(y) = \frac{1}{4\pi^2 \hbar^2 v_o^2} \int dK_z' \sum_{n \neq o} (E_n - E_o)$$

$$\cdot \, H'_{on}(0, y, K_z'-K) \, H'_{no}(0, y, K-K_z') \, \delta(k_z'-k_n), \qquad (12)$$

which coincides with the expression calculated by Kitagawa and
Ohtsuki [6] in different method.
    In Fig. 4, we show W(y) for 3 MeV He$^+$ ions in Au{100} channel
(curve a) with the local density approximation of Lindhard (curve b).

Concluding Remarks

    We note that our local transition probability is also used
for the star pattern, the Kikuchi pattern and the absorption coefficient. The detailed applications are discussed in our recent
publication [7].
    In most cases experiments are performed in low velocity region
in which the perturbation theory (golden rule) used in our derivation is not applicable. A general theory in low velocity region
[8] is presented in another publication.*

Acknowledgment

    The author thanks Prof. Ichinokawa and other members of chan-
neling study group in Waseda University for their useful discussions.
The work was supported partly by the Kawakami Foundation.

---

*For example, in low velocity region, the diffusion coefficient
 is not proportional to the velocity, but becomes a constant at
 lower velocity than a critical value [8].

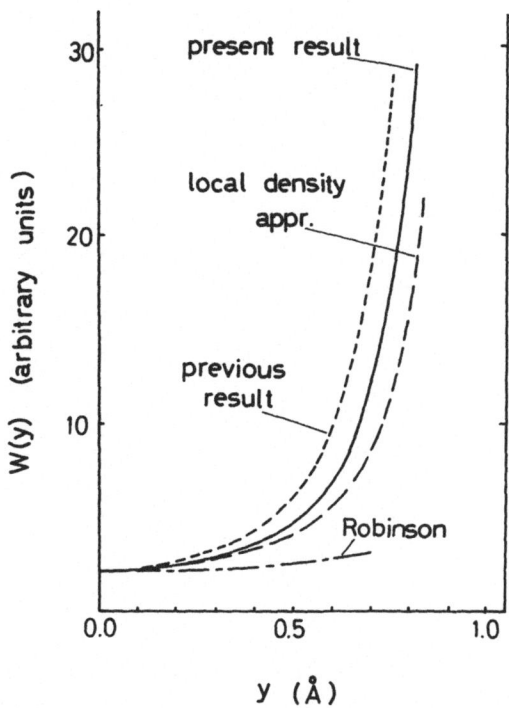

Fig. 4. Local stopping power for 3 MeV He ions in Au{100}
(curve a) and the local density approximation of Lindhard (curve b).

References

[1] Y. H. Ohtsuki, J. Phys. Soc. Japan 34, 473 (1973).
[2] M. Kitagawa and Y. H. Ohtsuki, Phys. Rev. B8, 3117 (1973).
[3] E. Bøgh, Rad. Effects 12, 13 (1972).
    J. C. Poizat and J. Remillieux, Thesis, Lyon, Sept. (1972).
    F. Abel, G. Amsel, M. Bruneaux and C. Cohen, Phys. Letters 42A,
    165 (1972).
[4] J. Lindhard, Kgl. Danske, Vid. Sels. Mat. Fys. 34, No. 14 (1965).
[5] T. Waho and Y. H. Ohtsuki, Rad. Effects 19 (1973).
[6] M. Kitagawa and Y. H. Ohtsuki, Phys. Rev. B5, 3418 (1972).
[7] Y. H. Ohtsuki, Phys. Status. Solidi (b) 59 (1973).
[8] M. Kitagawa and Y. H. Ohtsuki, Phys. Rev. B1 (in press).

CHANNELING IN A NON PERFECT CRYSTAL[*]

S. U. CAMPISANO, G. FOTI, F. GRASSO and E. RIMINI
*Istituto di Struttura della Materia dell'Università di Catania
Corso Italia 57 - I95129 Catania, Italy*

ABSTRACT

The present work concerns the extension of a dechanneling treatment, previously developed for perfect crystals and for those covered with amorphous layers, to the case of a damaged crystal. The minimum yield and the angular yield profiles have been measured in Si damaged by 0.3 MeV deuterons as a function of penetration depth by backscattering of MeV deuterons. In the case considered, i.e., low defect concentration, the aligned yield at a given depth results from wide-angle scattering on defects lying at the same depth, which adds to the dechanneled fraction. The last depends on the concentration of defects, which being spread inside the crystal contribute to multiple scattering in the channel as electrons and thermal vibrations do. The defect contribution should then be added to the electronic and nuclear reduced multiple scattering. The dechanneled fraction has been obtained then in the steady increase approximation as a function of the experimental parameters and of the defect concentration and profile. The comparison of the calculated minimum yield with experiments shows an extremely good agreement.

Introduction

In a perfect crystal the transition of channeled particles from the condition of correlated motion inside the crystal to the condition of ungoverned motion is due to the interaction of beam particles with crystal electrons and thermal vibrations of the

atomic rows. In this case dechanneling can be described analytically [1] in terms of elementary scattering processes [2] neglecting
diffusion [3] and accounting for a) the initial angular distributions
of particles just beneath the crystal surface, b) the interaction
with electrons and vibrating nuclei in the channel, and c) transition from the aligned to the random component of the beam.

In imperfect crystals where some of the lattice atoms are randomly located in the lattice, a channeled particle suffers small
angle scatterings which add to those experienced in a perfect crystal. The extension of the dechanneling treatment previously developed [1] for perfect crystals seems now of interest both as a test
of the applicability of the model and as a more correct method to
obtain lattice damage profiles.

Two extreme situations can be considered: i) very high defect
concentration localized in a well defined region, e.g., on the
crystal surface; ii) low defect concentration imbedded in the
crystal lattice. In the first case [4] the angular distribution of
particles impinging upon the crystal surface is drastically modified
with respect to the perfect surface because of the spreading experienced by the collimated beam in traversing the defect layer. The
calculated dechanneled fractions depend then on the treatment
adopted to describe the scattering in the surface imperfection,
i.e., on the resulting angular distribution (point a). In the
second case only the scattering experienced by a channeled particle
along its path inside the crystal are modified (point b); in fact
defects randomly imbedded in the crystal structure give rise both
to an increase in average of channeled particle transverse energy
with depth for large impact parameters, and to wide-angle scattering
beyond the critical angle for small-impact parameters.

In all previous works [5] which use channeling technique to
extract damage profiles or to characterize epitaxially grown single
crystals [6], the simultaneous occurence of small angular deflections due to the defects and to the nuclear and electronic scattering has not been taken into account.

It is then important to evaluate, even in an approximate way,
the influence of the term responsible for the increase in transverse
energy of a channeled particle. This has been achieved by a suitable
calculation of the mean spreading experienced by a channeled particle
scattered by a defect center, and by adding this contribution to the
other elementary scattering.

From the experimental point of view while the high defect concentration was obtained by deposition of a thin amorphous layer [4]
onto a Si single crystal surface, the low defect concentration has
been obtained by light particle bombardment of a Si crystal kept at
low temperature.

Experimental

    Backscattering measurements have been performed by using the
2.5 MeV Van de Graaff accelerator of Catania. Deuterons of energy
ranging between 0.3 and 1.5 MeV were collimated within 0.5 mrad
in 10^{-6} torr vacuum. Particles scattered at 150° were detected
by a 25 mm^2 surface-barrier detector 8 cm from the target, and
through standard electronics, pulses were fed to a 4096 channel
pulse height analyzer. The overall energy resolution of the system
was 10 keV FWHM for 1.0 MeV deuterons. The crystal target can be
oriented within 0.5 mrad in any space direction and its temperature
was maintained at 80°K or 300°K. A more detailed description of the
apparatus set-up has been previously reported [1].
    The low concentration of defects imbedded in the crystal
structure has been achieved by bombarding a Si crystal with doses
of the order 10^{15}-10^{16}ions/cm^2 of 0.3 MeV deuterons. During bombardment the Si crystal was randomly oriented and in thermal contact with a liquid nitrogen bath so to produce a detectable damage.
It has been recently shown [7] that at room temperature no damage
effect is observed, with channeling technique, for comparable doses
on Si. To avoid annealing effects the damaged samples were analyzed
at 80°K also. The analyzing beam spot of 1mm diameter was concentric with the 2mm diameter damaging beam spot.

Results and Calculations

    A typical experimental energy spectrum for a damaged crystal
is shown in Fig. 1. The peak in the aligned yield is caused by
Rutherford scattering of the aligned component of the beam on displaced atoms. This adds to the scattering of the random component
of the beam on all crystal atoms.
    This experimental energy spectrum can be described in a sche-
matic way as shown in Fig. 2. In the upper part are shown a few
typical trajectories for channeled particles. A particle in the
trajectory 1 is dechanneled because of the interaction with electrons
and vibrating nuclei (as in a perfect crystal) and therefore it
belongs to the random beam (component 1) and it can be backscattered
at any depth from all the atoms thus producing the contribution
marked as χ_1 in the spectrum (lower part of the figure). A particle
which follows the trajectory 2 is backscattered on defects and
gives to the spectrum the contribution marked as χ_2, the profile
of which is the defects profile. A particle in the trajectory 3
interacts with a defect at large distance and suffers an angular
deflection larger than the critical angle thus entering in the
random beam (component 3 and contribution χ_3). A particle in the
trajectory 4 which interacts with a defect at a distance larger
than 3 suffers a small angular deflection still remaining channeled
but with an increased transverse energy; it can be therefore dechanneled more easily by nuclear and electronic multiple scattering.
In a damaged crystal, because of the simultaneous occurrence of

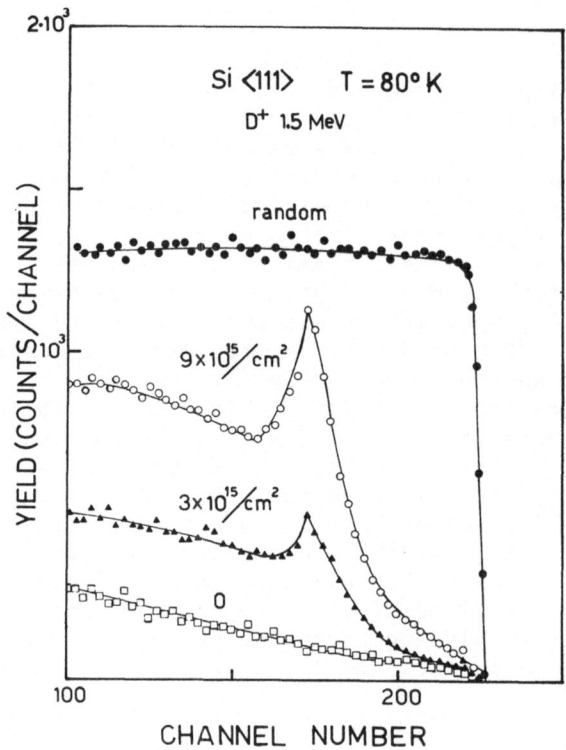

Fig. 1. Energy spectra for 1.5 MeV D$^+$ backscattered from a damaged silicon crystal at 80°K for incidence along a random and <111> directions: ☐ aligned virgin yield; Δ aligned yield after an 80°K bombardment with a dose of 3x10^{15}/cm^2 of 0.3 MeV D$^+$; 0 same as Δ but after 9x10^{15}/cm^2 of 0.3 MeV D$^+$.

multiple scattering on defects and on electrons and vibrating nuclei, the dechanneling processes cannot be separated. Particles following trajectories 1 and 4 produce then the contribution $\chi_{1.4}$ in the spectrum.

The last contribution has not been considered in previous works [5] and the concentration of defects there determined is then overestimated.

As said before our aim is to extract information on the damage profile from the depth dependence of the normalized aligned yield, then the energy scale of backscattered particles has been converted in a depth scale following the usual procedure [8] from reported stopping powers [9]. The same energy loss has been assumed for random and channeled trajectories.

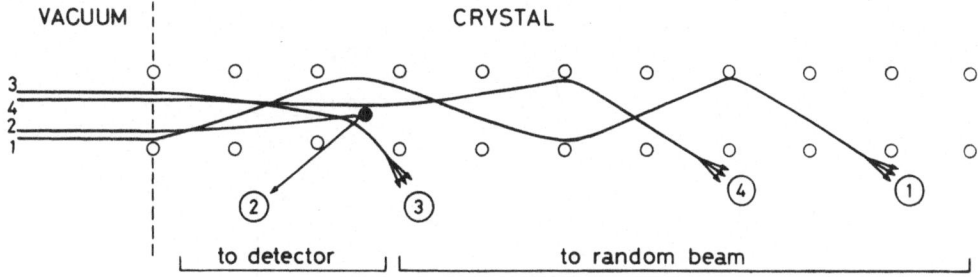

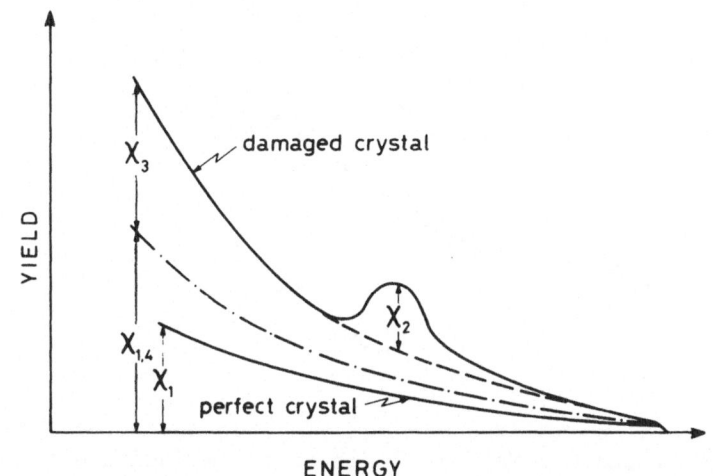

Fig. 2. Typical trajectories for channeled particles in a damaged crystal, upper part, and the resulting backscattered energy spectra, lower part. In particular trajectory 4 refers to a particle which suffers simultaneous multiple scattering on defect and on electrons and vibrating nuclei.

Fig. 3 shows the normalized minimum yield for a damaged crystal and for two different analyzing beam energies. For 1.5 MeV D^+ the backscattering depth range is higher than the total range of the damaging particles 0.3 MeV D^+. The damage peak appears then as a peak in the aligned yield. The $Y(0)$ value is quite independent from the damaging dose which increases instead the slope of the aligned yield vs depth.

Our aim is now to calculate the minimum yield $Y(z)$ vs depth z for a crystal containing defects diluted in the lattice. As

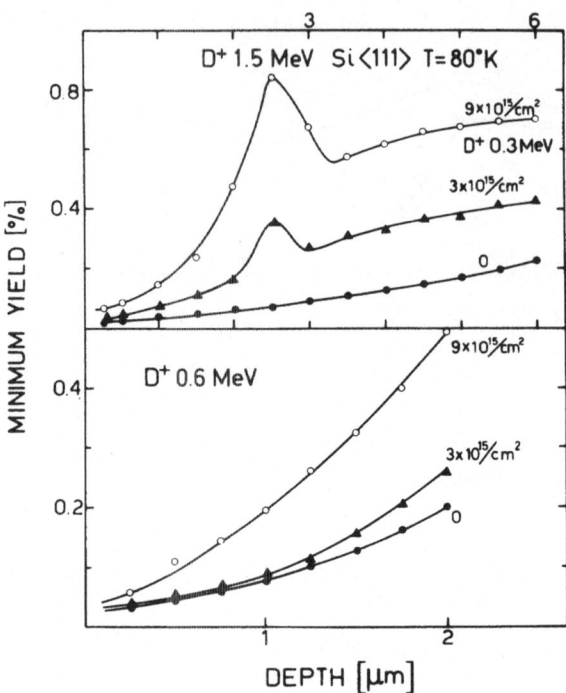

Fig. 3. Experimental normalized minimum yield vs depth for 1.5 MeV D$^+$ (upper part) and 0.6 MeV D$^+$ (lower part) impinging along the <111> axis of a silicon crystal damaged with different doses of 0.3 MeV D$^+$.

discussed in connection with Fig. 2 (contributions 1, 3 and 4) the minimum yield is partly due to dechanneling because of multiple scattering on electrons, vibrating nuclei and defects.

The calculation procedure of the dechanneled fraction $\chi_{1.4}$ in the steady increase approximation is sketched in Fig. 4. In the right upper part the integral distribution $P(\varepsilon_\perp)$ in reduced transverse energy for particles just beneath the crystal surface is shown. The $P(\varepsilon_\perp)$ value gives the fraction of particles entering the crystal with a transverse energy larger than $\varepsilon_\perp$. The reduced transverse energy is given by $\varepsilon_\perp = \dfrac{2E_\perp}{E\psi_1^2}$, where $E_\perp$ is the transverse energy, E the particle energy and ψ_1 the Lindhard critical angle. In the lower part of the function $z(\varepsilon_\perp)$ gives the depth z at which a particle having a well defined initial transverse energy $\varepsilon_\perp$ reaches the critical value $\varepsilon_\perp^* = \dfrac{2E\psi_{1/2}^2}{E\psi_1^2}$ to be dechanneled ($\psi_{1/2}$ is the

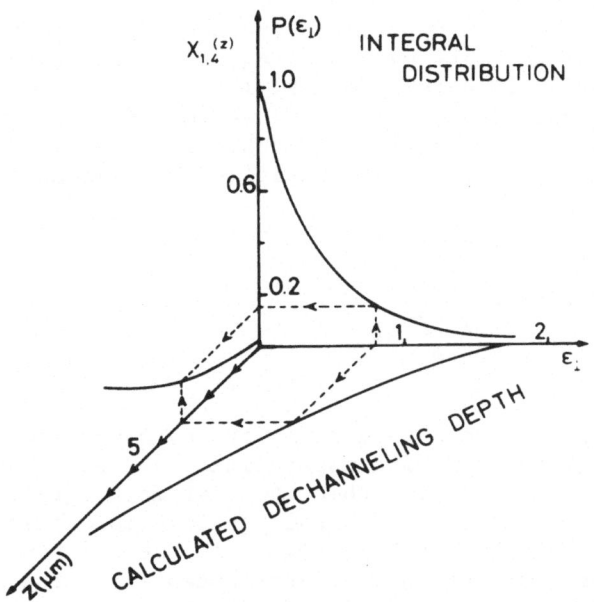

Fig. 4. Graphical determination of the dechanneled fraction from
the initial integral distribution and from the computed depth $z(\varepsilon_\perp)$
at which a particle of an initial reduced transverse energy $\varepsilon_\perp$

reaches the critical value $\varepsilon_\perp^* = \dfrac{2E\psi_{1/2}^2}{E\psi_1^2}$ to be dechanneled. Enter-

ing at a given $\varepsilon_\perp$ energy the resulting $X_{1,4}(z)$ value is equal to
the corresponding $P(\varepsilon_\perp)$ value while the corresponding z value is
obtained from the computed $z(\varepsilon_\perp)$ function (follow the arrows).

experimental critical angle). The calculated dechanneled fraction,
$X_{1,4}(z) \equiv P(\varepsilon_\perp(z))$, is found by eliminating $\varepsilon_\perp$ from the two above
functions as shown by the arrows.

To evaluate the $z(\varepsilon_\perp)$ function in a damaged crystal in addition
to the scattering on electrons and vibrating nuclei, as it occurs
in a perfect lattice, one should consider the small angular deflection ($< \psi_1$) experienced by the channeled particles on colliding
with displaced atoms. This extra contribution, which produces a
faster increase of dechanneling with depth, has been included
following the same treatment used for the multiple electronic and
nuclear reduced scattering. The increase with depth of a particle
transverse energy is then given by

$$\delta\varepsilon_\perp = [A_n(E,T)f_n(\varepsilon_\perp) + B_e(E)f_e(\varepsilon_\perp) + D(E,T)c(z)]\delta z \qquad (1)$$

where $A_n f_n, B_e f_e, Dc$ are respectively the nuclear, electronic and defect contributions. The values of A_n, B_e and the functions f_n and f_e are reported in detail in Ref. 1.

The defect contribtuion has been estimated as the mean angular spreading produced by a single defect $D(E,T)$ multiplied by the defect concentration c. In this description D is given by $2 \dfrac{<\Theta^2>}{\psi_1^2}$, where

$$<\Theta^2> \qquad \int_0^{\psi_1} \Theta^2 \frac{d\sigma}{d\omega} d\omega \qquad\qquad (2)$$

$\dfrac{d\sigma}{d\omega}$ is the differential cross section for the channeled particle-single defect scattering. The upper limit has been chosen to include only the scattering within the critical angle; in fact, for values of Θ larger than $\psi_{1/2}$ the particle is dechanneled directly by the defect (contribution 3 in Fig. 2). Table I reports $<\Theta^2>$ value calculated with the Thomas-Fermi potential. The use of a single scattering cross section in Eq. 2 gives the same result as the plural scattering treatment up to defect atomic concentrations of 20%. The analytic solution of Eq. 1 is shown in Fig. 5 for different defect concentrations. As said it represents the depth z at which a particle having a well defined initial transverse energy reaches the critical value $\varepsilon_\perp^*$, and it depends significantly on defect concentrations of the order of a few percent.

This defect contribution to dechanneling and then to minimum yield was neglected in the previous treatments [5].

Table I

Values of the mean square angular spreading $<\Theta^2>$ calculated by Thomas-Fermi potential for still channeled protons and for a single Si scattering center at 80° and 300°K crystal temperature.

E[MeV]	$10^{23}<\Theta^2>$ [rad^{2}xcm]	
	80°K	300°K
0.3	69.0	65.2
0.6	15.0	13.8
1.0	4.72	4.30
1.5	1.86	1.74
2.0	0.97	0.90

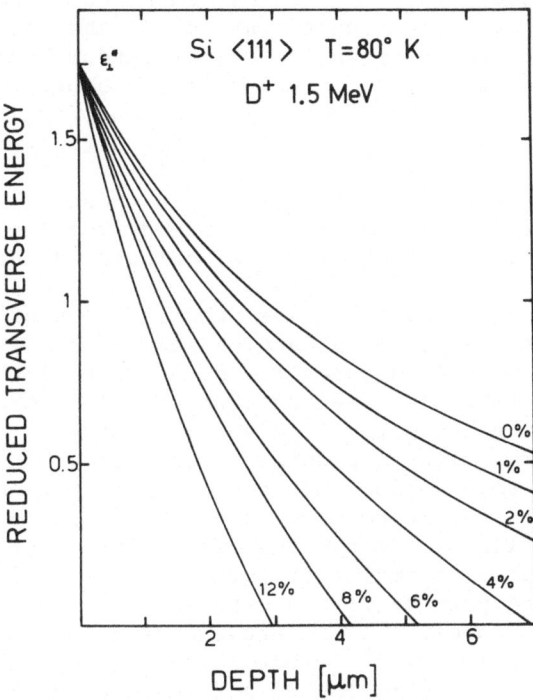

Fig. 5. Solution in the steady increase aoproximation of Eq. 1 for
different defect atomic concentrations in a Si crystal at 80°K for
1.5 MeV D$^+$.

The minimum yield $Y_{min}(z)$ in a non perfect crystal can be now
written, in terms of the contribution discussed in connection with
Fig. 2, as:

$$Y_{min}(z) = \chi_{1.4}(z) + \chi_3(z) + \chi_2(z)$$

$$= P(\varepsilon_\perp(z)) + [1-P(\varepsilon_\perp(z))] \int_{\Theta>\psi_{1/2}} f(\Theta,\tau)d\Omega + \left\{ 1-P(\varepsilon_\perp(z)) \right.$$

$$\left. -[1-P(\varepsilon_\perp(z))] \int_{\Theta>\psi_{1/2}} f(\Theta,\tau)d\Omega \right\} c(z) \qquad (3)$$

The first term $\chi_{1.4}$ is the contribution to the dechanneled
fraction due to the simultaneous occurrence of electronic and
nuclear reduced multiple scattering and of large impact parameter

scattering on a defect which, as discussed before, increases the
transverse energy of the channeled particles; the second term, χ_3,
is the component of the beam which becomes random because of suffered
angular deflection on defects larger than ψ_1 (contribution 3 of
Fig. 2); the last term comes from the fraction of the still channeled particles backscattered from the defects (contribution 2 in
Fig. 2).

The initial distribution $P(\varepsilon_{\perp})$ used in these calculations to
extract the defect profile has been obtained from the experimental
dechanneled fraction and from the $\varepsilon_{\perp}(z)$ calculated curve for a
virgin crystal.

To compute the second term, i.e. the fraction of channeled
particles which are deflected at angles larger than $\psi_{1,2}$ by scattering on defects we have assumed the occurrence of the plural
scattering regime [10,11]. The function $f(\theta,\tau)$ is the differential
angular distribution evaluated in Meyer's treatment [11], for
particles after traversing a reduced thickness τ. This τ value has
been evaluated by considering the defect concentration sampled by
channeled particles to the considered depth z. To obtain the defect
concentration profile we fit now the experimental data with the minimum yield calculated according to Eq. 3. The profile itself is
used as the fitting parameter. An example of the procedure is
shown in Fig. 6. Suppose that a particle is dechanneled after a
path length z_1 in a region of defect concentration c_1 and a path
length z_2 in a contiguous region of a defect concentration c_2. In
this case one must follow first the curve marked c_2 from $\varepsilon_{\perp}^*$ to
$\varepsilon_{\perp}(z_2)$ and from this ordinate value follow the curve marked c_1 for
an abscissa length z_1. The $\varepsilon_{\perp}$ value so obtained is the initial
transverse energy of the considered particle; one obtains then from
the initial-integral distribution calculated at this value the
first term of Eq. 3, i.e. $P(\varepsilon_{\perp}(z)) \equiv \chi_{1,4}(z)$. By iterative applications of this procedure one can fit the experimental data and then
extract the defect profile. An example is reported in Fig. 7; the
lower part shows the fits for two different analyzing energies while
the extracted profile is plotted as a solid line in the upper part
as an histogram.

The defect profiles so obtained nearly coincide in the common
sampled depth as shown by the dashed region. Moreover, the peak
of the defect concentration is located at a depth slightly smaller
than the L.S.S. [12] projected range, in accordance with previous
results [5].

In the upper part of the same Fig. 7 is plotted (dashed line)
the defect profile extracted according to ref. 5, i.e. neglecting
the simultaneous occurrence of small angle scattering on defects
and on electrons and vibrating nuclei, and using Meyer's [11] plural
scattering treatment. The profile so obtained overestimates the
total amount of defects by about 30% and moreover gives a tail
extending at depths larger than the damaging beam range.

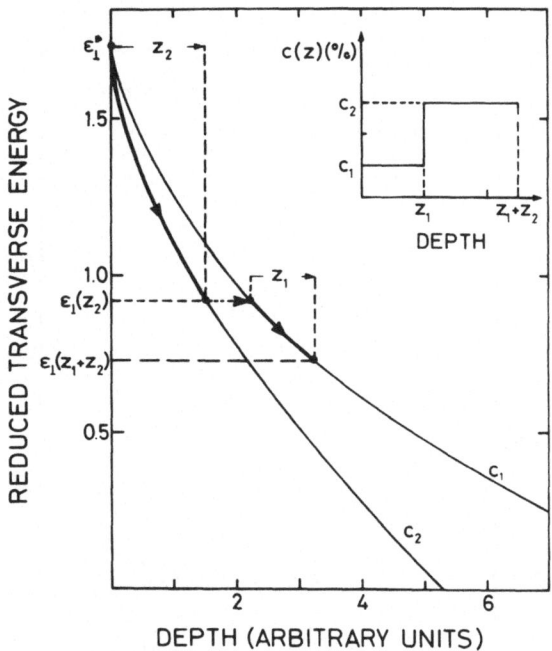

Fig. 6. Graphical determination of the initial reduced transverse
energy for particles dechanneled after traversing a depth z_1 in
a region of defect concentration c_1 and a depth z_2 in a region of
defect concentration c_2. The initial integral distribution $P(\varepsilon_\perp)$
calculated at the resulting ε_1 value gives the $\chi_{1.4}(z)$ contribution.

Discussion and Conclusion

     The dechanneling treatment developed for perfect crystals has
been applied with suitable modifications to the imperfect crystal,
i.e. for low concentration of defects imbedded in the crystal lattice.
     This treatment is based mainly on the steady-increase approxi-
mation in which the change in the mean value $\delta\langle\varepsilon_\perp\rangle$ of transverse
energy of channeled particle, having a well-defined initial value
(δ-function), with traversed depth δz is replaced by the change in
transverse energy $\delta\varepsilon_\perp$, neglecting smearing effects due to diffusion
processes.
     In this work we have included the spreading produced on the
channeled beam by small angle scattering with defects, so far neglected, in a consistent way with the other elementary scattering.
This has been done analytically while the correct diffusion approach
requires considerable computer effort.
     The defects in the above description have been assumed randomly
located in the crystal. The validity of this assumption has been

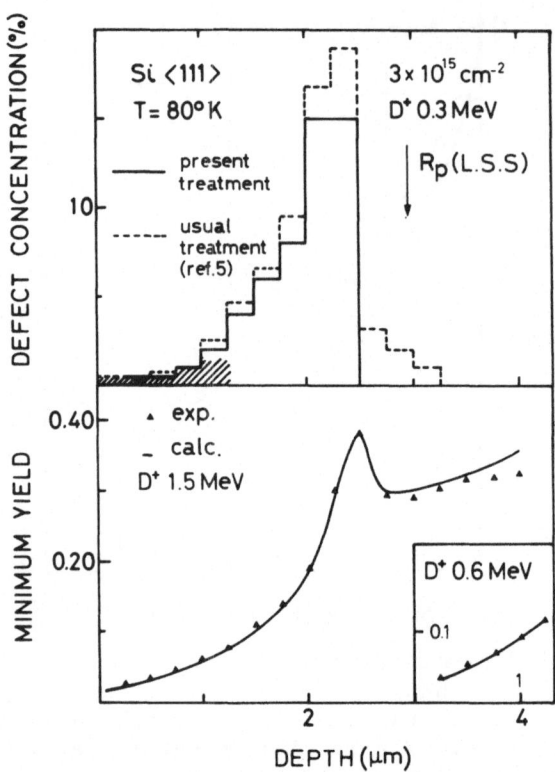

Fig. 7. Damage profile (upper part full lines) extracted by fitting the experimental dechanneled fractions with the calculated values. The obtained fit is shown in the lower part. The insert in the right hand low part shows the fit of data for 0.6 MeV D[+]. The corresponding extracted profile is reported in the dashed region and nearly coincides with that obtained by 1.5 MeV D[+] data. For comparison is also reported the profile obtained by the Bogh's procedure (dashed line) neglecting the effect of simultaneous scattering on defects and on electron and vibrating nuclei. The arrow represents the L.S.S. projected range.

tested by the absence of structure in the shape of measured angular yield profiles. This allowed one to neglect the distribution of beam particles within the channel, and to assume a defect contribution independent of the transverse energy.

Acknowledgment

The help of S. Pace, G. Panascì and A. Trovato in performing the measurements is gratefully acknowledged.

References

*Supported in part by Centro Siciliano di Fisica Nucleare e di Struttura della Materia and by Gruppo Nazionale di Struttura della Materia del Consiglio Nazionale delle Ricerche.

[1] G. Foti, F. Grasso, R. Quattrocchi and E. Rimini, Phys. Rev. B3, 2169 (1971).

[2] J. Lindhard, Kgl. Danske Videnskab. Selskab, Mat. Fys. Medd. 34, N.14 (1965).

[3] E. Bonderup, H. Esbensen, J. U. Andersen and H. E. Schiøtt, Rad. Effects 12, 261 (1972).

[4] S. U. Campisano, G. Foti, F. Grasso and E. Rimini, Phys. Rev. B8, 1811 (1973).

[5] E. Bøgh, Can. J. Phys. 46, 653 (1968);
L. C. Feldman and J. W. Rodgers, J. Appl. Phys. 41, 766 (1970);
J. E. Westmoreland, J. W. Mayer, F. H. Eisen and B. Welch, Rad. Effects 6, 161 (1970);
R. R. Hart, Rad. Eff. 6, 51 (1970);
J. F. Ziegler, J. Appl. Phys. 43, 2973 (1972).

[6] S. T. Picraux, J. Appl. Phys. 44, 583 (1973).

[7] S. U. Campisano, G. Foti, F. Grasso and E. Rimini, Appl. Phys. Letters 21, 425 (1972).

[8] J. A. Davies, J. Denhartog and J. L. Whitton, Phys. Rev. 165, 345 (1968).

[9] C. F. Williamson, J. P. Boujot, and J. Picard, Saclay Report N.R3042 1966) unpublished.

[10] E. Rimini, E. Lugujjo and J. W. Mayer, Phys. Rev. B6, 718 (1972);
E. Lugujjo and J. W. Mayer, Phys. Rev. B7, 1782 (1973).

[11] L. Meyer, Phys. Status Solidi B44, 253 (1971).

[12] J. Lindhard, M. Sharf and H. E. Schiøtt, Kgl. Danske Videnskab, Mat. Fys. Medd. 33, N.14 (1963).

TRANSMISSION ENERGY SPECTRA OF CHANNELED PROTONS SCATTERED IN THIN SILICON FILMS

F. H. EISEN and J. BØTTIGER*
*Science Center, Rockwell International
Thousand Oaks, California 91360, U.S.A.*

ABSTRACT

The transmission energy spectra of 0.2 and
0.4 MeV protons scattered from thin (∿1 μm) silicon crystals have been studied as a function of
sample orientation, with and without the introduction of a thin layer of disorder produced by
implanting krypton ions. From a comparison of
spectra observed for undamaged samples for random
penetration and for channeling along the <110>
axis with scattering angles of 10° and 15°, it is
concluded that the energy loss of dechanneled
protons is approximately equal to the random energy
loss. The energy spectra, for a 10° scattering
angle, from samples with a damaged layer on the
backside show enhanced scattering of protons with
low energy loss in comparison with undamaged
samples. By substracting the scattering yield
observed before the damage was introduced the
energy spectrum of the protons scattered from the
damage centers can be constructed. It is found
to be similar to the energy spectrum of the transmitted channeled protons. It is argued that the
protons observed in the defect spectrum, constructed
as described above, are scattered from interstitials
lying more than a few tenths of an angstrom off lattice sites. The number of such interstitials is
compared with the number of scattering centers observed in a standard backscattering–channeling effect
measurement on the same portion of the sample. It is
concluded that the channeling effect technique, using
backscattering, is primarily sensitive to such interstitials.

Introduction

The transmission energy spectra of channeled ions have sometimes been observed to be fairly wide compared to the width of the
random energy spectra for the same ions. For example, in silicon
the transmission energy spectrum for helium ions channeled along
the <1]0> axis is somewhat wider than the random spectrum [1].
The width of these channeled spectra is caused primarily by the
trajectory dependence of the channeled stopping power, since the
straggling in the channeled energy loss is small [1,2]. The trajectory dependence of channeled stopping can be understood in terms
of the electron densities sampled by ions having different transverse energies. Those with low transverse energies will be confined to regions of low electron density and so will have a relatively low electronic stopping. The channeled ions with the highest transverse energies will approach closest to the atomic rows
where the electron density is highest, and they will therefore
have the highest channeled stopping powers.

Because of this trajectory dependence of channeled stopping,
one would expect to find that the stopping power appropriate to different channeling experiments depends on the portion of the channeled beam detected in a given experiment. We have examined this
possibility experimentally for protons which are dechanneled in
silicon and for protons scattered in a disordered region produced
by ion implantation in silicon. The experiments were performed
by observing the transmission energy spectra of 0.2 or 0.4 MeV
protons scattered through angles of 0°, 10° or 15° in silicon samples about 1 μm thick using an electrostatic analyzer. The 10°
and 15° scattering angles are outside the proton multiple scattering
distribution at these energies, so that the protons which enter the
electrostatic analyzer are directed into it by a single scattering
event of about 10° or 15°. The energy loss of the protons in these
scattering events is negligible. Therefore, if a channeled proton
is scattered near the back surface of the sample, the energy with
which it enters the analyzer will primarily reflect the channeled
stopping power appropriate for that proton as it passed through
the crystal. As will become apparent later, in these experiments,
most of the channeled protons were scattered after they had penetrated more than half way through the sample. In the damage studies
in fact the damage layer constituted roughly about the last 5% of
the sample thickness. In addition to reaching conclusions concerning the energy loss of dechanneled protons and protons scattered
from damage sites, we have also been able to use channeling in a
new way to measure the number of displaced atoms in an ion implanted
sample. This method is compared with the conventional application
of the channeling effect technique to the measurement of disorder,
using backscattering.

Experimental

The experimental setup is illustrated schematically in Fig. 1.
The beam was provided by an 0.4 MeV Van de Graaff accelerator. The
beam was collimated to an angle of ±0.1° or less before striking the
thin silicon crystal which was mounted in a three-axis goniometer.
A phosphor screen could be inserted a few inches behind the sample
in order to observe the pattern of transmitted protons. This facilitated alignment of the sample, particularly for measurement of the
random spectrum. Random spectra were measured with the crystal
oriented so that no planar or axial channeling or blocking patterns
were visible within the portion of the beam entering the electrostatic analyzer. The energy spectra of the transmitted ions was
measured using the electrostatic analyzer (resolution about 0.5%)
shown in Fig. 1. The protons transmitted through the analyzer were
detected using either the movable Bendix magnetic electron multi-
plier, or surface barrier detector No. 4. When the surface barrier
detector was used, detector No. 2 in the sample chamber was used
for normalization. Detector No. 1 was used for standard backscattering measurements. It was also possible to measure the transmitted
energy spectrum using detector No. 3, or in some cases a detector
which could be positioned immediately behind the sample, so that
essentially all the transmitted protons were collected. The electrostatic analyzer could be tilted about the center of the sample
chamber so that it was possible to measure transmitted energy
spectra at any desired scattering angle. Further details of the
equipment, as well as sample preparation and mounting techniques
have been described in Ref. 3.

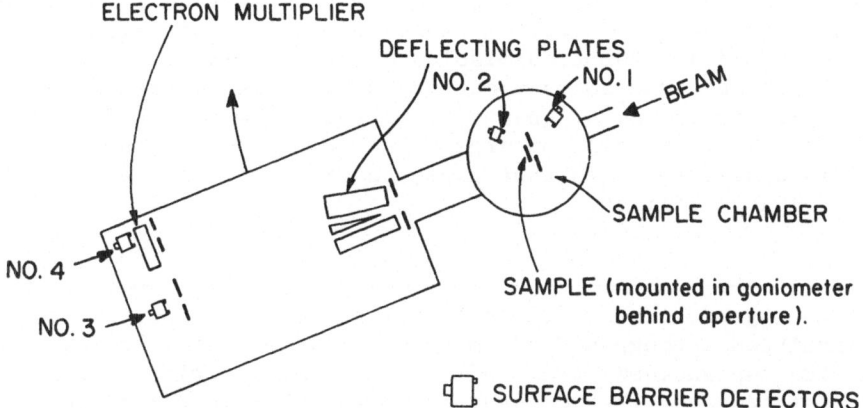

Fig. 1. Schematic drawing of the experimental apparatus.

Results and Discussion

Dechanneling

Transmission energy spectra for 0.2 MeV and 0.4 MeV protons incident on an 0.88 μm thick silicon crystal are shown in Fig. 2. These spectra were measured with the electrostatic analyzer set at an angle of 10° to the incident beam direction, with the beam oriented along the <110> axis of the sample or for random penetration of the incident beam. The channeled and random spectra are quite similar, though the spectra for the beam oriented along the <110> axis are somewhat wider than those for random penetration of the proton beam. This greater width of the <110> spectrum was observed for all samples for which this measurement was made. Similar results were observed with the analyzer oriented 15° from the incident beam direction.

The protons which were energy analyzed in the <110> spectra of Fig. 2 arose from three sources. They may have been part of the random beam on entry into the crystal, they may have been dechanneled while passing through the crystal, or they may have been scattered from defects in the crystal. The contribution from defect scattering should be negligible. From backscattering measurements we have found that more than half the counts in the <110> spectra of Fig. 2 come from dechanneled particles which were not dechanneled until they had penetrated more than half of the crystal thickness [2]. The similarity of the aligned and random spectra in Fig. 2 then indicates that the energy loss distribution for dechanneled protons is essentially the same as the energy loss distribution for random penetration of the protons.

Several factors may be responsible for the greater width of the aligned spectra as compared with the random spectra. On the low energy side the excess counts may be due to the type of high energy loss events discussed by Appleton, et al. [4]. On the high energy side of the <110> spectra protons were observed with energies in the range of the <110> spectrum measured at 0° scattering angle (i.e., the standard <110> transmission energy spectrum). This occurs in the 0.2 MeV spectrum above about 125 keV. Such counts can be caused by the scattering of channeled protons from the oxide layer on the backside of the silicon crystal. The number of counts observed is consistent with that expected from the oxide thickness normally found on silicon.

Fig. 2. Energy spectra of 0.2 MeV (top) and 0.4 MeV (bottom) protons transmitted through a 0.88 μm thick silicon single crystal aligned with the incident beam and with the sample oriented for random penetration of the incident protons. The electrostatic analyzer was tilted 10° away from the incident beam direction.

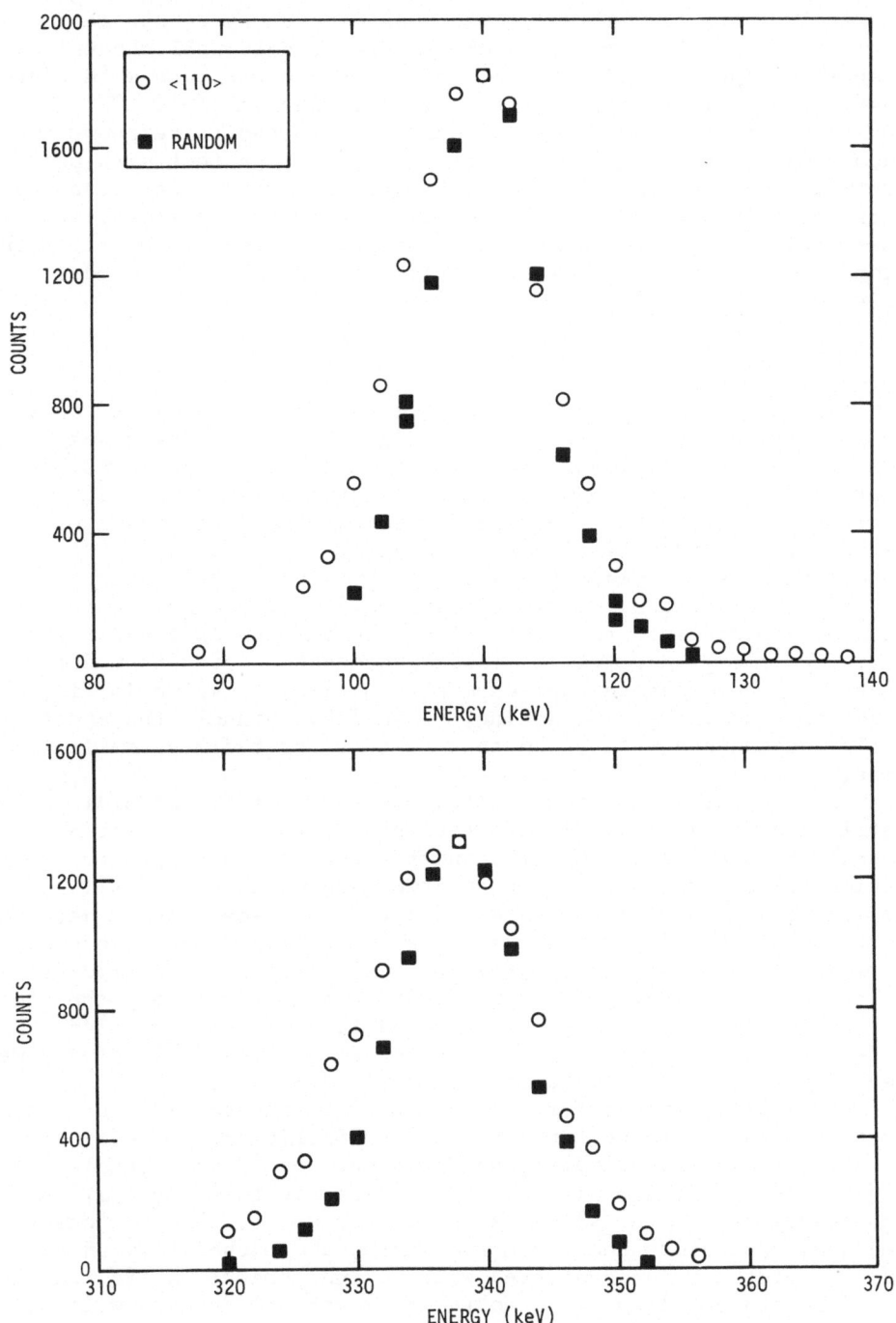

The greater width of the <110> spectrum in the energy region immediately above that at which the peak occurs may be due to the dechanneling of some protons which were relatively well channeled. For shallow depths in the crystal only channeled protons with high transverse energies would be expected to be dechanneled. The stopping powers of these protons would be expected to be about the random stopping power. As the depth increases, protons with progressively lower transverse energies (and therefore lower stopping powers) would be expected to become dechanneled. Therefore, in thicker crystals one would expect to observe a greater contribution to the <110> energy spectrum at energies higher than the random energy loss.

Damage studies

Transmission energy spectra of 0.4 MeV protons incident on a 1.00 μm silicon crystal are shown in Fig. 3. The analyzer was tilted 10° off the incident beam direction and the sample was oriented so the proton beam was incident parallel to the <110> direction in the crystal. Data are presented for the undamaged crystal, and after a thin layer of damage was introduced on the side of the crystal toward the analyzer by implanting krypton ions. The difference between the two spectra is due to the scattering of channeled protons into the analyzer by the defects in the damaged layer. The energy spectrum of these protons scattered from defects (referred to below as the defect spectrum) is also shown in Fig. 3. It was constructed by subtracting, channel by channel, the spectrum for the undamaged crystal from that taken after the krypton implantation.

Defect spectra were constructed for three different silicon crystals and for several different krypton doses. The position of the peak of these defect spectra and their width corresponded, within experimental error, to the peak position and width of the <110> transmitted energy spectrum measured with a surface barrier detector located so as to collect essentially all the transmitted protons. Therefore, the channeled stopping power for protons scattered from the defects in the damaged layers studied is approximately the same as the channeled stopping power measured for all the channeled protons (this is about 15% greater than the channeled stopping power measured with a well collimated detector) [2,3].

The result obtained for the energy position and width of the defect spectrum is to be expected if the distribution of silicon interstitials is approximately uniform over the <110> channels. In this case, since the distribution of interstitials in a given plane transverse to the <110> axis will be uniform and all channeled ions must cross the plane somewhere, the probability of a channeled proton being scattered will be independent of its trajectory. Therefore the energy spectrum of scattered protons will be about the same as that for the channeled protons transmitted through the crystal.

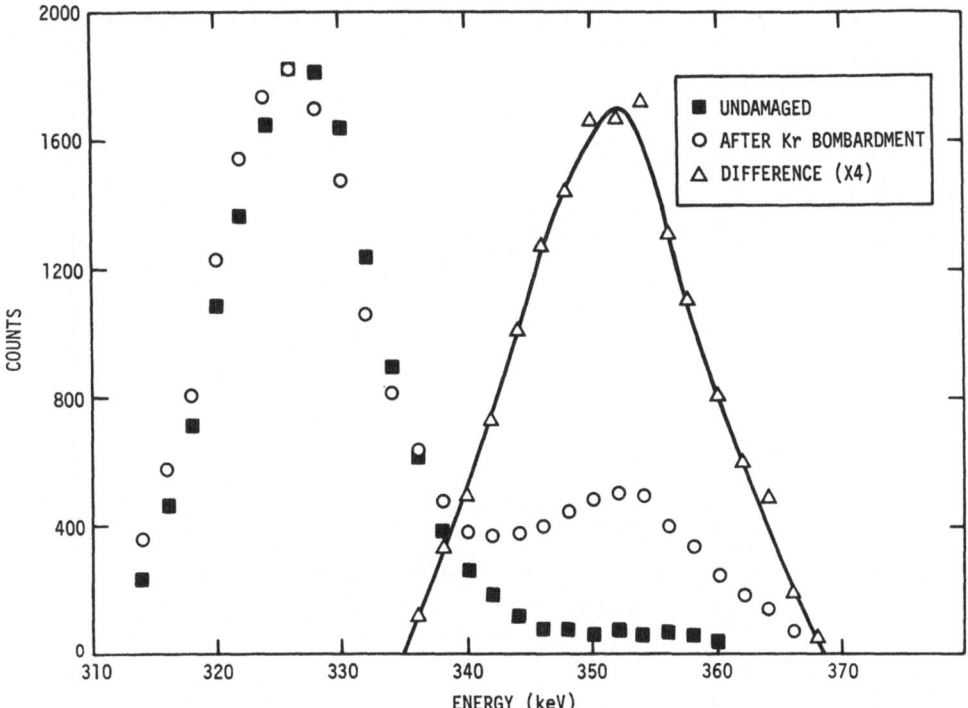

Fig. 3. Transmission energy spectra of 0.4 MeV protons incident on a 100 μm thick silicon single crystal. The <110> axis was aligned with the incident beam and the analyzer was tilted 10°. The two spectra were taken before (■) and after (o) implantation of 5×10^{13} ions/cm^2 of 100 keV Kr$^+$ into the side of the crystal closest to the analyzer.

The results discussed in the preceding paragraph suggest that the protons which contributed to the defect spectrum were scattered from "true" silicon interstitial atoms, with little contribution from strain effects which might produce displacements of silicon atoms by about 0.1 or 0.2Å off lattice sites. In fact, we suggest that the stopping power of channeled protons which could be scattered from such slightly displaced silicon atoms would be about the same as that observed for dechanneling, i.e., the random stopping power. Protons scattered from such slightly displaced atoms would therefore have little influence on the defect spectrum. By comparing the number of counts in the defect spectrum to the number of counts in the random spectrum we may, therefore, determine the number of displaced atoms (i.e., "true" interstitial atoms) in the damaged layer. The result of this comparison is 7.9×10^{16} displaced atoms/cm^2.

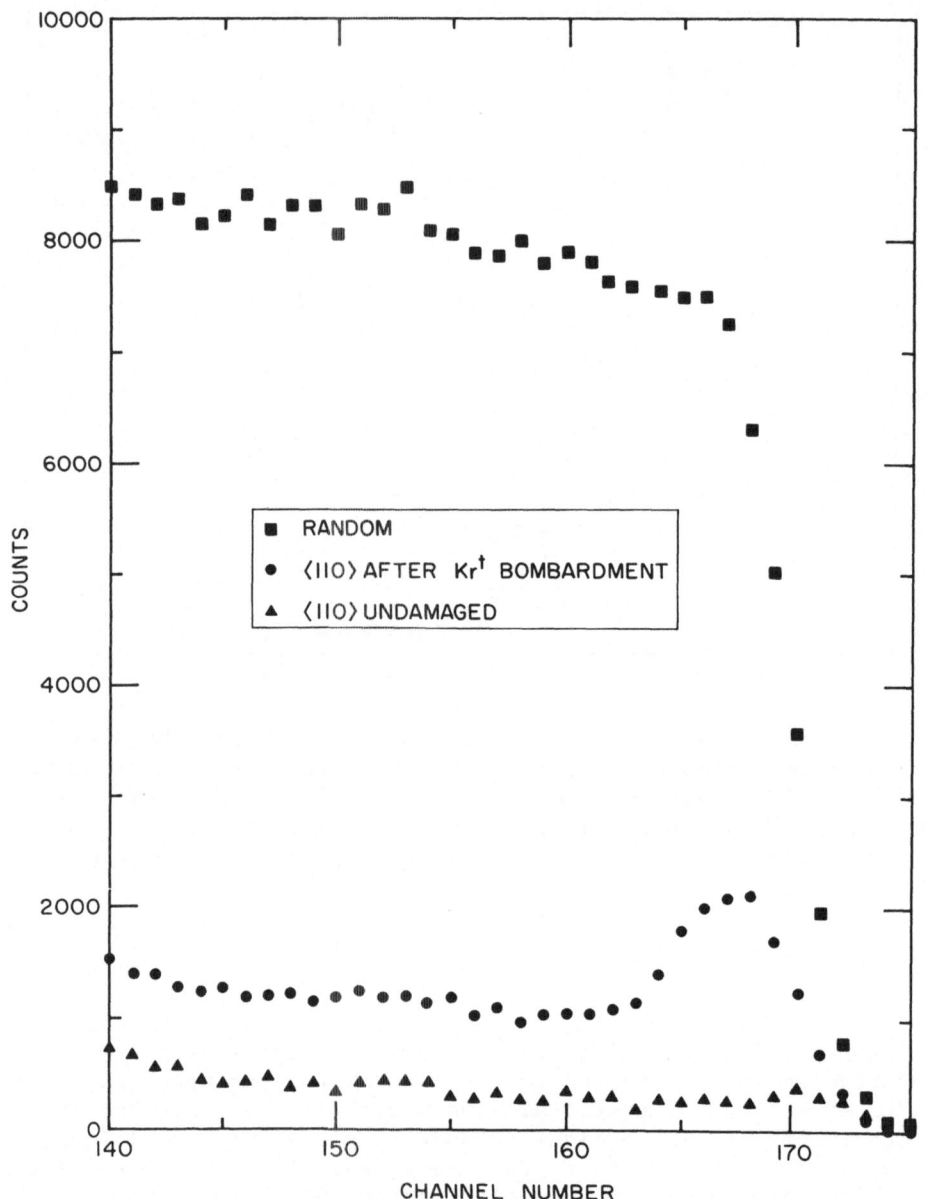

Fig. 4. Aligned and random energy spectra of 0.4 MeV protons back-scattered from the thin silicon crystal used to obtain the data in Fig. 3. The spectrum labeled "<110> after Kr+ bombardment" was taken with the damaged side toward the incident beam; i.e., tilted 180° from the orientation used to obtain the data in Fig. 3.

We have also made standard channeling effect measurements of the krypton induced damage using backscattering. Care was taken to insure that the backscattering measurement was made on exactly the same portion of the sample as the transmission measurements. The backscattering results are shown in Fig. 4. The number of displaced atoms obtained by standard analysis of these backscattering data is 9.5×10^{16} displaced atoms/cm^2. This is only about 20% greater than the number obtained from the transmission measurement. This result suggests that at least in the present case channeling effect measurements of disorder, using backscattering, are sensitive primarily to "true" interstitial atoms, and are only slightly sensitive to small displacements which might be produced by strain. Further work is required to determine whether or not this is true in general.

Finally, we wish to point out that it may be possible in some cases to obtain information on the location of defects from a defect spectrum constructed in the manner described above. For example, if the interstitial atoms were located close to the center of the channel the defect spectrum should peak at higher energies than in the present case. On the other hand, if interstitials were concentrated near the strings, the peak of the defect spectrum should occur at lower energies and there should be few, if any, counts observed at energies near the leading edge of the channeled energy spectrum observed at 0° scattering angle.

References

*Permanent address: Institute of Physics, University of Aarhus, DK-8000 Aarhus C, Denmark.
[1] F. H. Eisen, G. J. Clark, J. Bøttiger, and J. M. Poate, Rad. Eff. 13, 93 (1972).
[2] J. Bøttiger and F. H. Eisen, Thin Solid Films 19, 239 (1973).
[3] F. H. Eisen, Can. J. Phys. 46, 561 (1968).
[4] B. R. Appleton, C. Erginsoy, and W. M. Gibson, Phys. Rev. 161, 330 (1967).

INTERFERENCE DECREASE OF ELASTIC SCATTERING OF FAST CHARGED PARTICLES IN SINGLE CRYSTALS

N. P. KALASHNIKOV, E. A. KOPTELOV, and M. I. RIAZANOV
The Moscow Physical Engineering Institute
Moscow, U.S.S.R.

The elastic scattering of a fast particle in an atom at high energy may be treated as diffraction [1]. Therefore, there appears a diffraction shadow whose blackness depends on the particle interaction with the atom. If the scatterers are arranged in chain along the shadow, the interference of shadows from several "semitransparent" scatterers can result in a pronounced black shadow.

It follows from the above that in scattering fast particles in a single crystal the space orientation of the diffraction shadow region should be taken into account.

The problem was treated theoretically previously [2,3] only within the scope of the perturbation theory for a single-crystal potential and, thus, the shadowing effect was neglected. The condition for the applicability of the perturbation theory to the crystal as a whole is ($\beta = p/E$)

$$\frac{1}{\beta} \int dx U \simeq \frac{L}{a\beta} \int dx U_o << 1 \tag{1.1}$$

where a is the lattice constant, U_o is the atom potential; the axis x is directed along the particle motion, L is the crystal thickness along x. The thickness of the single crystal is greatly limited by the perturbation theory. Interacting with the atoms of a substance the fast particle gives the atom the transverse momentum of the order of the reciprocal screening radius $\kappa = me^2 Z^{1/3}$ (h=c=1) and, therefore, a typical deflection angle is $\theta \sim \kappa/p$. In the longitudinal direction the atom receives the momentum $\Delta p_{||} \sim p\theta^2 \sim (\kappa^2/p)$ which decreases as the particle energy increases. As a result, in scattering ultra-relativistic particles large longitudinal distances $(\Delta p_{||})^{-1} \sim p\kappa^{-2}$ are effective. At length $(p\theta^2)^{-1}$ the atoms start scattering coherently and, if $L < (p\theta^2)^{-1}$, the cross-section increases by a factor of L/a.

It follows that for crystals of some larger thickness the diffraction shadow from the chain of atoms will be essential. Below the interference scattering of fast particles in the single crystal whose thickness exceeds the maximum one allowing to apply the perturbation theory [4] is given.

The scattering amplitude on the single-crystal as a whole can be written as

$$f(\bar{p}_o, \bar{p}) = -\frac{E}{2\pi} \int d^3 r U(r) \psi_{\bar{p}_o}(r) e^{-i\bar{p}\bar{r}} \qquad (2.1)$$

where U is the total potential of the crystal atoms; satisfies the wave equation

$$(\Delta + p_o^2) \psi_{\bar{p}_o}(\bar{r}) = 2E\, U(\bar{r}) \psi_{\bar{p}_o}(\bar{r}) \ . \qquad (2.2)$$

Eq. (2.2) holds for the relativistic particles at $E = (p_o^2 + m^2)^{1/2} \gg U$. For the nonrelativistic particles it completely coincides with the Schrodinger equation. The diffractional approach of the scattering problem can be realized when on the right side of Eq. (2.1) the explicit solution $\psi_{\bar{p}_o}(\bar{r})$ is replaced by the approximate one

$$\psi(\bar{r}) = \exp\{ip_o x - \frac{i}{\beta} \int_{-\infty}^{x} dx' U(x', \bar{r}_\perp)\} \ . \qquad (2.3)$$

The condition of this approximation is the inequality

$$L < p\kappa^{-2} \qquad (2.4)$$

The procedure for (2.3) is similar to that of the geometrical optics type and the amplitude

$$f(\bar{p}_{o\perp}\bar{p}) = -\frac{E}{2\pi} \int d^3 r U(r) e^{i(p_o - p)r}$$

$$\exp\{ -\frac{i}{\beta} \int_{-\infty}^{x} dx' U(x', \bar{r}_\perp)\} \qquad (2.5)$$

takes into account small deviation from the geometrical optics. Integrating by parts it is convenient to reduce Eq. (2.5) to the form [5,6]

$$f(\bar{p}_{o\perp}\bar{p}) = \frac{ip}{2\pi} \int d^2 r_i e^{i(\bar{p}_o - \bar{p})_\perp \bar{r}_\perp}$$

$$\{1 - \exp[-\frac{i}{\beta} \int_{\infty}^{\infty} dx U(x, r_\perp)]\} \ . \qquad (2.6)$$

If the conditions (2.4 satisfy, the formula (2.6) is valid for an arbitrary potential.

The elastic scattering for the particle moving along the axis of the chain is of some interest. Let N atoms be in the points $\vec{R}_n = n\vec{a}(n=0,1,2...N-1)$ and $Na \ll p\kappa^{-2}$. Then Eq. (2.6) yields the expression for the scattering amplitude of a particle on the whole chain of atoms [5] ($\bar{q} = \bar{p}=\bar{p}_0$)

$$f_N(q) = ip \int_0^\infty J_0(qb)bdb\{1-\exp[-\frac{i}{\beta}\int_{-\infty}^\infty dxU(x,b)]\} \qquad (3.1)$$

where $J_0(x)$ is the Bessel function; $U(\vec{r}) = \sum_n U_a(\vec{r}-\vec{R}_n)$ and where the summation is carried out over all the atoms of the chain. The applicability of (3.1) is limited by the scattering angles

$$\theta < (\kappa/p)^{1/2} . \qquad (3.2)$$

It should be noted that the condition (3.2) allows to use (3.1) practically in the whole range of effective angles. The potential of the atom chain is

$$U(\vec{r})= \sum_{n=0}^{N-1} \frac{\alpha}{|\vec{r}-\vec{R}_n|} \exp\{-\kappa|\vec{r}-\vec{R}_n|\} \qquad (3.3)$$

where $\alpha \equiv Ze^2$. Substituting (3.3) into (3.1) and integrating yield

$$f_N(q) = ip \int_0^\infty J_0(qb)bdb\{1-\exp[-i\frac{2N\alpha}{\beta}K_0(b\kappa)] . \qquad (3.4)$$

Consider first the case when there is a sufficiently large chain for which

$$\alpha^{-1} \ll N \ll (p/\kappa^2 a^2) . \qquad (3.5)$$

In this case the assimptotic methods to calculate the integral can be used. In fact, by means of the stationary phase method one can get the explicit form of (3.4). Thus, at

$$\frac{a}{\kappa} \ln \frac{N\alpha}{\beta} < 1$$

we obtain

$$f_N(q) = \frac{ip}{\kappa} [\ln(2N\alpha\frac{\gamma\sqrt{\pi}}{\beta})+i\frac{\pi}{2}] \frac{J_1[\frac{9}{\kappa}(\ln(2N\alpha\frac{\gamma\sqrt{\pi}}{\beta})+i\frac{\pi}{2})]}{9} \qquad (3.6)$$

where $\gamma \approx 1.781$ is the Euler constant. In case $\frac{9}{\kappa} \ln \frac{N\alpha}{\beta} > 1$

$$f_N(q) = \frac{-ip}{\kappa} \sqrt{\ln(\frac{2N\alpha\kappa}{\beta 9})} \frac{\exp(-i\frac{9}{\kappa}\ln(\frac{2N\alpha\kappa}{\beta 9}))}{9} \qquad (3.7)$$

Using the optical theorem by means of either (3.4) or 3.6) one can get the total elastic scattering cross-section on a long chain of atoms (3.5)

$$\delta = \frac{2\pi}{\kappa^2} \{ (\ln A)^2 + \ln A \cdot [\frac{3}{4} + 2CI(1) - \frac{\sin A}{A}] \} \qquad (3.8)$$

where $Ci(x)$ is the integral cosine,

$$A = 2\frac{Ze^2}{\beta} N \sqrt{\frac{\pi}{2\ln(2N\alpha/\beta)}}$$

The total elastic scattering cross-section (3.8) can also be yielded by integrating the squared module of the scattering amplitude (3.6). The pattern of the formulae obtained is similar to those for the diffraction with a semitransparent body. The similarity confirms the interpretation of the results as being due to the superposition of shadows from several atoms. The "blackness" of the shadow (hence, the chain transparency) is determined by the parameter $N\alpha$. At $N\alpha \gg 1$ where (3.6)-(3.8) are valid the chain becomes almost unpenetrable for the particles parallel to its axis. The dependence of the scattering cross-section on the number of atoms in the chain is slight; from the case above it follows that the atoms lying before the black shadow take part in scattering. In other words, it is along the length

$$\ell \sim (a/\alpha) \qquad (3.9)$$

that the coherently scattering atoms form the potential with the effective interaction force of the order of unity.

It should be stressed that the results obtained for the total and differential scattering cross-section (3.6)-3.8) depend only upon the effective range of the potential κ^{-1}.

For the opposite limited case $N\alpha \ll 1$ the scattering cross-section on the atom chain can be obtained using the perturbation theory. Expanding the exponent in series in Eq. (3.4) yields the results that coincide with those of [2], i.e., they are proportional to $(L/a)^2$.

For the intermediate case $N\alpha \sim 1$ the integral (3.4) is calculated numerically. Applying (3.4) and the optical theorem total scattering cross-section can be expressed by

$$\delta = \frac{4\pi}{\kappa^2} \text{ Re} \int_0^\infty x dx \{1 - \exp(-i2A_0 K_0(x))\} \tag{3.10}$$

where $A_0 \equiv \frac{N\alpha}{\beta}$. By means of the integration methods the ratio of the total cross-section against the parameter A_0

$$\delta \kappa^2 / 4\pi = \phi(A_0)$$

has been obtained. The numerical values of the function $\phi(A_0)$ are equal to 0.995 10^{-2}, 0.390 10^{-1}, 0.292, 0.630, 1.41, 2.58, 4.15, 4.74, 6.86, 9.40, 13.4, 26.9; for A_0 = 0.1, 0.2, 0.6, 1, 2, 4, 8, 10, 20, 40, 10^2, 10^3, respectively.

From the values above it follows that at small A_0 the total cross-section is proportional to A_0^2 and at large A_0 it can be described by (3.8).

Consider the single-crystal bounded by the surfaces x = 0 and x = L, assuming $L \ll p\kappa^{-2}$. Depending on the particle direction the number of atoms in the chain that coincides with the direction will differ. Consider first the case when the particle falls along the crystallographic axis x. In this case the formulae for the cross-section in a single-crystal differ from (3.5)-(3.8) only by the factor equal to the number of atoms on the surfaces y, z, $N_\perp$; the latter is due to the fact that the scattering on various chains occurs independently. Therefore, from (3.6) and 3.8) yield

$$\frac{d\delta}{d\Omega} = \frac{N_i}{\kappa^2} \ln^2(2\frac{N\alpha}{\beta} \gamma\sqrt{\pi}) \frac{J_1^2[2\frac{P}{\kappa} \sin\frac{\theta}{2}\ln(2\frac{N\alpha}{\beta} \gamma\sqrt{\pi}]}{4 \sin^2 \frac{\theta}{2}} \tag{4.1}$$

and

$$\delta = 2N_i \frac{\pi}{\kappa^2} \ln^2(2\frac{N\alpha}{\beta} \gamma\sqrt{\pi}) \ . \tag{4.2}$$

It is worth noting that (4.1)-(4.2) are valid only if the effect of thermal vibration of atoms in a crystal lattice is neglected. It limits the longitudinal dimension of the single-crystal $L = Na_x$ by the condition

$$L \ll A_x \frac{\ln^2(2\frac{N\alpha}{\beta} \gamma\sqrt{\pi})}{\alpha^2 \kappa^2 \overline{u^2}} \tag{4.3}$$

where $\overline{u^2}$ is the mean square of the thermal atom displacement.

In changing the direction of the particle the numer of atoms along the direction $\vec{p}$, $N_{\vert\vert}$ depends upon the angle θ_0 formed by the particle velocity and the crystallographic axis x. For the limited case $N_{||}\alpha >> 1$ the cross-section slightly varies depending on the angle θ_0 while for $N_{||}\alpha << 1$ it radically changes according to the data [2]. If the single-crystal thickness increases the shadowing effect reduces the interference oscillation of scattering cross-section versus the angle θ_0. In its pure form the Ter-Mikayelyan interference effect is found only in fine single-crystals whose thickness is limited by the Born approximation

$$L << (a/Ze^2)$$

Finally, the treatment of the problem without the perturbation theory changes qualitatively the character of fast particle scattering in the single-crystal.

References

[1] A. I. Akhiezer and I. Ya. Pomeranchuk, Usp. Fiz. Nauk. 65, 593 (1958).
[2] M. L. Ter-Mikaleyan, Zh. Eksp. Teor. Fiz 25, 289 (1953).
[3] B. Ferretti, Nuovo Cim. 7, 118 (1950).
[4] N. P. Kalashnikov, E. A. Koptelov and M. I. Riazanov, ZhETF Pis. Red. 15, 120 (1972).
[5] L. I. Schiff, Phys. Rev. 103, 443 (1956).
[6] R. I. Glauber, Phys. Rev. 100, 242 (1955).
[7] A. Erdèlyi, Asymptotic Expansions, Dover Publ., 1960.

AUTHOR INDEX

Pages 1-478 are found in Volume 1, pages 481-934 in Volume 2

SUBJECT INDEX

Pages 1-478 are found in Volume 1, pages 481-934 in Volume 2

Alkali halides   47
Aluminum   219,423,429,430,431,
    440,514,765
Anti-particles   26
Argon   164,348,350,352,353,
    378,379,384,430,470,584
Auger spectra   414
Axial channeling   804,811
Axial dechanneling   843,863

Backscattering   77,142,177,215,
    251,315,569,819,825,864,898,
    908
  Ar from Cu   584
  F from Ni   83
  H from Au   332
  H from Be   320
  H from Cu   257,322,573
  H from Nb   322
  H from Mo   323
  H from Ta   324
  H from Si   333
  H from W   595
  H from V   320
  He from Cu   573
  Ne from Ni   600
  O from Au   83
  O from Ni   83
  S from Au   83
Back-sputtering   171
Beryllium   320
Bethe-Bloch formula   3,7,252
Bloch-wave channeling   547
Blocking   717,739,765
  H in Ge   735,745,752
  H in Si   752
  He in LiF   742

Blocking lifetimes   735
Boltzmann transport equation   125
Born approximation   25
Born-Mayer potential   92,605
Boron   812
Bragg's additivity rule   26
Brehmstrahlung   481,564
Broad band emission   511
Bromine   57,68

Carbon   59,263,385,389,402,470,
    493
Cauchois spectrometer   447
Channeling   4,57,779,804,811,825
  B in Ge   812
  B in Si   812
  D in Si   905
  H in Au   780
  H in Ge   812
  H in KCl   671
  H in LiF   671
  H in NaCl   671
  H in NaF   671
  H in TiO_x 791
  H in Si   75,812,919
  He in Au   780
  He in Fe   819
  He in Ge   812
  He in Si   812
  O in Ag   63,499
Channeling, electrons   529,531,
    539,547,550
Channeling, energy loss spectra
    919
Channeling, inelastic scattering
    in   897